Reviews in Modern Astronomy 22

Edited by
Regina von Berlepsch

The Series Reviews in Modern Astronomy

Vol. 21: Formation and Evolution of Cosmic Structures
2009
ISBN: 978-3-527-40910-5

Vol. 20: Cosmic Matter
2008
ISBN: 978-3-527-40820-7

Vol. 19: The Many Facets of the Universe - Revelations by New Instruments
2006
ISBN: 978-3-527-40662-3

Vol. 18: From Cosmological Structures to the Milky Way
2005
ISBN: 978-3-527-40608-1

Vol. 17: The Sun and Planetary Systems – Paradigms for the Universe
2004
ISBN: 978-3-527-40476-6

Vol. 16: The Cosmic Circuit of Matter
2003
ISBN: 978-3-527-40451-3

Vol. 15: Astronomy with Large Telescopes from Ground and Space
2002
ISBN: 978-3-527-40404-9

Reviews in Modern Astronomy
Vol. 22

Deciphering the Universe through Spectroscopy

Edited by
Regina von Berlepsch

WILEY-
VCH

WILEY-VCH Verlag GmbH & Co. KGaA

The Editor

Regina von Berlepsch
Astrophysical Institute
University of Potsdam
Germany
RBerlepsch@aip.de

All books published by **Wiley-VCH** are carefully produced. Nevertheless, authors, editors, and publisher do not warrant the information contained in these books, including this book, to be free of errors. Readers are advised to keep in mind that statements, data, illustrations, procedural details or other items may inadvertently be inaccurate.

Library of Congress Card No.:
applied for

British Library Cataloguing-in-Publication Data
A catalogue record for this book is available from the British Library.

Bibliographic information published by the Deutsche Nationalbibliothek
The Deutsche Nationalbibliothek lists this publication in the Deutsche Nationalbibliografie; detailed bibliographic data are available on the Internet at <http://dnb.d-nb.de>.

© 2010 Wiley-VCH Verlag & Co. KGaA, Boschstr.12, 69469 Weinheim, Germany

All rights reserved (including those of translation into other languages). No part of this book may be reproduced in any form – by photoprinting, microfilm, or any other means – nor transmitted or translated into a machine language without written permission from the publishers. Registered names, trademarks, etc. used in this book, even when not specifically marked as such, are not to be considered unprotected by law.

Composition Uwe Krieg, Berlin

Printing and Binding Strauss GmbH, Mörlenbach

Cover Design Schulz Grafik Design, Fußgönheim

Printed in the Federal Republic of Germany
Printed on acid-free paper

ISBN: 978-3-527-41055-2

Preface

The annual series *Reviews in Modern Astronomy* of the ASTRONOMISCHE GESELLSCHAFT was established in 1988 in order to bring the scientific events of the meetings of the Society to the attention of the worldwide astronomical community. *Reviews in Modern Astronomy* is devoted exclusively to the Karl Schwarzschild Lectures, the Ludwig Biermann Award Lectures, the invited reviews, and to the Highlight Contributions from leading scientists reporting on recent progress and scientific achievements at their respective research institutes.

The Karl Schwarzschild Lectures constitute a special series of invited reviews delivered by outstanding scientists who have been awarded the Karl Schwarzschild Medal of the Astronomische Gesellschaft, whereas excellent young astronomers are honoured by the Ludwig Biermann Prize.

Volume 22 continues the series with fifteen invited reviews and Highlight Contributions which were presented during the International Scientific Conference of the Society on "Deciphering the Universe through Spectroscopy" held in Potsdam, Germany, September 21 to 25, 2009 in conjunction with the fall meeting of the "Fachverband Extraterrestrische Physik" of the Deutsche Physikalische Gesellschaft and the "Arbeitsgemeinschaft Extraterrestrische Forschung e.V."

The Karl Schwarzschild medal 2009 was awarded to Professor Rolf-Peter Kudritzki, Hawaii. His lecture with the title "Dissecting galaxies with quantitative spectroscopy of the brightest stars in the Universe" opened the meeting.

The Ludwig Biermann Prize was awarded twice in 2009. The two winners are Anna Frebel, Cambridge, USA and Sonja Schuh, Göttingen, Germany. The title of Anna Frebel's talk was: "What the most metal-poor stars tell us about the early Universe" and Sonja Schuh gave a lecture on: "Pulsations and planets: the asteroseismology-extrasolar-planet connection".

Spectroscopy is a key method in modern astrophysical research throughout all wavelengths. Applications range from the measurements of magnetic fields on the surface of the Sun over detailed measurements of abundances in stellar atmospheres to the kinematics of the universe on its largest scales. The contributions to the meeting published in this volume discuss the complex of themes. A report on VLT-CRIRES with some highlights and recent results complete this volume.

The editor would like to thank the lecturers for their stimulating presentations. Thanks also to the local organizing committee from the Astrophysikalisches Institut Potsdam chaired by Hans Zinnecker.

Potsdam, Mai 2010 *Regina v. Berlepsch*

The ASTRONOMISCHE GESELLSCHAFT awards the **Karl Schwarzschild Medal**. Awarding of the medal is accompanied by the Karl Schwarzschild lecture held at the scientific annual meeting and the publication. Recipients of the Karl Schwarzschild Medal are

- 1959 Martin Schwarzschild:
  Die Theorien des inneren Aufbaus der Sterne.
  Mitteilungen der AG 12, 15
- 1963 Charles Fehrenbach:
  Die Bestimmung der Radialgeschwindigkeiten
  mit dem Objektivprisma.
  Mitteilungen der AG 17, 59
- 1968 Maarten Schmidt:
  Quasi-stellar sources.
  Mitteilungen der AG 25, 13
- 1969 Bengt Strömgren:
  Quantitative Spektralklassifikation und ihre Anwendung
  auf Probleme der Entwicklung der Sterne und der Milchstraße.
  Mitteilungen der AG 27, 15
- 1971 Antony Hewish:
  Three years with pulsars.
  Mitteilungen der AG 31, 15
- 1972 Jan H. Oort:
  On the problem of the origin of spiral structure.
  Mitteilungen der AG 32, 15
- 1974 Cornelis de Jager:
  Dynamik von Sternatmosphären.
  Mitteilungen der AG 36, 15
- 1975 Lyman Spitzer, jr.:
  Interstellar matter research with the Copernicus satellite.
  Mitteilungen der AG 38, 27
- 1977 Wilhelm Becker:
  Die galaktische Struktur aus optischen Beobachtungen.
  Mitteilungen der AG 43, 21
- 1978 George B. Field:
  Intergalactic matter and the evolution of galaxies.
  Mitteilungen der AG 47, 7
- 1980 Ludwig Biermann:
  Dreißig Jahre Kometenforschung.
  Mitteilungen der AG 51, 37
- 1981 Bohdan Paczynski:
  Thick accretion disks around black holes.
  Mitteilungen der AG 57, 27

1982	Jean Delhaye: Die Bewegungen der Sterne und ihre Bedeutung in der galaktischen Astronomie. Mitteilungen der AG 57, 123
1983	Donald Lynden-Bell: Mysterious mass in local group galaxies. Mitteilungen der AG 60, 23
1984	Daniel M. Popper: Some problems in the determination of fundamental stellar parameters from binary stars. Mitteilungen der AG 62, 19
1985	Edwin E. Salpeter: Galactic fountains, planetary nebulae, and warm H I. Mitteilungen der AG 63, 11
1986	Subrahmanyan Chandrasekhar: The aesthetic base of the general theory of relativity. Mitteilungen der AG 67, 19
1987	Lodewijk Woltjer: The future of European astronomy. Mitteilungen der AG 70, 21
1989	Sir Martin J. Rees: Is there a massive black hole in every galaxy. Reviews in Modern Astronomy 2, 1
1990	Eugene N. Parker: Convection, spontaneous discontinuities, and stellar winds and X-ray emission. Reviews in Modern Astronomy 4, 1
1992	Sir Fred Hoyle: The synthesis of the light elements. Reviews in Modern Astronomy 6, 1
1993	Raymond Wilson: Karl Schwarzschild and telescope optics. Reviews in Modern Astronomy 7, 1
1994	Joachim Trümper: X-rays from Neutron stars. Reviews in Modern Astronomy 8, 1
1995	Henk van de Hulst: Scaling laws in multiple light scattering under very small angles. Reviews in Modern Astronomy 9, 1
1996	Kip Thorne: Gravitational Radiation – A New Window Onto the Universe. Reviews in Modern Astronomy 10, 1
1997	Joseph H. Taylor: Binary Pulsars and Relativistic Gravity. not published

1998 Peter A. Strittmatter:
      Steps to the LBT – and Beyond.
      Reviews in Modern Astronomy 12, 1

1999 Jeremiah P. Ostriker:
      Historical Reflections
      on the Role of Numerical Modeling in Astrophysics.
      Reviews in Modern Astronomy 13, 1

2000 Sir Roger Penrose:
      The Schwarzschild Singularity:
      One Clue to Resolving the Quantum Measurement Paradox.
      Reviews in Modern Astronomy 14, 1

2001 Keiichi Kodaira:
      Macro- and Microscopic Views of Nearby Galaxies.
      Reviews in Modern Astronomy 15, 1

2002 Charles H. Townes:
      The Behavior of Stars Observed by Infrared Interferometry.
      Reviews in Modern Astronomy 16, 1

2003 Erika Boehm-Vitense:
      What Hyades F Stars tell us about Heating Mechanisms
      in the outer Stellar Atmospheres.
      Reviews in Modern Astronomy 17, 1

2004 Riccardo Giacconi:
      The Dawn of X-Ray Astronomy
      Reviews in Modern Astronomy 18, 1

2005 G. Andreas Tammann:
      The Ups and Downs of the Hubble Constant
      Reviews in Modern Astronomy 19, 1

2007 Rudolf Kippenhahn:
      Als die Computer die Astronomie eroberten
      Reviews in Modern Astronomy 20, 1

2008 Rashid Sunyaev:
      The Richness and Beauty of the Physics of Cosmological
      Recombination
      Reviews in Modern Astronomy 21, 1

2009 Rolf-Peter Kudritzki:
      Dissecting Galaxies with Quantitative Spectroscopy
      of the Brightest Stars in the Universe
      Reviews in Modern Astronomy 22, 1

The **Ludwig Biermann Award** was established in 1988 by the ASTRONOMISCHE GESELLSCHAFT to be awarded in recognition of an outstanding young astronomer. The award consists of financing a scientific stay at an institution of the recipient's choice. Recipients of the Ludwig Biermann Award are

- 1989 Dr. Norbert Langer (Göttingen),
- 1990 Dr. Reinhard W. Hanuschik (Bochum),
- 1992 Dr. Joachim Puls (München),
- 1993 Dr. Andreas Burkert (Garching),
- 1994 Dr. Christoph W. Keller (Tucson, Arizona, USA),
- 1995 Dr. Karl Mannheim (Göttingen),
- 1996 Dr. Eva K. Grebel (Würzburg) and
        Dr. Matthias L. Bartelmann (Garching),
- 1997 Dr. Ralf Napiwotzki (Bamberg),
- 1998 Dr. Ralph Neuhäuser (Garching),
- 1999 Dr. Markus Kissler-Patig (Garching),
- 2000 Dr. Heino Falcke (Bonn),
- 2001 Dr. Stefanie Komossa (Garching),
- 2002 Dr. Ralf S. Klessen (Potsdam),
- 2003 Dr. Luis R. Bellot Rubio (Freiburg im Breisgau),
- 2004 Dr. Falk Herwig (Los Alamos, USA),
- 2005 Dr. Philipp Richter (Bonn),
- 2007 Dr. Henrik Beuther (Heidelberg) and
        Dr. Ansgar Reiners (Göttingen)
- 2008 Dr. Andreas Koch (Los Angeles)
- 2009 Dr. Anna Frebel (Cambridge, USA) and
        Dr. Sonja Schuh (Göttingen)

Contents

Karl Schwarzschild Lecture:
Dissecting galaxies with quantitative spectroscopy of the
brightest stars in the Universe
By Rolf-Peter Kudritzki (With 20 Figures) 1

Ludwig Biermann Award Lecture:
Pulsations and planets: The asteroseismology-extrasolar-planet
connection
By Sonja Schuh (With 4 Figures) 29

Ludwig Biermann Award Lecture:
Stellar archaeology: Exploring the Universe with
metal-poor stars
By Anna Frebel (With 5 Figures) 53

Quantitative solar spectroscopy
By Klaus Wilhelm (With 7 Figures) 81

Metallicity and kinematical clues
To the formation of the Local Group
By Rosemary Wyse (With 5 Figures) 99

Probing dark matter, galaxies and the expansion history
of the Universe with Lyα in absorption and emission
By Martin Haehnelt (With 14 Figures) 117

Hypervelocity stars in the Galactic halo
By Holger Baumgardt 133

Schwarzschild modelling of elliptical galaxies
and their black holes
By Jens Thomas (With 4 Figures) 143

Star and protoplanetary disk properties in Orion's suburbs
By Roy van Boekel, Min Fang, Wei Wang, Andrès Carmona,
Aurora Sicilia-Aguilar, and Thomas Henning
(With 4 Figures) 155

Molecular gas at high redshift
By Fabian Walter, Chris Carilli, and Emanuele Daddi (With 1 Figure) 167

X-ray spectroscopy and mass analysis of galaxy clusters
By Robert W. Schmidt (With 5 Figures) 179

High-fidelity spectroscopy at the highest resolution
By Dainis Dravins (With 4 Figures) 191

Spectroscopy of solar neutrinos
By Michael Wurm, Franz von Feilitzsch, Marianne Göger-Neff,
Tobias Lachenmaier, Timo Lewke, Qurin Meindl, Randolph Möllenberger
Lothar Oberauer, Walter Potzel, Marc Tippmann, Christoph Traunsteiner, and
Jürgen Winter (With 5 Figures) 203

Open clusters and the galactic disk
By Siegfried Röser, Nina V. Kharchenko, Anatoly E. Piskunov,
Elena Schilbach, Ralf-Dieter Scholz, and Hans Zinnecker (With 7 Figures) ... 215

VLT-CRIRES: "Good Vibrations"
Rotational-vibrational molecular spectroscopy in astronomy
By Hans Ulrich Käufl (With 4 Figures) 229

Index of Contributors 241

General Table of Contents 243

General Index of Contributors 258

Karl Schwarzschild Lecture

Dissecting galaxies with quantitative spectroscopy of the brightest stars in the Universe

Rolf-Peter Kudritzki

Institute for Astronomy, University of Hawaii
2680 Woodlawn Dr., Honolulu, Hawaii 96822, USA
kud@ifa.hawaii.edu

Abstract

Measuring distances to galaxies, determining their chemical composition, investigating the nature of their stellar populations and the absorbing properties of their interstellar medium are fundamental activities in modern extragalactic astronomy helping to understand the evolution of galaxies and the expanding universe. The optically brightest stars in the universe, blue supergiants of spectral A and B, are unique tools for these purposes. With absolute visual magnitudes up to $M_V \cong -9.5$ they are ideal to obtain accurate quantitative information about galaxies through the powerful modern methods of quantitative stellar spectroscopy. The spectral analysis of individual blue supergiant targets provides invaluable information about chemical abundances and abundance gradients, which is more comprehensive than the one obtained from H II regions, as it includes additional atomic species, and which is also more accurate, since it avoids the systematic uncertainties inherent in the strong line studies usually applied to the H II regions of spiral galaxies beyond the Local Group. Simultaneously, the spectral analysis yields stellar parameters and interstellar extinction for each individual supergiant target, which provides an alternative very accurate way to determine extragalactic distances through a newly developed method, called the Flux-weighted Gravity–Luminosity Relationship (FGLR). With the present generation of 10 m-class telescopes these spectroscopic studies can reach out to distances of 10 Mpc. The new generation of 30 m-class telescopes will allow to extend this work out to 30 Mpc, a substantial volume of the local universe.

1 Introduction

To measure distances to galaxies, to determine their chemical composition, and to investigate the nature of their stellar populations and the absorbing properties of their interstellar medium are fundamental activities in modern extragalactic astronomy.

They are crucial to understand the evolution of galaxies and of the expanding universe and to constrain the history of cosmic chemical enrichment, from the metal-free universe to the present-day chemically diversified structure. However, while stars are the major constituents of galaxies, little of this activity is based on the quantitative spectroscopy of individual stars, a technique which over the last fifty years has proven to be one of the most accurate diagnostic tools in modern astrophysics. Given the distances to galaxies beyond the Local Group individual stars seem to be too faint for quantitative spectroscopy and, thus, astronomers have settled to restrict themselves to the photometric investigation of resolved stellar populations, the population synthesis spectroscopy of integrated stellar populations or the investigation of H II region-emission lines. Of course, color-magnitude diagrams and the study of nebular emission lines have an impressive diagnostic power, but they are also limited in many ways and subject to substantial systematic uncertainties, as we will show later in the course of this lecture.

Thus, is it really out of the question to apply the methods of quantitative spectroscopy of individual stars as a most powerful complementary tool to understand the evolution of galaxies beyond the Local Group? The answer is, no, it is not. In the era of 10 m-class telescopes with most efficient spectrographs it is indeed possible to quantitatively analyze the spectra of individual stars in galaxies as distant as 10 Mpc and to obtain invaluable information about chemical composition and composition gradients, interstellar extinction and extinction laws as well as accurate extragalactic distances. With the even larger and more powerful next generation of telescopes such as the TMT and the E-ELT we will be able to extend such studies out to distances as large as 30 Mpc. All one has to do is to choose the right type of stellar objects and to apply the extremely powerful tools of NLTE spectral diagnostics, which have already been successfully tested with high resolution, high signal-to-noise spectra of similar objects in the Milky Way and the Magellanic Clouds.

Of course, now when studying objects beyond the Local Group the analysis methods need to be modified towards medium resolution spectra with somewhat reduced signal. For a stellar spectroscopist, who is usually trained to believe that only the highest resolution can give you an important answer, this requires some courage and boldness (or maybe naive optimism). But as we will see in the course of this lecture, once one has done this step, a whole new universe is opening up in u the true sense of the word.

2 Choosing the right objects: A and B supergiants

It has long been the dream of stellar astronomers to study individual stellar objects in distant galaxies to obtain detailed spectroscopic information about the star formation history and chemodynamical evolution of galaxies and to determine accurate distances based on the determination of stellar parameters and interstellar reddening and extinction. At the first glance, one might think that the most massive and, therefore, most luminous stars with masses higher than 50 $M_\odot$ are ideal for this purpose. However, because of their very strong stellar winds and mass-loss these objects keep very hot atmospheric temperatures throughout their life and, thus, waste most of their precious photons in the extreme ultraviolet. As we all know, most of these UV photons are killed by dust absorption in the star forming regions, where these stars are born, and the few which make it to the earth can only be observed with tiny UV telescopes in space such as the HST or FUSE and are not accessible to the giant telescopes on the ground.

Thus, one learns quickly that the most promising objects for such studies are massive stars in a mass range between 15 to 40 $M_\odot$ in the short-lived evolutionary phase, when they leave the hydrogen main-sequence and cross the HRD in a few thousand to ten thousand years as blue supergiants of B and early A spectral type. Because of the strongly reduced absolute value of bolometric correction when evolving towards smaller temperature these objects increase their brightness in visual light and become the optically brightest "normal" stars in the universe with absolute visual magnitudes up to $M_V \cong -9.5$ rivaling with the integrated light brightness of globular clusters and dwarf spheroidal galaxies. These are the ideal stellar objects to obtain accurate quantitative information about galaxies.

The optical spectra of B- and A-type supergiants are rich in metal absorption lines from several elements (C, N, O, Mg, Al, S, Si, Ti, Fe, among others). As young objects they represent probes of the current composition of the interstellar medium. Abundance determinations of these objects can therefore be used to trace the present abundance patterns in galaxies, with the ultimate goal of recovering their chemical and dynamical evolution history. In addition to the α-elements, key for a comparison with H II region results and to understand the chemical evolutionary status of each individual star, the stellar analysis of blue supergiants provides the only accurate way to obtain information about the spatial distribution of Fe-group element abundances in external star-forming galaxies. So far, beyond the Local Group most of the information about the chemical properties of spiral galaxies has been obtained through the study of H II regions oxygen emission lines using so-called strong-line methods, which, as we will show, have huge systematic uncertainties arising from their calibrations. Direct stellar abundance studies of blue supergiants open a completely new and more accurate way to investigate the chemical evolution of galaxies.

In addition, because of their enormous intrinsic brightness, blue supergiants are also ideal distance indicators. As first demonstrated by Kudritzki, Bresolin & Przybilla (2003) there is a very simple and compelling way to use them for distance determinations. Massive stars with masses in the range from 12 $M_\odot$ to 40 $M_\odot$ evolve through the B and A supergiant stage at roughly constant luminosity. In addition, since the evolutionary timescale is very short when crossing through the B and A supergiant domain, the amount of mass lost in this stage is small. As a consequence, the evolution proceeds at constant mass and constant luminosity. This has a very simple, but very important consequence for the relationship between gravity and effective temperature along each evolutionary track, namely that the flux-weighted gravity, $g_F = g/T_{\text{eff}}^4$, stays constant. As shown in detail by Kudritzki et al. (2008) and explained again further below this immediately leads to the *'flux-weighted gravity–luminosity relationship' (FGLR)*, which has most recently proven to be an extremely powerful to determine extragalactic distances with an accuracy rivalling the Cepheid- and the TRGB-method.

3 Spectral diagnostics and studies in the Milky Way and Local Group

The quantitative analysis of the spectra of these objects is not trivial. NLTE effects and but also the influence of stellar winds and atmospheric geometrical extension are of crucial importance. However, over the past decades with an effort of hundreds of person-years sophisticated model atmosphere codes for massive stars have been developed including the hydrodynamics of stellar winds and the accurate NLTE opacities of millions of spectral lines. Detailed tests have been carried out reproducing the observed spectra of Milky Way stars from the UV to the IR and constraining stellar parameters with unprecedented accuracy; see reviews by Kudritzki (1998), Kudritzki & Puls (2000), and Kudritzki & Urbaneja (2007). For instance, the most recent work on A-type supergiants by Przybilla et al. (2006, 2008), Schiller & Przybilla (2008), and also by Przybilla et al. (2000), Przybilla et al. (2001a), Przybilla et al. (2001b), Przybilla & Butler (2001), and Przybilla (2002) demonstrates that with high resolution and very high signal-to-noise spectra stellar parameters and chemical abundances can be determined with hitherto unknown precision (T_{eff} to $\leq 2\%$, $\log g$ to ~ 0.05 dex, individual metal abundances to ~ 0.05 dex).

At the same time, utilizing the power of the new 8 m to 10 m class telescopes, high resolution studies of A supergiants in many Local Group galaxies were carried out by Venn (1999) (SMC), McCarthy et al. (1995) (M33), McCarthy et al. (1997) (M31), Venn et al. (2000) (M31), Venn et al. (2001) (NGC 6822), Venn et al. (2003) (WLM), and Kaufer et al. (2004) (Sextans A) yielding invaluable information about the stellar chemical composition in these galaxies. In the research field of massive stars, these studies have so far provided the most accurate and most comprehensive information about chemical composition and have been used to constrain stellar evolution and the chemical evolution of their host galaxies.

4 The challenging step beyond the Local Group

The concept to go beyond the Local Group and to study A supergiants by means of quantitative spectroscopy in galaxies out to the Virgo cluster has been first presented by Kudritzki, Lennon & Puls (1995) and Kudritzki (1998). Following-up on this idea, Bresolin et al. (2001) and Bresolin et al. (2002) used the VLT and FORS at 5 Å resolution for a first investigation of blue supergiants in NGC 3621 (6.7 Mpc) and NGC 300 (1.9 Mpc). They were able to demonstrate that for these distances and at this resolution spectra of sufficient S/N can be obtained allowing for the quantitative determination of stellar parameters and metallicities. Kudritzki, Bresolin & Przybilla (2003) extended this work and showed that stellar gravities and temperatures determined from the spectral analysis can be used to determine distances to galaxies by using the correlation between absolute bolometric magnitude and flux weighted gravity $g_F = g/T_{\text{eff}}^4$ (FGLR). However, while these were encouraging steps towards the use of A supergiants as quantitative diagnostic tools of galaxies beyond the Local Group, the work presented in these papers had still a fundamental deficiency. At the low resolution of 5 Å it is not possible to use ionization equilibria for the determination of $T_{\rm eff}$ in the same way as in the high resolution work mentioned in the previous paragraph. Instead, spectral types were determined and a simple spectral type-temperature relationship as obtained for the Milky Way was used to determine effective temperatures and then gravities and metallicities. Since the spectral type-$T_{\rm eff}$ relationship must depend on metallicity (and also gravities), the method becomes inconsistent as soon as the metallicity is significantly different from solar (or the gravities are larger than for luminosity class Ia) and may lead to inaccurate stellar parameters. As shown by Evans & Howarth (2003), the uncertainties introduced in this way could be significant and would make it impossible to use the FGLR for distance determinations. In addition, the metallicities derived might be unreliable. This posed a serious problem for the the low resolution study of A supergiants in distant galaxies.

This problem was overcome only very recently by Kudritzki et al. (2008) (hereafter KUBGP), who provided the first self-consistent determination of stellar parameters and metallicities for A supergiants in galaxies beyond the Local Group based on the detailed quantitative model atmosphere analysis of low resolution spectra. They applied their new method on 24 supergiants of spectral type B8 to A5 in the Scultor Group spiral galaxy NGC 300 (at 1.9 Mpc distance) and obtained temperatures, gravities, metallicities, radii, luminosities and masses. The spectroscopic observations were obtained with FORS1 at the ESO VLT in multiobject spectroscopy mode. In addition, ESO/MPI 2.2 m WFI and HST/ACS photometry was used. The observations were carried out within the framework of the Araucaria Project (Gieren et al. 2005b). In the following we discuss the analysis method and the results of this pilot study.

5 A pilot study in NGC300: the analysis method

For the quantitative analysis of the spectra KUBGP use the same combination of line blanketed model atmospheres and very detailed NLTE line formation calculations as Przybilla et al. (2006) in their high signal-to-noise and high spectral resolution study of galactic A-supergiants, which reproduce the observed normalized spectra and the spectral energy distribution, including the Balmer jump, extremely well. They calcuate an extensive, comprehensive and dense grid of model atmospheres and NLTE line formation covering the potential full parameter range of all the objects in gravity ($\log g = 0.8$ to 2.5), effective temperature ($T_{\rm eff} = 8300$ to 15000K) and metallicity ($[Z] = \log Z/Z_\odot = -1.30$ to 0.3). The total grid comprises more than 6000 models.

The analysis of the each of the 24 targets in NGC 300 proceeds in three steps. First, the stellar parameters ($T_{\rm eff}$ and $\log g$) are determined together with interstellar reddening and extinction, then the metallicity is determined and finally, assuming a distance to NGC 300, stellar radii, luminosities and masses are obtained. For the first step, a well established method to obtain the stellar parameters of supergiants of late B to early A spectral type is to use ionization equilibria of weak metal lines (O I/II, Mg I/II, N I/II, etc.) for the determination of effective temperature $T_{\rm eff}$ and the Balmer lines for the gravities $\log g$. However, at the low resolution of 5 Å the weak spectral lines of the neutral species disappear in the noise of the spectra and an alternative

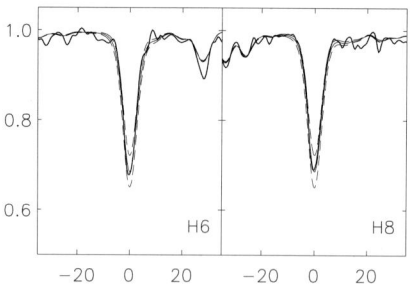

Figure 1: Model atmosphere fit of two observed Balmer lines of NGC300 target No. 21 of KUBGP for $T_{\rm eff} = 10000$ K and $\log g = 1.55$ (solid). Two additional models with same $T_{\rm eff}$ but $\log g = 1.45$ and 1.65, respectively, are also shown (dashed).

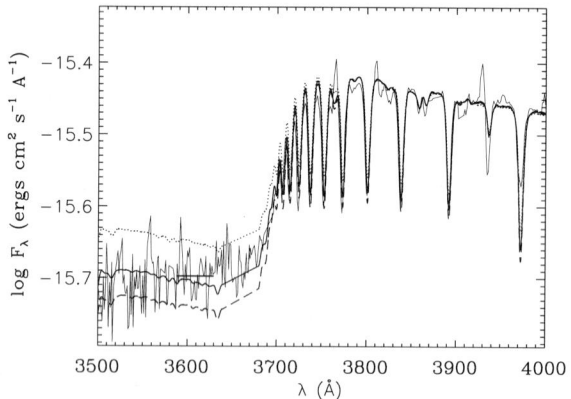

Figure 2: Model atmosphere fit of the observed Balmer jump of the same target as in Fig. 1 for $T_{\rm eff} = 10000$ K and $\log g = 1.55$ (solid). Two additional models with the same $\log g$ but $T_{\rm eff} = 9750$ K (dashed) and 10500 K (dotted) are also shown. The horizontal bar at 3600 Å represents the average of the flux logarithm over this wavelength interval, which is used to measure D_B.

technique is required to obtain temperature information. KUBGP confirm the result by Evans & Howarth (2003) that a simple application of a spectral type-effective temperature relationship does not work because of the degeneracy of such a relationship with metallicity. Fortunately, a way out of this dilemma is the use of the spectral energy distributions (SEDs) and here, in particular of the Balmer jump D_B. While the observed photometry from B-band to I-band is used to constrain the interstellar reddening, D_B turns out to be a reliable temperature diagnostic. A simultaneous fit of the Balmer lines and the Balmer jump allows to constrain effective temperature and gravity independent of assumptions on metallicity. Figures 1 and 2 demonstrate the sensitivity of the Balmer lines and the Balmer jump to gravity and effective temperature, respectively. The accuracy obtained by this method is $\leq 4\%$ for T_{eff} and ~ 0.05 dex for $\log g$.

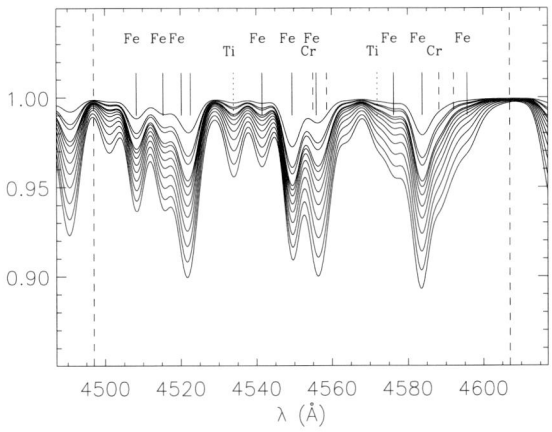

Figure 3: Synthetic metal line spectra calculated for the stellar parameters of target No. 21 as a function of metallicity in the spectral window from 4497 Å to 4607 Å. Metallicities range from $[Z] = -1.30$ to 0.30, as described in the text. The dashed vertical lines give the edges of the spectral window as used for a determination of metallicity.

Knowing the stellar atmospheric parameters T_{eff} and $\log g$ KUBGP are able to determine stellar metallicities by fitting the metal lines with their comprehensive grid of line formation calculations. The fit procedure proceeds again in several steps. First, spectral windows are defined, for which a good definition of the continuum is possible and which are relatively undisturbed by flaws in the spectrum (for instance caused by cosmic events) or interstellar emission and absorption. A typical spectral window used for all targets is the wavelength interval $4497\,\text{Å} \leq \lambda \leq 4607\,\text{Å}$. Figure 3 shows the synthetic spectrum calculated for the atmospheric parameters of the same target as analyzed in Figs. 1 and 2 for all the metallicities of the grid ranging from $-1.30 \leq [Z] \leq 0.30$. It is very obvious that the strengths of the metal line features are a strong function of metallicity.

In Fig. 4 the observed spectrum of the target in this spectral window is shown overplotted by the synthetic spectrum for each metallicity. Separate plots are used for each metallicity, because the optimal relative normalization of the observed and calculated spectra is obviously metallicity dependent. This problem is addressed by renormalizing the observed spectrum for each metallicity so that the synthetic spectrum always intersects the observations at the same value at the two edges of the spectral window (indicated by the dashed vertical lines). The next step is a pixel-by-pixel comparison of calculated and normalized observed fluxes for each metallicity and a calculation of a χ^2-value. The minimum $\chi([Z])$ as a function of $[Z]$ is then used to determine the metallicity. This is shown in Fig. 5. Application of the same

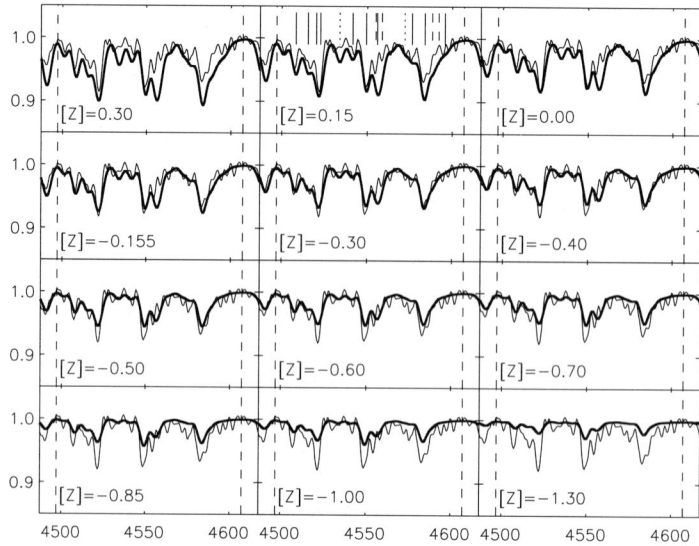

Figure 4: Observed spectrum of the same target as in Figs. 1 and 2 in the same spectral window as Fig. 3 but now the synthetic spectra for each metallicity overplotted separately.

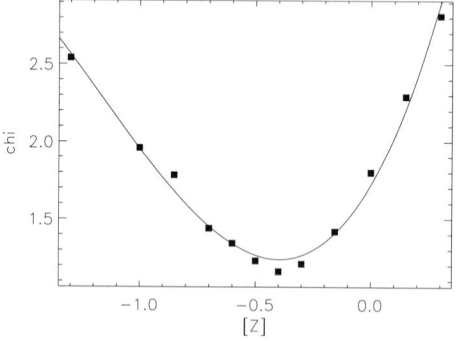

Figure 5: $\chi([Z])$ as obtained from the comparison of observed and calculated spectra. The solid curve is a third order polynomial fit.

method on different spectral windows provides additional independent information on metallicity and allows to determine the average metallicity obtained from all windows. A value of $[Z] = -0.39$ is found with a very small dispersion of only 0.02 dex. However, one also need to consider the effects of the stellar parameter uncertainties on the metallicity determination. This is done by applying the same correlation method for $[Z]$ for models at the extremes of the error box for $T_{\rm eff}$ and $\log g$. This increases the uncertainty of $[Z]$ to ± 0.15 dex, still a very reasonable accuracy of the abundance determination.

After the determination of T_{eff}, $\log g$, and $[Z]$, the model atmosphere SED is used to determine interstellar reddening $E(B-V)$ and extinction $A_V = 3.1 E(B-V)$. Simultaneously, the fit also yields the stellar angular diameter, which provides the stellar radius, if a distance is adopted. Gieren et al. (2005a) in their multi-wavelength study of a large sample of Cepheids in NGC 300 including the near-IR have determined a new distance modulus $m - M = 26.37$ mag, which corresponds to a distance of 1.88 Mpc. KUBGP have adopted these values to obtain the radii and absolute magnitudes.

6 A pilot study in NGC300: results

As a first result, the quantitative spectroscopic method yields interstellar reddening and extinction as a by-product of the analysis process. For objects embedded in the dusty disk of a star forming spiral galaxy one expects a wide range of interstellar reddening $E(B-V)$ and, indeed, a range from $E(B-V) = 0.07$ mag up to 0.24 mag was found (see Fig. 6). The individual reddening values are significantly larger than the value of 0.03 mag adopted in the HST distance scale key project study of Cepheids by Freedman et al. (2001) and demonstrate the need for a reliable reddening determination for stellar distance indicators, at least as long the study is restricted to optical wavelengths. The average over the observed sample is $\langle E(B-V) \rangle = 0.12$ mag in close agreement with the value of 0.1 mag found by Gieren et al. (2005a) in their optical to near-IR study of Cepheids in NGC 300. While Cepheids have somewhat lower masses than the A supergiants of our study and are consequently somewhat older, they nonetheless belong to the same population and are found at

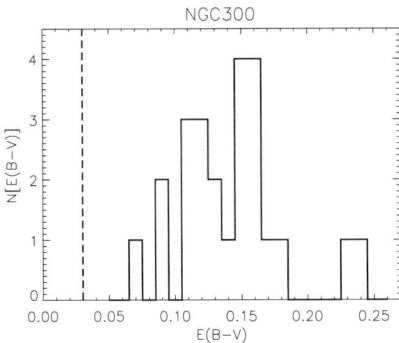

Figure 6: Histogram of the reddening distribution of blue supergiants in NGC 300 determined from spectral analysis and photometry. The dashed lines show the $E(B-V)$ value adopted by the HST Key Project. Data from Kudritzki et al. (2008).

similar sites. Thus, one expects them to be affected by interstellar reddening in the same way as A supergiants.

Note that a difference of 0.1 mag in reddening corresponds to 0.3 mag in the distance modulus. It is, thus, not surprising that Gieren et al. (2005a) found a significantly shorter distance modulus for NGC 300 than the Key Project. Another illustrative example of the importance of an accurate determination of interstellar extinction is given in Fig. 7, which compares the reddening distribution obtained from the quantitative spectral analysis of blue supergiants in M 33 with the value adopted by the Key Project. Again a wide distribution in reddening is found, but this time the values are significantly smaller than the Key Project value (see discussion in Sect. 9).

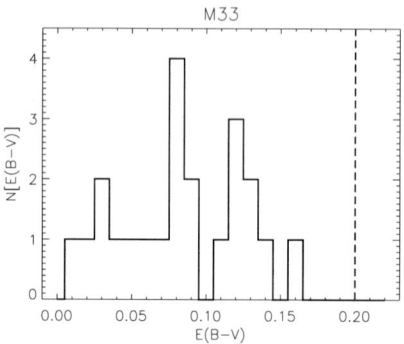

Figure 7: Histograms of the reddening distribution of blue supergiants in M33. The dashed lines show the $E(B-V)$ value adopted by the HST Key Project. Data from U et al. (2009).

Figures 8 and 9 show the location of all the observed targets in the ($\log g$, $\log T_{\text{eff}}$) plane and in the HRD. The diagrams include the early B-supergiants studied by Urbaneja et al. (2005b). The comparison with evolutionary tracks gives a first indication of the stellar masses in a range from 10 $M_\odot$ to 40 $M_\odot$. Three A supergiant targets have obviously higher masses than the rest of the sample and seem to be on a similar evolutionary track as the objects studied by Urbaneja et al. (2005b). The evolutionary information obtained from the two diagrams appears to be consistent. The B-supergiants seem to be more massive than most of the A supergiants. The same three A supergiants apparently more massive than the rest because of their lower gravities are also the most luminous objects. This confirms that quantitative spectroscopy is – at least qualitatively – capable to retrieve the information about absolute luminosities. Note that the fact that all the B supergiants studied by Urbaneja et al. (2005b) are more massive is simply a selection effect of the V magnitude limited spectroscopic survey by Bresolin et al. (2002). At similar V magnitude as the A supergiants those objects have higher bolometric corrections because of their higher effective temperatures and are, therefore, more luminous and massive.

Figure 10 shows the stellar metallicities and the metallicity gradient as a function of angular galactocentric distance, expressed in terms of the isophotal radius, ρ/ρ_0

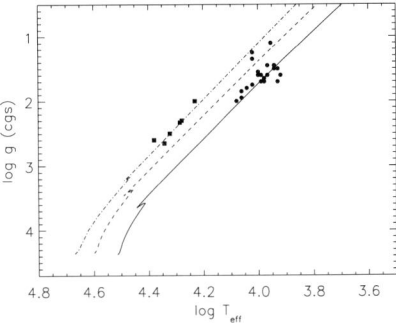

Figure 8: NGC 300 A supergiants (filled circles) and early B supergiants (filled squares) in the (log g, log $T_{\rm eff}$) plane compared with evolutionary tracks by Meynet & Maeder (2005) of stars with 15 M$_\odot$ (solid), 25 M$_\odot$ (dashed), and 40 M$_\odot$ (dashed-dotted), respectively.

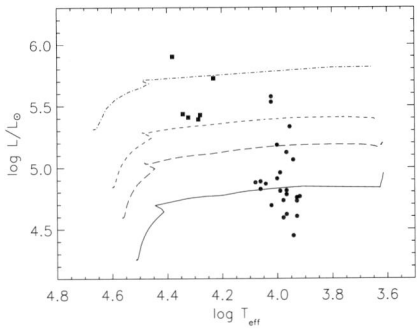

Figure 9: NGC 300 A and early B supergiants in the HRD compared with evolutionary tracks for stars with 15 M$_\odot$ (solid), 20 M$_\odot$ (long-dashed), 25 M$_\odot$ (short-dashed), and 40 M$_\odot$ (dashed-dotted), respectively.

(ρ_0 corresponds to 5.33 kpc). Despite the scatter caused by the metallicity uncertainties of the individual stars the metallicity gradient of the young disk population in NGC 300 is very clearly visible. A linear regression for the combined A- and B-supergiant sample yields (d in kpc, see also Bresolin et al. 2009)

$$[Z] = -0.07 \pm 0.05 - (0.081 \pm 0.011)\, d. \tag{1}$$

Note that the metallicities of the B supergiants refer to oxygen only with a value of $\log(N({\rm O})/N({\rm H})) = -3.35$ adopted for the Sun (Asplund et al. 2005) On the other hand, the A supergiant metallicities reflect the abundances of a variety of heavy elements such as Ti, Fe, Cr, Si, S, and Mg. KUBGP discuss the few outliers in Fig. 10 and claim that these metallicities seem to be real. Their argument is that the expec-

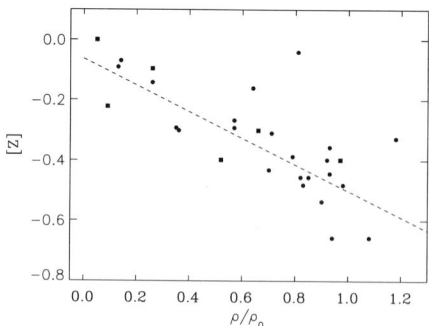

Figure 10: Metallicity $[Z]$ as a function of angular galacto-centric distance ρ/ρ_0 for the A supergiants (filled circles) and early B-supergiants (filled squares). Note that for the latter metallicity refers to oxygen only. The dashed curve represents the regression discussed in the text.

tation of homogeneous azimuthal metallicity in patchy star forming galaxies seems to be naive. Future work on other galaxies will show whether cases like this are common or not.

It is important to compare these results with metallicity measurements obtained from H II region emission lines. Beyond the Local Group and for metal rich galaxies H II region metallicities are mostly restricted to oxygen and measured through the so-called 'strong-line method', which uses the fluxes of the strongest forbidden lines of [O II] and [O III] relative to Hβ. Unfortunately, abundances obtained with the strong-line method depend heavily on the calibration used. Consequently, for the comparison of stellar with H II region metallicities KUBGP (extending the discussion started by Urbana et al. 2005b) used line fluxes published by Deharveng et al. (1988) and applied various different published strong line method calibrations to determine nebular oxygen abundances, which could then be used to obtain the similar regressions as above.

As a very disturbing result, the different strong line method calibrations lead to significant differences in the central metallicity as well as in the abundance gradient. The calibrations by Dopita & Evans (1986) and Zaritsky et al. (1994) predict a metallicity significantly supersolar in the center of NGC 300 contrary to the other calibrations. On the other hand, the work by KUBGP yields a central metallicity slightly smaller than solar in good agreement with Denicolo et al. (2002) and marginally agreeing with Kobulnicky et al. (1999), Pilyugin (2001), and Pettini & Pagel (2004). At the isophotal radius, 5.3 kpc away from the center of NGC 300, KUBGP obtain an average metallicity significantly smaller than solar $[Z] = -0.50$, close to the average metallicity in the SMC. The calibrations by Dopita & Evans (1986), Zaritsky et al. (1994), and Kobulnicky et al. (1999) do not reach these small values for oxygen in the H II regions either because their central metallicity values are too high or the metallicity are gradients too shallow.

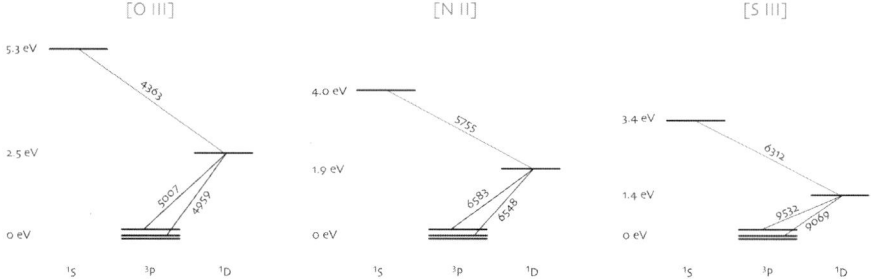

Figure 11: Energy level diagrams for the strong and auroral nebular forbidden lines. The auroral lines arise from the higher excited levels. (Bresolin, priv. comm.).

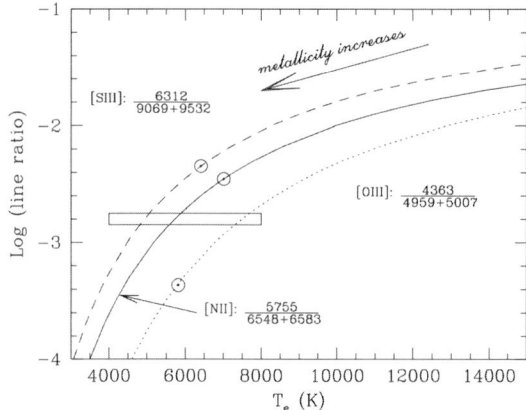

Figure 12: Diagnostic of H II regions. Ratio of auroral to strong line fluxes as a function of nebular electron temperature. The horizontal box indicates the observational limit for 10 m-class telescopes. (Bresolin, priv. comm.)

Thus, which of the metallicities determined for NGC 300 are correct? Those obtained from the blue supergiant study or the ones obtained from H II regions with one of the strong line calibrations? One possible way out of this problem is the use of the faint auroral emission lines (e.g. [O III] λ4363) of H II regions, instead of relying solely on strong lines, for nebular metallicity determinations (see Fig. 11). This requires a substantially larger observational effort at 10 m-class telescopes. Bresolin et al. (2009a) have recently used FORS at the VLT and studied 28 H II regions in NGC 300, for all of which they were able to detect the auroral lines and to use them to constrain nebular electron temperatures (see Fig. 12). This allowed for a much more accurate determination of nebular oxygen abundances, avoiding the calibration uncertainty intrinsic to the strong-line method. The result of this work compared to the blue supergiant metallicities is shown in Fig. 13. The agreement between the stars and the H II regions is excellent, confirming independently the quality of the supergiant work and ruling out many of the strong line calibrations.

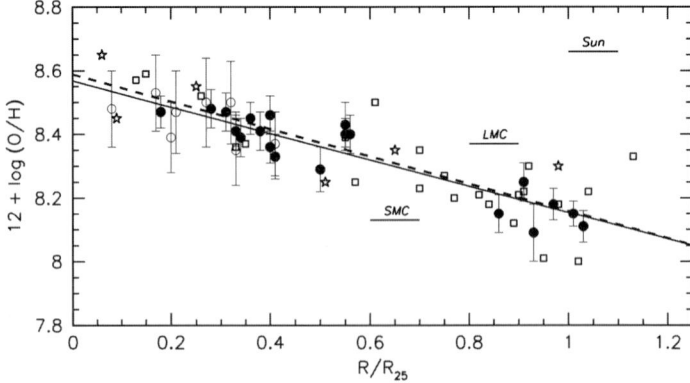

Figure 13: Radial oxygen abundance gradient obtained from H II regions (circles) and blue supergiants (star symbols: B supergiants; open squares: A supergiants). The regression to the H II region data is shown by the continuous line. The dashed line represents the regression to the BA supergiant star data. For reference, the oxygen abundances of the Magellanic Clouds (LMC, SMC) and the solar photosphere are marked. From Bresolin et al. (2009a).

Bresolin et al. (2009a) also perform a very educative experiment. They use their new and very accurate measurements of strong line fluxes and simply apply a set of different more recent strong line calibrations to obtain oxygen abundances without using the information from the auroral lines. The result of this experiment compared with the oxygen abundances using the auroral lines is shown in Fig. 14. The comparison is again shocking. The abundance offsets introduced by the application of inappropriate strong-line calibrations can be as large as 0.6 dex, putting the whole business of constraining galaxy evolution through the measurement of nebular metallicities and metallicity gradients into jeopardy.

Table 1: Central metallicity $[Z]$ and metallicity gradient (dex/kpc) in NGC 300.

Source	Central Abund.	Gradient	Comments
Dopita & Evans 1986	0.29±0.17	−0.118±0.019	H II, oxygen
Zaritsky et al. 1994	0.32±0.04	−0.101±0.017	H II, oxygen
Kobulnicky et al. 1999	0.09±0.04	−0.051±0.017	H II, oxygen
Denicolo et al. 2002	−0.05±0.05	−0.086±0.019	H II, oxygen
Pilyugin 2001	−0.14±0.06	−0.053±0.023	H II, oxygen
Pettini & Pagel 2004	−0.16±0.04	−0.068±0.015	H II, oxygen
KUBGP	−0.07±0.09	−0.081±0.011	Stars, metals

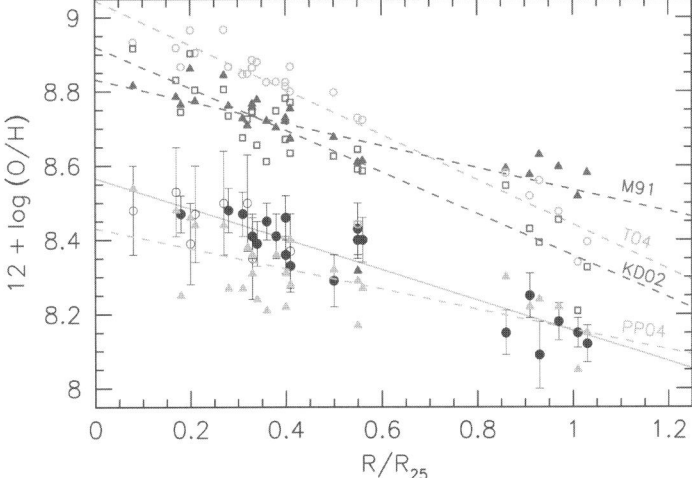

Figure 14: H II region galactocentric oxygen abundance gradients in NGC 300 obtained from our dataset but different strong line calibrations: McGaugh (1991) = M91, Tremonti et al. (2004) = T04, Kewley & Dopita (2002) = KD02, and Pettini & Pagel (2004) = PP04, as shown by the labels to the corresponding least squares fits. The auroral line-based abundances determined by Bresolin et al. (2009a) are shown by the full and open circle symbols, and the corresponding linear fit is shown by the continuous line.

7 The metallicities of galaxies

As is well known, the metallicity of the young stellar population of spiral galaxies has the potential to provide important constraints on galactic evolution and the chemical evolution history of the universe. For instance, the relationship between central metallicity and galactic mass appears to be a Rosetta stone to understand chemical evolution and galaxy formation; see Lequeux et al. (1979), Tremonti et al. (2004), and Maiolino et al. (2008). In addition, the observed metallicity gradients in spiral galaxies, apparently large for spirals of lower mass and shallow for high mass galaxies, are, according to Garnett et al. (1997), Skillman (1998), and Garnett (2004), the result of a complex interplay of star formation history, initial mass function and matter infall into the disks of spirals and allow in principle to trace the evolutionary history of spiral galaxies. However, as intriguing the observations of the mass-metallicity relationship and the metallicity gradients of galaxies are, the published results are highly uncertain in a quantitative sense, since they reflect the intrinsic uncertainty of the calibration of the strong line method applied. As a striking example, Kewley & Ellison (2008) have demonstrated that the quantitative shape of the mass-metallicity relationship of galaxies can change from very steep to almost flat depending on the calibration used (Fig. 15). In the same way and as demonstrated by Fig. 14 and Table 1, metallicity gradients of spiral galaxies can change from steep to flat simply as the result of the calibration used. Obviously, the much larger effort of either stellar spectroscopy of blue supergiants or the observation of faint nebular au-

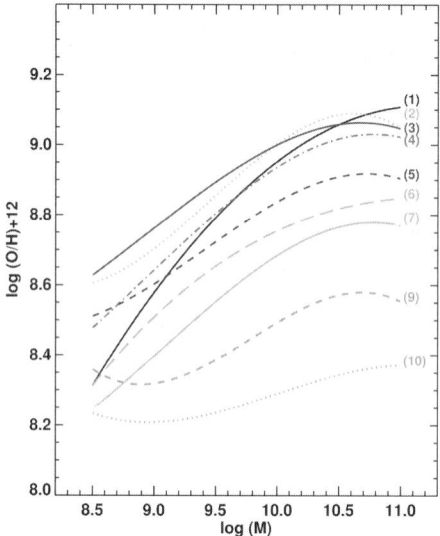

Figure 15: The mass-metallicity relationship of star forming galaxies in the nearby universe obtained by applying several widely used empirical metallicity calibrations based on different strong line ratios. This figure illustrates that there is an effect not only on the absolute scale, but also on the relative shape of this relationship. Adapted from Kewley & Ellison (2008).

roral lines will be needed in the future to observationally constrain the metallicities and metallicity gradients of spiral galaxies and their evolutionary history. We note that the use of blue supergiants provides the additional information about chemical abundances of iron-group elements as an important additional constraint of galaxy evolution history.

8 Flux weighted gravity: luminosity relationship (FGLR)

Massive stars with masses in the range from 12 M$_\odot$ to 40 M$_\odot$ evolve through the B and A supergiant stage at roughly constant luminosity (see Fig. 5). In addition, since the evolutionary timescale is very short when crossing through the B and A supergiant domain, the amount of mass lost in this stage is small. This means that the evolution proceeds at constant mass and constant luminosity. This has a very simple, but very important consequence for the relationship of gravity and effective temperature along each evolutionary track. From

$$L \propto R^2 T_{\text{eff}}^4 = \text{const.}, \quad M = \text{const.}, \tag{2}$$

follows immediately that

$$M \propto g\,R^2 \propto L\,(g/T_{\text{eff}}^4) = L\,g_{\text{F}} = \text{const.} \tag{3}$$

Thus, along the evolution through the B and A supergiant domain the *"flux-weighted gravity"* $g_F = g/T_{\text{eff}}^4$ should remain constant. This means each evolutionary track of different luminosity in this domain is characterized by a specific value of g_F. This value is determined by the relationship between stellar mass and luminosity, which in a first approximation is a power law

$$L \propto M^x \tag{4}$$

and leads to a relationship between luminosity and flux-weighted gravity

$$L^{1-x} \propto (g/T_{\text{eff}}^4)^x. \tag{5}$$

With the definition of bolometric magnitude $M_{\text{bol}} \propto -2.5 \log L$ one then derives

$$-M_{\text{bol}} = a_{\text{FGLR}} (\log g_F - 1.5) + b_{\text{FGLR}}. \tag{6}$$

This is the *"flux-weighted gravity-luminosity relationship"* (FGLR) of blue supergiants. Note that the proportionality constant a_{FGLR} is given by the exponent of the mass-luminosity power law through

$$a_{\text{FGLR}} = 2.5x/(1-x). \tag{7}$$

For instance, for $x = 3$, one obtains $a_{\text{FGLR}} = -3.75$. Note that the zero point of the relationship is chosen at a flux weighted gravity of 1.5, which is in the middle of the range encountered for blue supergiant stars.

KUBGP use the mass-luminosity relationships of different evolutionary tracks (with and without rotation, for Milky Way and SMC metallicity) to calculate the FGLRs predicted by stellar evolution. Very interestingly, while different evolutionary model types yield somewhat different FGLRs, the differences are rather small.

Kudritzki, Bresolin & Przybilla (2003) were the first to realize that the FGLR has a very interesting potential as a purely spectroscopic distance indicator, as it relates two spectroscopically well defined quantities, effective temperature and gravity, to the absolute magnitude. Compiling a large data set of spectroscopic high resolution studies of A supergiants in the Local Group and with an approximate analysis of low resolution data of a few targets in galaxies beyond the Local Group (see discussion in previous chapters) they were able to prove the existence of an observational FGLR rather similar to the theoretically predicted one.

With the improved analysis technique of low resolution spectra of A supergiants and with the much larger sample studied for NGC 300 KUBGP resumed the investigation of the FGLR. The result is shown in Fig. 16, which for NGC 300 reveals a clear and rather tight relationship of flux weighted gravity $\log g_F$ with bolometric magnitude M_{bol}. A simple linear regression yields $b_{\text{FGLR}} = 8.11$ for the zero point and $a_{\text{FGLR}} = -3.52$ for the slope. The standard deviation from this relationship is $\sigma = 0.34$ mag. Within the uncertainties the observed FGLR appears to be in agreement with the theory.

In their first investigation of the empirical FGLR Kudritzki, Bresolin & Przybilla (2003) have added A supergiants from six Local Group galaxies with stellar parameters obtained from quantitative studies of high resolution spectra (Milky Way, LMC,

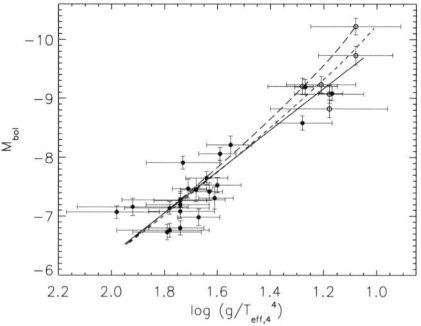

Figure 16: The FGLR of A (solid circles) and B (open circles) supergiants in NGC 300 and the linear regression (solid). The stellar evolution FGLRs for models with rotation are also overplotted (dashed: Milky Way metallicity, long-dashed: SMC metallicity).

SMC, M31, M33, NGC 6822) to their results for NGC 300 to obtain a larger sample. They also added 4 objects from the spiral galaxy NGC 3621 (at 6.7 Mpc) which were studied at low resolution. KUBGP added exactly the same data set to their new enlarged NGC 300 sample, however, with a few minor modifications. For the Milky Way they included the latest results from Przybilla et al. (2006) and Schiller & Przybilla (2008) and for the two objects in M31 we use the new stellar parameters obtained by Przybilla et al. (2008a). For the objects in NGC 3621 they applied new HST photometry. They also re-analyzed the LMC objects using ionization equilibria for the temperature determination.

Figure 17 shows bolometric magnitudes and flux-weighted gravities for the full sample of eight galaxies again revealing a tight relationship over one order of magnitude in flux-weighted gravity. The linear regression coefficients are $a_{\mathrm{FGLR}} = -3.41 \pm 0.16$ and $b_{\mathrm{FGLR}} = 8.02 \pm 0.04$, very similar to the NGC 300 sample alone. The standard deviation is $\sigma = 0.32$ mag. The stellar evolution FGLR for Milky Way metallicity provides a fit of almost similar quality with a standard deviation of $\sigma = 0.31$ mag.

9 First distances using the FGLR-method

With a relatively small residual scatter of $\sigma \sim 0.3$ mag the observed FGLR with the calibrated values of a_{FGLR} and b_{FGLR} is an excellent tool to determine accurate spectroscopic distance to galaxies. It requires multicolor photometry and low resolution (5 Å) spectroscopy to determine effective temperature and gravity and, thus, flux-weighed gravity directly from the spectrum. With effective temperature, gravity and metallicity determined one also knows the bolometric correction, which is particularly small for A supergiants. This means that errors in the stellar parameters do not largely affect the determination of bolometric magnitudes. Moreover, one knows the intrinsic stellar SED and, therefore, can determine interstellar reddening

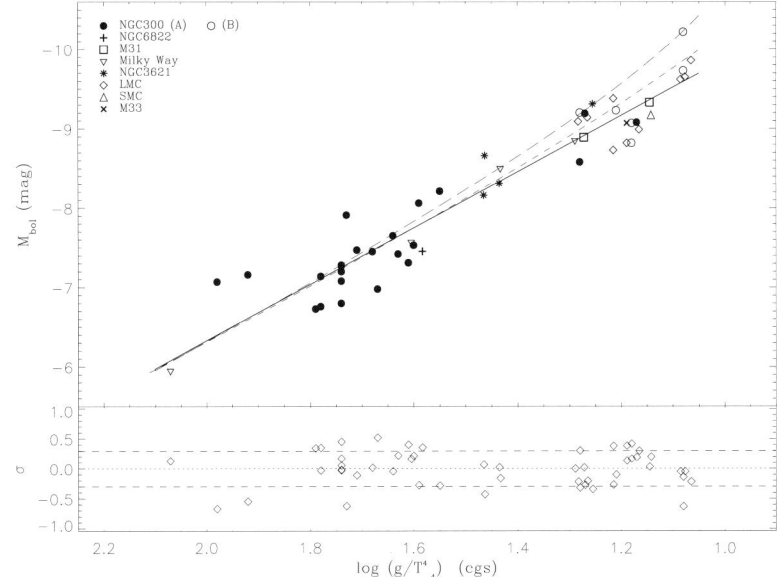

Figure 17: The FGLR of A (solid circles) and B (open circles) supergiants in 8 galaxies including NGC 300 and the linear regression (solid). The stellar evolution FGLRs for models with rotation are again overplotted.

and extinction from the multicolor photometry, which then allows for the accurate determination of the reddening-free apparent bolometric magnitude. The application of the FGLR then yields absolute magnitudes and, thus, the distance modulus.

The first distance determination of this type has been carried out by Urbaneja et al. (2008) who studied blue supergiants in WLM, one of faintest dwarf irregular galaxies in the Local Group. The quantitative spectral analysis of VLT FORS spectra yields an extremely low metallicity of the young stellar population in this galaxy with an average of -0.9 dex below the solar value. The interstellar extinction is again extremely patchy ranging from 0.03 to 0.30 mag in $E(B-V)$ (note that the foreground value given by Schlegel et al. (1998) is 0.037 mag). The individually de-reddened FGLR – in apparent bolometric magnitude – is shown in Fig. 18. Using the FGLR calibration by Kudritzki et al. (2008) and minimizing the residuals Urbaneja et al. (2008) determined a distance modulus of 24.99 ± 0.10 mag (995 ± 46 kpc). This value is in good agreement with the TRGB distance by Rizzi et al. (2007) and the K-band Cepheid distance by Gieren et al. (2008), albeit 0.07 mag larger.

Most recently, U et al. (2009) have analyzed blue supergiant spectra obtained with DEIMOS and ESI at the Keck telescopes to determine a distance to the triangulum galaxy M 33. The case of M 33 is particularly interesting, since many independent distance determinations have been carried out for this galaxy during the last decade using a variety of techniques, including Cepheids, RR Lyrae, TRGB, red clump stars, planetary nebulae, horizontal branch stars and long-period variables.

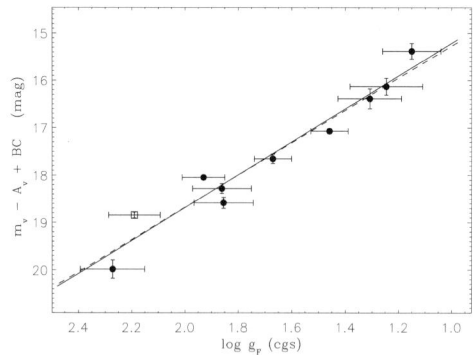

Figure 18: The FGLR of the Local Group dwarf irregular galaxy WLM, based on apparent bolometric magnitudes ($m_{\rm bol} = m_V - A_V + {\rm BC}$). The solid line corresponds to the FGLR calibration. The distance is determined by using this calibration through a minimization of the residuals. From Urbaneja et al. (2008).

The surprising result of all of these studies has been that the distance moduli obtained with these different methods differ by as much as 0.6 mag, which is more than 30% in the linear distance. At the same time, also the metallicity and the metallicity gradient of the young stellar population in the disk of M 33 are still heavily disputed. Different published results obtained from H II region work range from no gradient at all (Roslowsky & Simon 2008) to a very steep gradient of −0.11 dex/kpc (Garnett et al. 1997) and everywhere in between (Willner & Nelson-Patel 2002; Magrini et al. 2007) including a bimodal break with a very steep inner gradient (Vilchez et al. 1988; Magrini et al. 2007).

The blue supergiant spectroscopy by U et al. (2009) supports a moderate metallicity gradient of 0.07dex/kpc without any indication of a bimodal break (see Fig. 19). The FGLR-method (Fig. 20) yields a long distance modulus for M33 of 24.93 ± 0.11 mag, in basic agreement with a TRGB distance of 24.84 ± 0.10 mag obtained by the same authors from HST ACS imaging. The long distance modulus agrees also very well with the eclipsing binary distance obtained by Bonanos et al. (2006). U et al. (2009) relate the difference between their result and the published cepheid distances to the difference in accounting for interstellar reddening (see Fig. 7).

These successfull first applications of the FGLR-method for distance determinations indicate the very promising potential of the method to provide an independent constraint on the extragalactic distance scale.

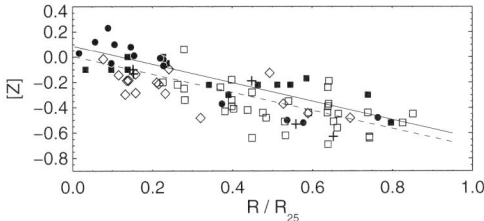

Figure 19: Metallicity of blue supergiants, H II regions and Cepheids as a function of dimensionless angular galactocentric distance in the triangulum galaxy M33. A supergiants (circles) and B supergiants (squares) are shown as solid symbols. Logarithmic oxygen abundances of H II regions in units of the solar value as published by Magrini et al. (2007) are plotted as open squares. Logarithmic neon abundances of H II regions normalized to the value for B stars in the solar neighbourhood and as obtained from Rubin et al. (2008) are shown as large open diamonds. The metallicity [Z] for beat Cepheids as determined by Beaulieu et al. (2006) are given as crosses. The solid line is the regression for the supergiants only, whereas the dashed lines is the regression for all objects.

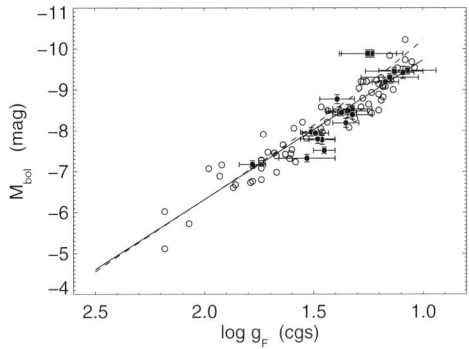

Figure 20: FGLR fits of the blue supergiants in M 33. Solid circles are late B and A supergiants and solid squares are early B supergiants in M 33. The solid line is a linear fit as described in the text. In addition to the M 33 targets, objects from nine other galaxies investigated in the studies by K08 and Urbaneja et al. (2008) are also shown. The solid line is the regression FGLR from K08. The dashed curve is the stellar evolution FGLR for Milky Way metallicity.

10 The potential of FGLR-method for extragalactic distance determinations

One of the most challenging chapters of modern astrophysics is the effort to establish the extragalactic distance scale with with sufficient accuracy. Over the past years, substantial improvement has been made by the HST Key Project on the Extragalactic

Distance Scale (Freeman et al. 2001) and by the SN Ia HST Calibration Program (Saha et al. 2001; Sandage et al. 2006), in which Cepheid-based distances to galaxies permitted the calibration of far-reaching secondary indicators. In addition, the 'Tip of the Red Giant Branch' (TRGB) method has become an additional very reliable and effective extragalactic distance indicator (Rizzi et al. 2007), which can also be used to calibrate secondary indicators (Mould & Sakai 2008, 2009ab). However, in spite of this progress, there are still a number of systematic uncertainties which affect both, the Cepheids and secondary methods of distance measurement, and which do not yet allow to obtain extragalactic distances, and thus the Hubble constant, with the high accuracy desired and needed by cosmologists, i.e. below the current 10% uncertainty.

For instance, the results on the Hubble constant H_0 obtained by Freedman et al. (2001) and Sandage et al. (2006) differ by 20% and the agreement with the TRGB-based calibration and the HST Key Project is at the margin of 10%. As is well known (see the discussion in Macri et al. 2006, Sect. 4.6), the determination of cosmological parameters from the cosmic microwave background is affected by degeneracies in parameter space and cannot provide strong constraints on the value of H_0 (Spergel 2006; Tegmark et al. 2004). Only, if assumptions are made, for instance that the universe is flat, H_0 can be predicted with high precision (i.e. 2%) from the observations of the cosmic microwave background, baryonic acoustic oscillations and type I high redshift supernovae. If these assumptions are relaxed, then much larger uncertainties are introduced (Spergel et al. 2007; Komatsu 2009). As an example, the uncertainty of the determination of the dark energy equation-of-state parameter $w = p/(c^2\rho)$ is related to the uncertainty of the Hubble constant through $\delta w/w \sim 2\,\delta H_0/H_0$. Thus, an independent determination of H_0 with an accuracy of 5% will allow to reduce the 1σ uncertainty of the cosmological equation of state parameter w to ± 0.1. A combination with other independent measurements constraining the cosmological parameters (large scale structure, SN Ia) will then allow for even tighter constraints. A very promising step forward in this regard has been made most recently by Riess et al. (2009a,b), but it is clear that the complexity of their approach requires additional and independent tests.

Among the remaining uncertainties affecting the extragalactic distance scale, probably the most important one is interstellar reddening. As young stars, Cepheids tend to be embedded in dusty regions which produce a significant *internal* extinction, in addition to the galactic foreground extinction. It seems likely that most of the Cepheid distances to galaxies determined from optical photometry *alone* are affected by sizeable systematic errors, due to a flawed determination of the appropriate reddening correction. The examples for NGC 300 and M33 give in the sections above clearly illustrate the problem. IR photometry (J, H, and K-band) of Cepheids is a promising way to address the issue, as has been shown in the Araucaria collaboration (Gieren et al. 2005b) or by Riess et al. (2009b). However, photometry at these wavelengths is still subject to important systematic effects, thus an entirely independent and complementary approach to address the issue of reddening is highly desirable.

The TRGB method is, in principle, not free from reddening errors, too. Usually, fields in the outskirts of the galaxies investigated are observed and Galactic foreground reddening values obtained from the interpolation of published maps are used to apply an extinction correction. While this seems to be a reasonable assumption, it has also been demonstrated that for a few galaxies (NGC 300, Vlajic et al. 2009; M83, Bresolin et al. 2009b) the stellar disks extend much further out than previously assumed and what the intrinsic reddening is in these very faint extended disks is completely unexplored. In the application of the TRGB method, colors are used to constrain metallicity and cannot, therefore, provide information about reddening.

An equally important uncertainty in the use of Cepheids as distance indicators is the dependence of the period-luminosity (P-L) relationship on metallicity. Work by Kennicutt et al. (1998) on M101 and by Macri et al. (2006) on the maser galaxy NGC 4258 using metallicity gradient information from H II region oxygen emission lines (Cepheids beyond the Local Group are too faint for a determination of their metallicity directly from spectra) indicates an *increase of Cepheid brightness with metallicity*. This agrees with Sakai et al. (2004), who related the difference between TRGB and Cepheid distances to the galactic H II region metallicities (not taking into account metallicity gradients, though) and derived a similar P-L dependence on metallicity.

However, these results are highly uncertain. Rizzi et al. (2007) have argued that many of the TRGB distances used by Sakai et al. (2004) need to be revised. With the Rizzi et al. (2007) TRGB distances the Sakai et al. (2004) dependence of the P-L relationship on metallicity disappears and the results are in much closer agreement with stellar pulsation theory (Fiorentino et al. 2002; Marconi et al. 2005; Bono et al. 2008), which predicts a small *decrease of Cepheid brightness with metallicity*. Moreover, all the H II region (oxygen) metallicities adopted when comparing Cepheid distances with TRGB distances or when using metallicity gradients are highly uncertain. They result from the application of the "strong-line method", using the calibration by Zaritsky et al. (1994). As we have shown above, this calibration gives metallicities and metallicity gradients that are not in agreement with results obtained from blue supergiants or from H II regions, when the more accurate method involving auroral lines is used. We point out, following the discussion in Macri et al. (2006), that even small changes in the P-L metallicity dependence can have an effect of several percent on the determination of H_0. We also note that the most recent work by Riess et al. (2009b) makes use of this calibration.

Also the TRGB method has a metallicity dependence. Usually, the metallicity of the old metal-poor population used for the method is obtained from the $V - I$ color, assuming that only foreground reddening is important. Then, a calibration of the TRGB magnitude as a function of metallicity is used. As shown in the careful work by Mager et al. (2008) on the TRGB distance to the maser galaxy NGC 4258, this metallicity correction introduces a systematic uncertainty of 0.12 mag in the distance modulus.

Very obviously, in order to improve the determination of extragalactic distances of star-forming spiral and irregular galaxies in the local universe an independent and complementary method is desirable, which can overcome the problems of interstellar extinction and variations of chemical composition. The FGLR-method presented in the two previous sections is such a method. The tremendous advantage of this technique is that individual reddening and extinction values, together with metallicity, can be determined for each supergiant target directly from spectroscopy combined with photometry. This reduces significantly the uncertainties affecting competing methods such as cepheids and the TRGB.

The FGLR technique is robust. Bresolin et al. (2004, 2006) have shown, from observations in NGC 300 and WLM, that the photometric variability of blue supergiants has negligible effect on the distances determined through the FGLR. Moreover, the study by Bresolin et al. (2005) also confirms that with HST photometry the FGLR method is not affected by crowding out to distances of at least 30 Mpc. This is the consequence of the enormous intrinsic brightness of these objects, which are 3 to 6 magnitudes brighter than Cepheids.

The current calibration of the FGLR rests on the 5 Å resolution stellar spectra obtained in NGC 300 and in seven additional galaxies (in some cases with higher resolution). Ideally, a large number of stars, observed in a single galaxy with a well-established distance, should be used. Obviously, a very natural step to improve the precision of the method by recalibrating in the LMC. The LMC currently defines the zero point for many classic photometric distance indicators, and its distance is presently well constrained, e.g. from eclipsing binaries ($m - M = 18.50 \pm 0.06$, Pietrzynski et al. 2009). This work is presently under way. In addition, the analysis of a large sample of SMC blue supergiants will provide additional information about the metallicity dependence of the method, but also about the geometrical depth of the SMC. New spectra in IC 1613 (10% solar metallicity) and M31 ($\sim$ solar metallicity) will help to constrain how the FLRG depends on metallicity, even though the results already obtained on WLM (10% solar metallicity, Urbaneja et al. 2008) does not indicate a strong effect.

An alternative and certainly more ambitious approach will be to by-pass the LMC as a distance scale anchor point and to use the maser galaxy NGC 4258 (Humphreys et al. 2008) as the ultimate calibrator. While challenging, such a step would be entirely feasible and provide an independent test of the approach taken by Riess et al. (2009a,b). Last but not least, the GAIA mission will provide a very accurate determination of the distances of blue supergiants in the Milky Way, which will allow for an accurate local calibration of the FGLR.

11 Perspectives of future work

It is evident that the type of work described in this paper can be in a straightforward way extended to the many spiral galaxies in the local volume at distances in the 4 to 12 Mpc range. Bresolin et al. (2001) have already studied A supergiants in NGC 3621 at a distance of 7 Mpc. Pushing the method we estimate that with present day 8 m to 10 m class telescopes and the existing very efficient multi-object spectrographs one can reach down with sufficient S/N to $V = 22.5$ mag in two nights of observing time under very good conditions. For objects brighter than $M_V = -8$ mag this means metallicities and distances can be determined out to distances of 12 Mpc ($m - M = 30.5$ mag). This opens up a substantial volume of the local universe for metallicity and galactic evolution studies and independent distance determinations complementary to the existing methods. With the next generation of extremely large telescopes such as the TMT, GMT or the E-ELT the limiting magnitude can be pushed to $V = 24.5$ equivalent to distances of 30 Mpc ($m - M = 32.5$ mag).

Acknowledgements

The work presented here is the result of an ongoing collaborative effort over many years. I wish to thank my colleagues in Hawaii Miguel Urbaneja, Fabio Bresolin and Vivian U for their tremendous dedication and skillful contributions to help me to make this project happen. It is more than just a science collaboration. It is a joyful endeavor with a lot of challenges but also a lot of fun. My colleagues Norbert Przybilla and Florian Schiller from Bamberg Observatory have made crucial contributions without which this project would not have been possible. I hope that the good time we had together in Hawaii has been an adequate compensation for their efforts. I also want to thank Wolfgang Gieren and his team from Universidad de Concepcion for inviting us to be part of the Araucaria collaboration. This has given a much wider perspective to our project. It has also created the new spirit of a "trans-Pacific" collaboration.

References

Allende Prieto, C., Lambert, D.L., Asplund, M.: 2001, ApJ 556, L63

Asplund, M., Grevesse, N., Sauval, A.J.: 2005, in: D.L. Lambert (ed.) *Cosmic Abundances as Records of Stellar Evolution and Nucleosynthesis in honor of David L. Lambert*, ASPC 336, p. 25

Beaulieu, J.P., Buchler, J.R., Marquette, J.B., Hartman, J.D., Schwarzenberg-Czerny, A.: 2006, ApJ653, L101

Bonanos, A.Z., Stanek, K.Z., Kudritzki, R.P., et al.: 2006, ApJ652, 313

Bono, G., Caputo, F., Fiorentino, G., Marconi, M., Musella, I.: 2008, ApJ684, 102

Bresolin, F., Kudritzki, R.-P., Méndez, R.H., Przybilla, N.: 2001, ApJ 548, L159

Bresolin, F., Gieren, W., Kudritzki, R.-P., Pietrzyński, G., Przybilla, N.: 2002, ApJ 567, 277

Bresolin, F., Pietrzynski, G., Gieren, W., Kudritzki, R.P., Przybilla, N., Fouque, P.: 2004, ApJ600, 182

Bresolin, F., Pietrzynski, G., Gieren, W., Kudritzki, R.P.: 2005, ApJ634, 1020

Bresolin, F., Gieren, W., Kudritzki, R.P., Pietrzynski, G., Urbaneja, M., Carraro, G.: 2009a, ApJ 700, 1141

Bresolin, F., Ryan-Weber, E., Kennicutt, R.C., Goddard, Q.: 2009b, ApJ695, 580

Deharveng, L., Caplan, J., Lequeux, J., Azzopardi, M., Breysacher, J., Tarenghi, M., Westerlund, B.: 1988, A&AS 73, 407

Evans, C.J.,, Howarth, I.D.: 2003, MNRAS 345, 1223

Denicolo, G., Terlevich, R., Terlevich, E.: 2002, MNRAS 330, 69

Dopita, M.A., Evans, I.N.: 1986, ApJ 307, 431

Fiorentino, G., Caputo, F., Marconi, M., Musella, I.: 2002, ApJ576, 402

Freedman et al. 2001, *ApJ*, 553, 47

Garnett, D.R.: 2004, in: C. Esteban et al. (eds.), *Cosmochemistry. The melting pot of the elements. XIII Canary Islands Winter School of Astrophysics*, p. 171

Garnett, D.R., Shields, G.A., Skillman, E.D., Sagan, S.P., Dufour, R.J.: 1997, ApJ489, 63

Gieren, W., Pietrzyński, G., Soszynski, I., Bresolin, F., Kudritzki, R.P., Miniti, D., Storm, J.: 2005a, ApJ 628, 695

Gieren, W., et al.: 2005b, ESO Messenger 121, 23

Gieren, W., Pietrzyński, G., Szewczyk, O., Soszynski, I., Bresolin, F., Kudritzki, Urbaneja, M.A., Storm, J., Miniti, D.: 2008, ApJ 683, 611

Humphreys, E.M.L., Reid, M.J., Greenhill, L.J., Moran. J.M., Argon, A.L.: 2008, ApJ672, 800

Kaufer, A., Venn, K.A., Tolstoy, E., Pinte, C., Kudritzki, R.P.: 2004, AJ 127, 2723

Kennicutt, R.C., Stetson, P.B., Saha, A., et al.: 1998, ApJ498, 181

Kewley, L.J., Ellison, S.L.: 2008, ApJ681, 1183

Kewley, L.J., Dopita, M.A.: 2002, ApJS 142, 35

Kobulnicky, H.A., Kennicutt, R.C., Pizagno, J.L.: 1999, ApJ 514, 544

Komatsu, E., et al.: 2009, ApJS 180, 330

Kudritzki, R.P.: 1998, in: A. Aparicio, A. Herrero, F. Sanchez (eds.), *Stellar Physics for the Local Group*, p. 149

Kudritzki, R.P., Puls, J.: 2000, ARA&A 38, 613

Kudritzki, R.P., Urbaneja, M.: 2007, in: M. Livio, E. Villaver (eds.), *Massive Stars: From Pop III and GRB to the Milky Way*, STSci Symp. 20, p. 126

Kudritzki, R.-P., Lennon, D.J., Puls, J.: 1995, in: J.P. Welsh, I.J. Danziger (eds.), *Science with the VLT*, p. 246

Kudritzki, R.-P., Bresolin, F., Przybilla, N.: 2003, ApJ 582, L83

Kudritzki, R.-P., Urbaneja, M.A., Bresolin, F., Przybilla, N., Gieren, W., Pietrzynski, G.: 2008, ApJ 681, 269

Lequeux, J., Peimbert, M., Rayo, J.F., Serrano, A., Torres-Peimbert, S.: 1979, A&A 80, 155

Macri, L.M., Stanek, K.Z., Bersier, D., Greenhill, L.J., Reid, M.J.: 2006, ApJ652, 1133

Mager, V.A., Madore, B.F., Freedman, W.L.: 2008, ApJ689, 721

Magrini, L., Vilchez, J.M., Mampaso, A., Corradi, R.L.M., Leisy, P.: 2007, A&A 470, 865

Maiolino, R., et al.: 2008, A&A 488, 463

Marconi, M., Musella, I., Fiorentino, G.: 2005, ApJ632, 590

McCarthy, J.K., Lennon, D.J., Venn, K.A., Kudritzki, R.P., Puls, J., Najarro, F.: 1995, ApJ 455, L135

McCarthy, J.K., Kudritzki, R.P., Lennon, D.J., Venn, K.A., Puls, J.: 1997, ApJ 482, 757

McGaugh, S.: 1991, ApJ 380, 140

Meynet, G., Maeder, A.: 2005, A&A 429, 581

Mould, J., Sakai, S.: 2008, ApJ 686, L75

Mould, J., Sakai, S.: 2009a, ApJ 694, 1331

Mould, J., Sakai, S.: 2009b, ApJ 697, 996

Pettini, M., Pagel, B.E.J.: 2004, MNRAS 348, L59

Pietrzynski, G., Thompson, I.B., Graczyk, D., et al.: 2009, ApJ 689, 862

Pilyugin, L.S.: 2001, A&A 369, 594

Przybilla, N.: 2002, PhD Thesis, Ludwig-Maximilian University, Munich

Przybilla, N., Butler, K.: 2001, A&A 379, 955

Przybilla, N., Butler, K., Becker, S.R., Kudritzki, R.P., Venn, K.A.: 2000, A&A 359, 1085

Przybilla, N., Butler, K., Becker, S.R., Kudritzki, R.P.: 2001a, A&A 369, 1009

Przybilla, N., Butler, K., Kudritzki, R.P.: 2001b, A&A 379, 936

Przybilla, N., Butler, K., Becker, S.R., Kudritzki, R.P.: 2006, A&A 445, 1099

Przybilla, N., Butler, K., Kudritzki, R.P.: 2008a, in: G. Israelian, G. Meynet (eds.), *The Metal-Rich Universe*, p. 332

Przybilla, N., Nieva, M., Butler, K.: 2008b, ApJ 688, L103

Riess, A.G., Macri, L., Li, W., et al.: 2009a, ApJS 183, 109

Riess, A.G., Macri, L., Casertano, S., et al.: 2009b, ApJ 699, 539

Rizzi, L., Tully, R.B., Makarov, D., Makarova, L., Dolphin, A.E., Sakai, S., Shaya, E.J.: 2007, ApJ 661, 815

Rosolowsky, E., Simon, J.D.: 2008, ApJ 675, 1213

Rubin, R.H., Simpson, J.P., Colgan, S.W.J., et al.: 2008, MNRAS 387, 45

Saha, A., Sandage, A., Tammann, G.A., Dolphin, A.E., Christensen, J., Panagia, N., Macchetto, F.D.: 2001, ApJ 562, 314

Sakai, S., Ferrarese, L., Kennicutt, R.C., Jr., Saha, A.: 2004, ApJ 608, 42

Sandage, A., Tammann, G.A., Saha, A., Reindl, B., Macchetto, F.D., Panagia, N.: 2006, ApJ 653, 843

Schiller, F., Przybilla, N.: 2008, A&A 479, 849

Schlegel, D.J., Finkbeiner, D.P., Davis, M.: 1998, ApJ 500, 525

Skillman, E.D.: 1998, in: A. Aparicio, A. Herrero, F. Sanchez (eds.), *Stellar Astrophysics for the Local Group: VIII Canary Islands Winter School of Astrophysics*, p. 457

Spergel, D.: 2006, APS April Meeting, abstract C5.002

Spergel, D., et al.: 2007, ApJS 170, 377

Tegmark, M., et al.: 2004, Phys. Rev. D 69, 103501

Tremonti, C.A., Heckman, T.M., Kauffmann, G., et al.: 2004, ApJ 613, 898

U, V., Urbaneja, M.A., Kudritzki, R.-P., Jacobs, B.A., Bresolin, F., Przybilla, N.: 2009, ApJ 704, 1120

Urbaneja, M.A., Herrero, A., Kudritzki, R.-P., et al.: 2005a, ApJ 635, 311

Urbaneja, M.A., Herrero, A.J., Bresolin, F., et al.: 2005b, ApJ 622, 877

Urbaneja, M.A., Kudritzki, R.P., Bresolin, F., Przybilla, N., Gieren, W., Pietrzynski, G.: 2008, ApJ 684, 118

Venn, K.A.: 1999, ApJ 518, 405

Venn, K.A., McCarthy, J.K., Lennon, D.J., Przybilla, N., Kudritzki, R.P., Lemke, M.: 2000, ApJ 541, 610

Venn, K.A., Lennon, D.J., Kaufer, A., et al.: 2001, ApJ 547, 765

Venn, K.A., Tolstoy, E., Kaufer, A., et al.: 2003, AJ 126, 1326

Vilchez, J.M., Pagel, B.E., Diaz, A.I., Terlevich, E., Edmunds, M.G.: 1988, MNRAS 235, 633

Vlajic, M., Bland-Hawthorn, J., Freeman, K.C.: 2009 ApJ 697, 361

Willner, S.P., Nelson-Patel, K.: 2002, ApJ 568, 679

Zaritski, D., Kennicutt, R.C., Huchra, J.P.: 1994, ApJ 420, 87

Pulsations and planets:
The asteroseismology-extrasolar-planet connection

Sonja Schuh

TEA Visiting Professor, Eberhard-Karls-Universität Tübingen
Kepler Center for Astro and Particle Physics
Institut für Astronomie und Astrophysik
Sand 1, 72076 Tübingen, Germany

Georg-August-Universität Göttingen, Institut für Astrophysik
Friedrich-Hund-Platz 1, 37077 Göttingen, Germany
schuh@astro.physik.uni-goettingen.de

Abstract

The disciplines of asteroseismology and extrasolar planet science overlap methodically in the branch of high-precision photometric time series observations. Light curves are, amongst others, useful to measure intrinsic stellar variability due to oscillations, as well as to discover and characterize those extrasolar planets that transit in front of their host stars, periodically causing shallow dips in the observed brightness. Both fields ultimately derive fundamental parameters of stellar and planetary objects, allowing to study for example the physics of various classes of pulsating stars, or the variety of planetary systems, in the overall context of stellar and planetary system formation and evolution. Both methods typically also require extensive spectroscopic follow-up to fully explore the dynamic characteristics of the processes under investigation. In particularly interesting cases, a combination of observed pulsations and signatures of a planet allows to characterize a system's components to a very high degree of completeness by combining complementary information. The planning of the relevant space missions has consequently converged with respect to science cases, where at the outset there was primarily a coincidence in instrumentation and techniques. Whether space- or ground-based, a specific type of stellar pulsations can themselves be used in an innovative way to search for extrasolar planets. Results from this additional method at the interface of stellar pulsation studies and exoplanet hunts in a beyond-mainstream area are presented.

1 Introduction

The (transiting) extrasolar planet fields and the asteroseismology field see a convergence of instrumentation that culminates in the insight that beyond this purely technical level, a much more fundamental connection exists in the shared desire for

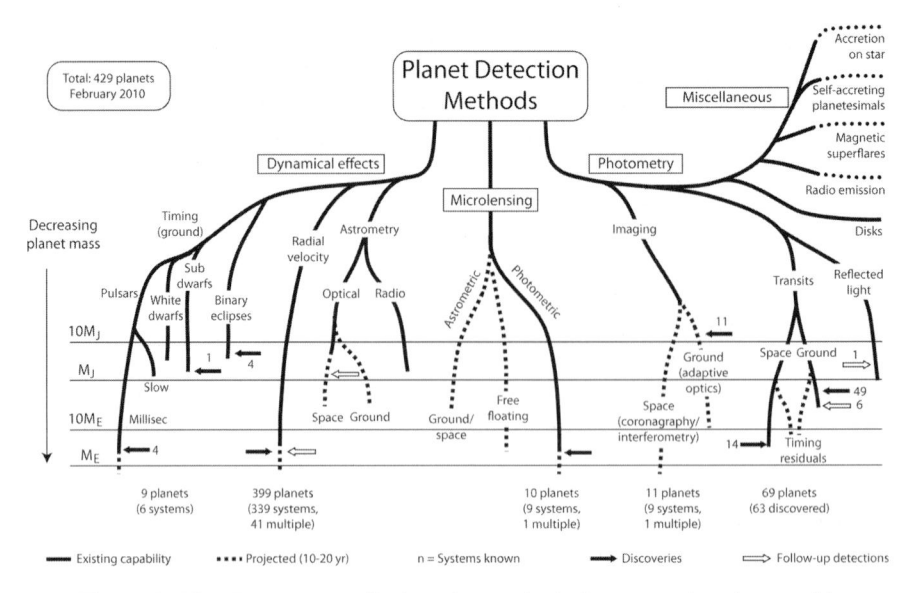

Figure 1: The "Perryman tree": detection methods for extra-solar planets with detectable masses on a (logarithmic) mass scale. Adopted from Fig. 1 in Perryman (2000), updated to include recent detections up to February 2010 (courtesy of M. A. C. Perryman).

the most exhaustive characterization of stellar and planetary systems at all possible with the available diagnostics. I will first sum up the relevant current context with an emphasis on planets around evolved stars, and then specifically address the topic of oscillation timing as a means to detect planets.

In terms of successes to detect planets around evolved stars, the subdwarf B stars as host stars stand out as a group. This class of evolved objects will be introduced, and the difficulties in explaining their sheer existence mentioned. Incidentally, the asteroseismology of subdwarf B stars is also a very active field. The EXOTIME planet searching program will be presented, which takes advantage of the long-term behaviour of these pulsations. Current ideas on subdwarf B evolution, and the potentially crucial role of planets, will be discussed, frequently resorting to the V391 Pegasi system.

2 Extrasolar planet detection methods

2.1 Overview

The first extrasolar planet candidate was observed in 1989, without however being claimed as such at the time. Latham et al. (1989) instead suggested that the substellar companion they had found around HD 114762 probably was a brown dwarf: a class of intensely searched-for objects, yet mostly elusive, at that period, and found to be intrinsically rare in the role of companions to normal stars today.

The first detection of an extrasolar planet around a solar-type star properly published as a "Jupiter-mass companion" was by Mayor & Queloz (1995). This companion to 51 Pegasi was immediately confirmed by Marcy & Butler (1995). 51 Pegasi b constitutes the prototype of the hot Jupiters, but it can also be more generally regarded as the prototype for all extrasolar planets discovered with the radial velocity method. While the radial velocity method still is to be credited with the top score in terms of number of planets (and multiple planetary systems) discovered, other discovery methods gain their importance from the fact that the observational biases involved can be significantly different.

Direct imaging, for instance, is obviously biased towards large semi-major axes (e.g. Kalas et al. 2008; Marois et al. 2008; and probably Lagrange et al. 2009), whereas radial velocity and transit measurements are biased towards the detection of planets on orbits with small semi-major axes (and large masses resp. radii). This correspondence between the radial velocity and transit methods is in a sense a good thing, since transit detections always need to be confirmed by radial velocity measurements in order to secure a planet discovery, while planet candidates from radial velocities detections usually remain candidates as long as the inclination cannot be constrained. For ground-based surveys, the micro-lensing technique is most sensitive in the vicinity of the Einstein radius at 2–3 AU (Bennett et al. 2009; actual detections exist for semi-major axes in the range of 0.6–5.1 AU).

The timing method, or more precisely, the various timing methods, also constitute indirect methods. While exploiting the same stellar "wobbling" effect induced by an unseen companion that is put to use in the radial velocity method, it measures the varying light travel time from the star(s) to the observer in the course of a "wobbling" cycle. Its amplitude (a direct measure of the projected semi-major axis of the orbit followed by the central object) increases with large companion masses, but also with large separations. Despite this increasing sensitivity towards wider orbits, the fact that the observational time base required for a detection increases for longer orbital periods also must be factored into the overall detection probability. Exactly as in the case of detections from radial velocity variations, candidates discovered with any of the timing methods suffer from a systematic uncertainty in the mass determination whenever the orbital inclination remains unknown.

A more detailed overview of the available planet detection methods with an emphasis on their respective sensitivity to masses has been given by Perryman (2000); an updated version with planet detection counts up to early 2010 is given in Fig. 1.

2.2 Planetary systems around evolved stars

The majority of extrasolar planets known today were found around solar-like stars. On the one hand, this has practical reasons – both the number of lines in the optical and their sharpness decreases towards earlier spectral types, while later spectral types tend to show augmented spectral variability due to activity, making small periodic radial velocity signals harder to detect. On the other hand, any dedicated searches for solar system analogues and Earth-like planets will obviously primarily target solar-like stars. In the quest to understand how our own solar system including the Earth has formed, it seems plausible to assume that both the knowledge of the frequency of similar systems as well as an overview of just how differently planet formation has proceeded elsewhere are important ingredients.

For stars evolving off the main sequence, the radial velocity method remains applicable in the red giant regime. It has turned up a total of 27 detections around G and K giants that add to the diversity of systems known. Among the initial discoveries were those by Frink et al. (2002), Hatzes et al. (2005), Hatzes et al. (2006), and Döllinger et al. (2007). For all stars evolved beyond the first red giant branch, however, the method that has most successfully been applied to detect extrasolar planets so far is the timing method.

2.3 Timing methods

The timing method actually comes in a variety of flavours. In all cases, a mechanism intrinsic to the central object provides a stable clock the time signal of which reaches the observer with a delay or in advance to the mean arrival times when a further body causes a cyclic displacement (wobble) of the "clock". As a side note, the associated change in period of the clock due to the Doppler effect is typically much smaller than the light travel time delays.

The most prominent example for a central object's clock are the pulses from rapidly rotating neutron stars. Pulsar planets were discovered in this way to orbit PSR 1257+12 by Wolszczan & Frail (1992), and confirmed by Wolszczan (1994). A further detection was reported for PSR B1620-26 by Backer (1993), Thorsett et al. (1993), and Backer et al. (1993). This system in the globular cluster M4 is known for its assortment of components, with a white dwarf in a tight orbit around the pulsar (Thorsett et al. 1999; imaged using *HST* by Richer et al. 2003), so the planet may be referred to as a circumbinary planet.[1]

In a further approach, the timing method specifically targets circumbinary planets by design. The clock in this case are frequent, sharp eclipses in a close central binary system. Eclipse timing has uncovered sub-stellar companions to subdwarf B stars (sdB, see section 4) in HW Vir-like binaries, in (pre-)cataclysmic variables and possibly also in W UMa systems. Two planetary companions have been accepted as confirmed around the prototype system HW Vir (Lee et al., 2009), while the tertiary component to the binary HS 0705+6700 (Qian et al., 2009d) probably lies in the mass range for brown dwarfs. Further HW Vir systems with preliminary detections by Qian et al. (2010c) include HS 2231+2441 and NSVS 14256825. For the (pre-)cataclysmic variable central systems, tertiary detections have been reported for the polar DP Leo (Qian et al., 2010a), for the DA+dme binary QS Vir (Qian et al., 2010b), and perhaps for the nova NN Ser (Qian et al. 2009a, status of the planet unconfirmed).[2]

---

[1] As a side note, a different example for a hierarchical system is the planet in the binary γ Cephei AB, where the planet orbits the primary K subgiant component (early speculations from radial velocity measurements by Campbell et al. 1988 were first confirmed by Hatzes et al. 2003, and refined through a direct detection of the secondary M dwarf component by Neuhäuser et al. 2007).

[2] Provisional reports on more detections are also out on the W UMa-type eclipsing binaries NY Lyr and DD Mon (Qian et al., 2009b,c).

A variant of this method is to look for timing residuals in known planetary transits in order to uncover additional planets in the system.

The third possibility is to resort to stellar oscillations as a clock. This lead to the first discovery of a planet around a subdwarf B star, V391 Pegasi, by Silvotti et al. (2007). The existence of this system suggested the possibility that a planet in an orbit similar to that of the Earth may have survived the red giant expansion of its presumably single host star, and has triggered follow-up searches for comparable systems. I will come back to this application in Sect. 3.3, and in more detail in Sect. 5.

3 Stellar oscillation – extrasolar planet links

An obvious benefit of high-precision photometric time series of planet-hosting stars (such as those obtained in ambitious transit searches) is that the host star can additionally be characterized in detail with asteroseismic methods if it oscillates.

3.1 Asteroseismology

For small perturbations to a spherical equilibrium solution, an infinite series of non-radial modes, with the radial modes included as a special case, can occur (which can be described by nodal planes in two angular directions, and spherical nodal surfaces in the radial direction), leading to multi-periodic frequency spectra. The equilibrium is restored by the actions of pressure and buoyancy, with one of the two usually dominating in a particular region of the star. The number of actually measurable frequencies depends on whether the corresponding modes are excited (i.e., if a driving mechanism is converting radiative energy or, temporarily, convective movement into pulsational kinetic energy), and on whether they are observable as photometric or radial velocity variations (i.e., if the mode geometry yields a detectable net effect in the integrated observables).

Given appropriate modelling capabilities, the density structure of a star can be inferred from observations of a sufficiently large number of excited eigenmodes that probe the interior conditions differentially. While the exact analysis approach can vary depending on the class of variables considered, important fundamental parameters that can be derived from this exercise are the stellar mass, radius and (depth-dependent) chemical composition.

In solar-like pulsators, oscillatory eigenmodes are excited stochastically by convection. The known pulsations are p modes (acoustic modes, with displaced material restored by the action of pressure), although g modes (gravity modes, action of buoyancy) are also thought to exist in the deep interior of the Sun. Considering that derivatives of the local equilibrium gravity and radius can be neglected in this case, approximations for high radial order can be found. For acoustic modes described within the asymptotic theory, one finds a regular frequency spacing for modes of the same low angular degree corresponding to subsequent radial orders. The large frequency separations, along with modifications introduced by considering the effect of different degrees leading to additional small frequency separations, together allow the definition of valuable diagnostic tools. This includes the famous échelle diagrams, as well as the "asteroseismic HR diagram" which allows to relate the two observables large and small frequency separation directly to the masses and ages of solar-like stars.

3.2 Connection with solar-like oscillations

The precise stellar parameters available through asteroseismic investigations contribute to answering a number of key questions in extrasolar planet research. One of the early noteworthy examples was the idea to investigate the scenarios for the origin of the enhanced metal content of μ Arae (HD 160691), host to a system of at least four planets, with asteroseismic methods. From ground-based radial velocity observations (Bouchy et al., 2005), Bazot et al. (2005) attempted to decide whether the overmetallicity was limited to the outer layers (accretion scenario) or was present throughout the star (resulting from enhanced metallicity in the original proto-stellar cloud). The latter scenario with enhanced planet formation rates in intrinsically metal-richer star-forming clouds has now been generally accepted, a result incorporated into the more recent analysis by Soriano & Vauclair (2009) that has in addition uncovered a high helium abundance. It will remain to be seen if contributions from asteroseismology will also be able to help solve the possibly related mystery of enhanced lithium depletion in planet-hosting stars (Israelian et al., 2009).

Fundamental links between the two programmes of the space mission Corot, the exoplanet search and the asteroseismology programme, have been pointed out by Vauclair et al. (2006); see also Soriano et al. (2007) and Vauclair (2008).

In Corot these two programs are conducted with separate pairs of detectors operated with different instrumental setups: an on-focus setup with dispersion through a bi-prism sampled every 512 s in the exoplanet field, and a highly out-of-focus setup sampled every 1 s in the seismology field.

While a smaller technical dichotomy continues to exist with the long (30 min) and short (1 min) cadence readout modes for the individual apertures on the Kepler satellite's detectors, the boundaries start to dissolve here since the sampling is at best loosely associated with the classification of a target as belonging to the planet-hunting core program or a program such as the asteroseismic investigation, and can simply be re-assigned. Kepler has demonstrated its capabilities early on in the mission schedule with observations of the known transiting exoplanet host HAT-P-7 (Borucki et al., 2009). HAT-P-7 has in the meantime been further characterized through an analysis of its simultaneously discovered solar-like oscillations (Christensen-Dalsgaard et al., 2010).

Kepler has now also delivered its first five genuine extrasolar planet discoveries (Borucki et al., 2010). The routine analysis of solar-like oscillations in newly-discovered planet-hosting stars, as well as the analysis of a large variety of asteroseismology targets of interest to the pulsation science community, have been institutionalized in the Kepler Asteroseismic Investigation (KAI) and the Kepler Asteroseismic Science Consortium (KASC, see Christensen-Dalsgaard et al. 2008, and in particular the first results in Gilliland et al. 2010 and Christensen-Dalsgaard et al. 2010). The importance of determining the stellar radius as accurately as possible stems from the circumstance that a transit light curve yields the radius *ratio* between star and planet. In order to derive the transiting planet's absolute radius the value of the stellar radius must be known. Together with the planet mass determined from the confirmation radial velocity curve (again, the stellar mass must be known), the planet's mean density is found.

In the proposed Plato mission (e.g. Catala & ESA Plato Science Study Team, 2009), the high-precision determination of planet host star radii and other fundamental stellar parameters from asteroseismology is an integral part of the planet hunting and characterization concept.

3.3 Connection with coherent oscillations

The above considerations could in principle with the same rationale be extended to planet-hosting stars that exhibit pulsations other than solar-like. Well beyond this, the striking capabilities in particular of Kepler evidently open many possibilities for genuine asteroseismological applications that target questions in many areas of stellar astrophysics. This includes the classical pulsators which, instead of being stochastically excited as the solar-like stars, exhibit unstable modes driven by the κ mechanism, with topical applications extensively described by Gilliland et al. (2010).

The standard asteroseismic exercise derives the instantaneous structure of a star, often relying on model structures from full evolutionary calculations in the process. As an extension, period changes in the oscillatory eigenfrequencies due to evolutionary effects can be considered as an additional constraint to find the best solution in parameter space. Given a series of evolutionary models already subjected to a stability analysis, the rate of change $\dot{P}$ of modes in a specific model can be determined without too much trouble. Due to the typically very long evolutionary time scales involved, measuring the secular evolution in real stars is observationally expensive and somewhat more complex.

An example of a class of objects where pulsators can be found that prove to be coherent and stable on time scales of many years are the ZZ Ceti variables on the DA white dwarf cooling track. The pulsations in these objects are due to the recombination of hydrogen in a narrow temperature range, leading, via the associated increase of the opacity in the outer layers, to the manifestation of low-degree low-order g modes. In contrast to the millisecond pulses in spun-up pulsars, and also in comparison to the pulse duration in normal "slow" pulsars, the pulsation periods in white dwarfs are much longer: of the order of a few minutes. The precision that can be reached in determining the clock rate is hence accordingly lower, and long-term changes are more readily analyzed using $O - C$ techniques, instead of determining $\dot{P}$ directly as the derivative of a series of quasi-instantaneous P measurements over time.

Measurements of $\dot{P}$ and its interpretation in the context of cooling times exist for a small number of suitable pulsating white dwarfs: PG 1159−035 (Costa et al. 1999; Costa & Kepler 2008), G117-B15A (Kepler et al. 1991; Kepler et al. 2000; Kepler et al. 2005), and a larger sample of a total of 15 objects investigated by Mullally et al. (2008).

It was quickly recognized (e.g. by Provencal, 1997) that the influence of a possible unseen companion can be measured as a side effect from the same data. Applying the ideas of the timing method (Sect. 2.3) has given the long-term photometric monitoring of pulsations in white dwarfs and related objects a new spin as a means to search for planets around evolved stars. With respect to previous work, the efforts by Mullally et al. (2008) show a shift of focus to that effect. Their proposed planet candidate around GD 66 has however remained unconfirmed.

The same measurements are possible for a different group of compact oscillators, the subdwarf B stars (see below), a quantitatively uncommon feeder channel for white dwarfs. The rapid pulsations in subdwarf B stars, of the order of minutes just as in the ZZ Ceti white dwarfs, are generated via a κ mechanism, providing potentially suitable conditions for reasonable long-term coherence. Yet, results have only been published for one pulsating subdwarf B star so far, V391 Pegasi (aka HS 2201+2610). As stated in Sect. 2.3, these measurements, simultaneously to $\dot{P}$, revealed the presence of a companion with a planetary mass. The possible reasons for this initially unexpected, instant success, sometimes jokingly referred to as "100% discovery rate", are worth a more in-depth investigation.

4 Hot subdwarf stars

4.1 Evolution

Subdwarf B stars (sdBs) are subluminous hot stars that are found in an effective temperature range from 20 000 K to 40 000 K at surface gravities between about 5.0 and 6.2 in $\log(g/\mathrm{cm\,s}^{-2})$, and that can in many cases be identified with evolved stellar models on the extreme horizontal branch (EHB). Their masses are expected to peak around the value for the He core flash at $0.46\,\mathrm{M}_\odot$. As is true for all horizontal branch stars, extreme horizontal branch stars have a He burning core but, due to previous significant mass loss, no H-shell burning in their thin hydrogen shells. The thinness of the shell leads to their blue appearance, so that the flux from hot subdwarfs (including the hotter sdOs) contributes significantly to the UV excess observed in galaxy bulges and elliptical galaxies. A great recent review of the observational properties of hot subdwarfs including binarity, kinematics, as well as current modelling capabilities to describe their atmospheres, interiors, and evolutionary history, can be found in Heber (2009).

The most puzzling question about hot subdwarfs remains what the precise evolutionary status of these core-helium burning or even more evolved objects really is. Here I only focus on how to possibly produce the sdB type. The basic problem in the standard single-star evolutionary scenario is to shed the hydrogen envelope almost entirely just before or at the moment of the He flash. This would require *ad hoc* strong mass loss through stellar winds for a certain fraction of stars upon reaching the tip of the first giant branch. Explaining the formation of subdwarf B stars in the context of binary evolution is therefore now generally favoured over single-star scenarios (Han et al., 2002, 2003).

Following the overview by Podsiadlowski et al. (2008), three genuine binary formation scenarios, and additionally the merger scenario, can be distinguished. All three of the binary scenarios involve Roche-Lobe overflow (RLOF), each at different stages and under different conditions.

In the "stable RLOF + CE" channel, the system initially goes through a first mass-transfer phase with stable RLOF that turns the evolving component into a He white dwarf. When the second component, the future subdwarf B star, then evolves to become a red giant, the second mass-transfer phase can happen dynamically. This unstable RLOF, where the matter transferred cannot all be accreted by the He white dwarf, leads to the formation of a common envelope (CE). After spiral-in and envelope ejection, the resulting system consists of the He white dwarf and the sdB – the core of the giant with its envelope removed – in a short-period binary (<10 days).

In the "CE only" channel, unstable RLOF occurs when the future subdwarf B star starts transferring matter to a lower-mass main-sequence companion near the tip of the first giant branch. The ensuing common-envelope phase again leads to a closer final configuration of the resulting system, which will consist of a low-mass main sequence star and the sdB in a short-period binary (<10 days).

Besides these two common-envelope channels (involving unstable RLOF), the "stable RLOF" channel can also produce sdBs. Stable Roche-Lobe overflow can occur when the future subdwarf B star starts transferring matter to a main sequence companion at mass ratio below ~ 1.2. As before, this happens near the tip of the red giant branch, but this time the system more likely widens due to the mass transfer. The resulting system consists of a main sequence or subgiant star and the sdB in a wide binary ($\gtrsim 10$ days).

The problem with all of the above binary scenarios is that (apparently?) single sdBs also exist. In addition to the single star scenario involving variable mass loss as mentioned above, a further possibility to produce these are mergers. In the "merger" scenario, two He white dwarfs in a close system, produced while undergoing one or two common envelope phases, spiral towards each other due to angular momentum loss via gravitational radiation until the less massive one gets disrupted and its matter accreted onto the more massive component. At a critical mass, the accretor can ignite helium fusion and the merger product would hence indeed turn into a single sdB.

In the confrontation with observations, both the types of companion found in close systems (white dwarf or low-mass main-sequence stars) as well as the close binary frequency are roughly as expected when the above merger channel is considered. Yet observationally, the mass spectrum for the companions is broader than expected from the standard formation scenarios for sdB+WD, sdB+dM and single sdBs. On the low-mass end, it can be argued to comprise the V391 Pegasi planet and, possibly, further secondary sub-stellar companions. Four companions with unusually high masses have also been reported; according to Geier et al. (2008, 2009b, 2010), the massive compact companions found from radial velocity variations must at least be heavy white dwarfs or in two cases even neutron stars or black holes.

While the problem of sdB formation has not been fully solved, the subsequent evolution of a subdwarf B star towards the white dwarf cooling sequence is more straightforward. One interesting characteristic is that this evolution, due to the thinness of the outer layers, bypasses the asymptotic giant branch, as was noted early on by Dorman et al. (1993). Another aspect is that the sdB+dM close systems constitute potential progenitors for cataclysmic variables. Furthermore, given the variety of configurations in which sdBs are found as one component, the diversity of possible progeny systems is not restricted to pre-CVs.

All of the above, while far from understood in the details, implies that the sdB's stellar core as it looks like after the first giant branch – or after a merger – is *almost* laid bare. It is hence very well accessible to asteroseismological methods.

4.2 Pulsating subdwarf B stars

Only a small fraction of the sdBs show pulsational variations, with non-pulsators also populating the region in the HRD where the pulsators are found. There are p- (pressure-) mode and g- (gravity-)mode types of pulsation.

The rapidly pulsating subdwarf B stars (sdBV$_r$) were discovered observationally by Kilkenny et al. (1997). The short periods of these p-mode pulsators are of the order of minutes and have amplitudes of a few tens mmag. These pulsations were independently predicted to exist by Charpinet et al. (1997), who in this initial and subsequent research papers explain the driving to result from a κ mechanism due to a Z-opacity bump accumulated by radiative levitation.

The longer periods in the slowly pulsating subdwarf B stars (sdBV$_s$) range from 30 to 80 min at even lower amplitudes of a few mmag. This group of g-mode pulsators was discovered by Green et al. (2003) and has been explained to pulsate due to the same κ mechanism as the sdBV$_r$ type by Fontaine et al. (2003).

A number of objects are known that show both types of modes simultaneously and are referred to as hybrid pulsators, or sdBV$_{rs}$ in the nomenclature of Kilkenny et al. (2010) that is used here. These hybrids lie in the overlapping region between the hotter short-period and the cooler long-period pulsators. The prototype and a prominent example is HS 0702+6043. Its rapid pulsations were discovered by Dreizler et al. (2002), and subsequently the slower ones by Schuh et al. (2005, 2006). Further examples can be found in Schuh (2008) and references therein. It should however be noted here that the planet host star V391 Pegasi (HS 2201+2610) is also among the hybrids. It has five p modes (Østensen et al. 2001b; Silvotti et al. 2002a) and at least one g mode (Lutz et al., 2008a, 2009a).

Exploiting the sdBV$_r$ asteroseismologically has the potential to test diffusion processes and has lead to a first mass distribution for subluminous B stars (Charpinet et al. 2008; Fontaine et al. 2008; Østensen 2009 and references therein.) On the other hand, the extent of the instability region for the sdBV$_s$, but in particular the existence of the sdBV$_{rs}$, have challenged details of the input physics for the models. The actual composition of the "Z" in the Z bump (iron, nickel) as well as the role of opacities have been discussed in this context (Jeffery & Saio, 2006, 2007). While, quite similarly to the findings for the sdB variables, using updated Opacity Project (OP) instead of OPAL opacities also improves the situation in β Cep and [SPB] variables, the opposite is the case in the sun with helioseismology. The most appropriate opacities are therefore still under discussion.

Further open questions in this field are the co-existence of pulsators and non-pulsators, and also the origin of the amplitude variability observed in a number of sdBVs (Kilkenny, 2010).

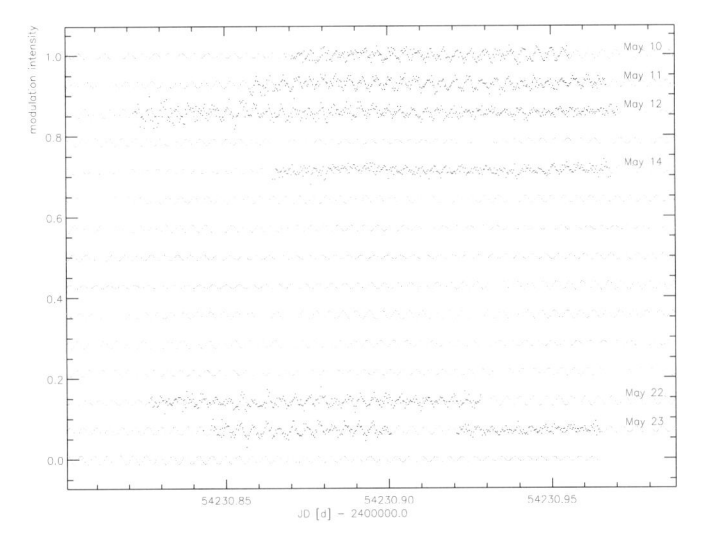

Figure 2: Photometric data obtained with the 1.2 m MONET/North telescope during a period of 2 weeks in May 2005, showing the pulsations in the light curve of V391 Pegasi. Starting from a mean intensity of unity on the first night, subsequent observations are shifted downward by a fixed offset each night. The actual data points (black dots) are overlaid with a model for the pulsations in red. One such run spanning several nights is required to derive one $O - C$ point by comparing the current phasing to that of a mean model.

4.3 Sub-stellar companions of subdwarf B stars

When a companion with planetary mass was found around the hybrid pulsating subdwarf B star V391 Pegasi with the timing method, this indicated that a previously undiscovered population of sub-stellar companions to apparently single subdwarf B stars might exist (Silvotti et al., 2002a, 2007).

As for low-mass tertiary bodies around HW Vir-like systems detected by eclipse timing, Lee et al. (2009) have reported two planetary companions around the system HW Vir itself. Qian et al. (2009d) put the mass of the tertiary body in the HS 0705+6700 system in the range for brown dwarfs. Still unconfirmed detections by Qian et al. (2010c) exist for HS 2231+2441 and NSVS 14256825.

A different type of detection has been put forward by Geier et al. (2009a, 2010), who report measurements of radial velocity variations in HD 149382 indicative of a planetary companion, variations which if real are indicative of a close sub-stellar companion in a 2.4 d orbit. Based on the above and this finding (that yet has to be confirmed), these authors also argue for a decisive influence of sub-stellar companions on the late stages of stellar evolution. The EXOTIME program aims to increase the empirical data available on which to base such discussions.

5 The EXOTIME program

5.1 The planet-hosting pulsating sdB V391 Pegasi

V391 Pegasi (HS 2201+2610) was first discovered to be a rapidly pulsating subdwarf B star by Østensen et al. (2001b). Additional slow pulsations were subsequently reported by Lutz et al. (2009a).

Silvotti et al. (2002a, 2007) were able to derive $\dot{P}$ values for the two strongest pulsation modes, and found an additional pattern in the observed–calculated ($O - C$) diagrams that revealed the presence of a giant planet in a 3.2 year orbit. The fact that this cyclic variation has been measured independently from two frequencies considerably strengthens the credibility of this discovery. Actually, Silvotti et al. (2007) detected parabolic and sinusoidal variations in the $O - C$ diagram constructed for the two main pulsation frequencies at 349.5 s and 354.2 s over the observing period of seven years. The sinusoidal component with its 3.2 year periodicity is attributed to the presence of the very low-mass companion V391 Pegasi b at $m \sin i = 3.2 \pm 0.7 \, M_{\rm Jup}$. The scenarios proposed for the origin of this planet are discussed in Sect. 6.2.

5.2 Further characterization of V391 Pegasi system

The determination of the true mass of the "asteroseismic planet" V391 Pegasi b requires to find a constraint on the orbital inclination. Besides the orbital inclination, the orbit eccentricity has also not been well determined so far. The constraints on possible further planets are weak and currently allow for a second planet in the system massive enough to be detected. Continued photometric monitoring (on-going, see for example Fig. 2, Silvotti et al., in prep.) will be able to: check if the $O - C$ evolves as predicted from the orbital solution, investigate the eccentricity, investigate a possible multiplicity, attempt another independent re-detection of the planet from $\dot{P}_3$, and search for rotational splitting in the pulsations (see below).

As a first step to derive the mass of the known companion object with follow-up observations, Schuh et al. (2009) have attempted to determine the orbital inclination from spectroscopy. The approach suggested to achieve this is to use the stellar inclination as a primer for the orbital orientation. "Stellar inclination" can refer to the rotational or the pulsational axis, which as a further necessary simplification are assumed to be aligned, and can in turn then be derived by combining measurements of $v_{\rm rot}$ and $v_{\rm rot} \sin i$.

The value for $v_{\rm rot}$ is in principle accessible through rotational splitting in the photometric frequency spectrum (which has however not been found for V391 Pegasi yet), while the projected rotational velocity $v_{\rm rot} \sin i$ can be measured from the rotational broadening of spectral lines. This rotational broadening must be deconvolved from the additional pulsational broadening caused by the surface radial velocity variation in high S/N phase averaged spectra.

Both phase averaged and phase resolved high resolution échelle spectra were obtained in May and September 2007 with the Hobby-Eberly Telescope (HET), and one phase averaged spectrum in May 2008 with the Keck 1 telescope, in order to

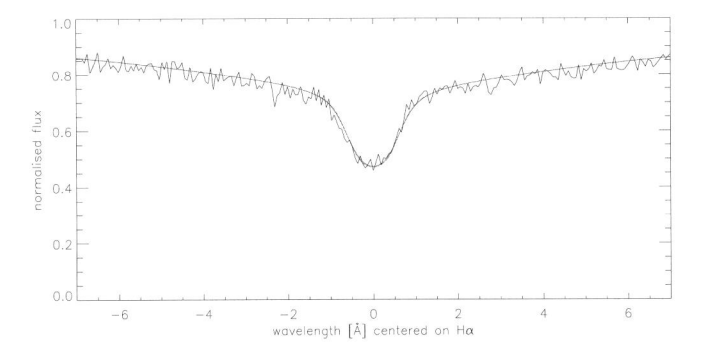

Figure 3: Detail of the observed Hα core in a pulsation-integrated Keck spectrum of V391 Pegasi. The data (shown as a black line) are plotted centered on 6562.798 Å. While measurements on the Hα core may be subject to systematic effects, this line is the only one available for such investigations in the spectral range covered by the observation. The flux was normalized; it does not reach unity at the edges of the plot as these regions are still well within the line wings. Overlaid are 1) a model spectrum in red calculated from a NLTE pure H atmosphere model at $T_{\text{eff}} = 30\,000$ K and $\log(g/\text{cm s}^{-2}) = 5.50$, and 2) a model spectrum in blue from the same atmosphere model as before but rotationally broadened with 0.2 Å corresponding to $9\,\text{km s}^{-1}$.

put limits on the pulsational radial velocities. Échelle spectra of V391 Pegasi were taken during May and September 2007 with the HRS ($R = 15\,000$) of the HET at the McDonald Observatory, and with HIRES ($R = 31\,000$) at the Keck 1 telescope atop Mauna Kea in May 2008.

Using standard data reduction procedures, the individual échelle orders were merged and the final spectra carefully normalized and finally summed. This results in a set of individual spectra (S/N ≈ 3), in particular two September 2007 high time resolution series, and summed spectra for May and September 2007 (Kruspe et al., 2008).

In an attempt to "clean" the relevant rotational broadening from pulsational effects, the spectra in September obtained in time resolved mode were combined to a series of ten phase resolved averaged spectra (S/N ≈ 9) for the main pulsation period of 349.5 s (similar to Tillich et al. 2007).

The cross-correlation of this series of averaged spectra with a pure hydrogen NLTE model spectrum at $T_{\text{eff}} = 30\,000$ K and $\log(g/\text{cm s}^{-2}) = 5.5$ as a template yields pulsational radial velocity measurements for the different Balmer lines. The maximum amplitude of a sinusoidal curve (fixed at the expected period) that could be accommodated in comparison to the weighted means of the Balmer lines reveals that any pulsational radial velocity amplitude is smaller than the accuracy of our measurements and confirms the upper limit of $16\,\text{km s}^{-1}$ given by Kruspe et al. (2008).

The resolution of the model template matches the spectral resolution of the (pulsation-averaged) Keck spectrum. A comparison of the Hα NLTE line core shape

Table 1: Overview of the EXOTIME targets, see Sect. 5 for details.

Target	Coordinates (Equinox 2000)		m_B	Status	
HS 2201+2610	22:04:12.0	+26:25:07	14.3	Collecting data	aka V391 Pegasi, planet candidate, $\sin i$ unknown
HS 0702+6043	07:07:09.8	+60:38:50	14.7	Collecting data,	see Schuh et al. 2010
HS 0444+0458	04:47:18.6	+05:03:35	15.2	Collecting data,	see Schuh et al. 2010
EC 09582−1137	10:00:41.8	−11:51:35	15.0	Collecting data,	
PG 1325+101	13:27:48.6	+09:54:52	13.8	Collecting data	

(see Fig. 3) yields a more stringent upper limit for the *combined* broadening effect of pulsation and rotation of at most 9 km s^{-1}, meaning better spectral resolution and signal to noise data will be necessary to measure $v_{\rm puls}$ and $v_{\rm rot} \sin i$.

5.3 Exoplanet Search with the Timing Method

Following the serendipitous discovery of V391 Pegasi b, the EXOTIME[3] monitoring program was set up to search for similar systems. EXOTIME monitors the pulsations of a number of selected rapidly pulsating subdwarf B stars on time-scales of several years with the immediate observational goals of a) determining $\dot{P}$ of the pulsational periods P and b) searching for signatures of sub-stellar companions in $O - C$ residuals due to periodic light travel time variations, which would be tracking the central star's companion-induced wobble around the centre of mass.

The long-term data sets should therefore on the one hand be useful to provide extra constraints for classical asteroseismological exercises from the $\dot{P}$ (comparison with "local" evolutionary models), and on the other hand allow to investigate the preceding evolution of apparently single sdB targets in terms of possible "binary" evolution by extending the otherwise unsuccessful search for companions to potentially very low masses.

As noted before, timing pulsations to search for companions samples a different range of orbital parameters, inaccessible through orbital photometric effects or the radial velocity method: the latter favours massive close-in companions, whereas the timing method becomes increasingly more sensitive towards wider separations. A further advantage of timing versus radial velocities is that the former, although observationally expensive, is easier to measure than the latter. In fact, it is very hard to achieve the required accuracy in radial velocity measurements from the few and broad lines in hot subdwarf stars (a notable exception is the publication of small RV variations in the bright sdB HD 149382 as reported by Geier et al. 2009a.)

The targets selected for monitoring in the EXOTIME program are listed in Table 1. The target selection criteria applied to compile this list over time have been described by Schuh et al. (2010).

V391 Pegasi (HS 2201+2610) appears as the first entry, as the monitoring of this system is on-going (see 5.1). The rapid pulsations in HS 2201+2610 were discovered

---
[3] http://www.na.astro.it/~silvotti/exotime/

by Østensen et al. (2001b), and additional slow pulsations by Lutz et al. (2009a). An asteroseismology analysis of the star is included in Silvotti et al. (2002a, 2007).

Rapid oscillations were discovered in the second target on the list, HS 0702+6043, by Dreizler et al. (2002), and simultaneous slow oscillations were reported by Schuh et al. (2006). The on-going EXOTIME observations for HS 0702+6043 have also previously been summarized by Lutz et al. (2008b, 2009b) and include a significant contribution of data by Francœur et al. (2010).

The target HS 0444+0458 was first discovered to pulsate by Østensen et al. (2001a), and has been further characterized by Reed et al. (2007). EXOTIME has followed it regularly since 2008.

Kilkenny et al. (2006) discovered rapid pulsations in EC 09582−1137 which Randall et al. (2009) subjected to an asteroseismology analysis. EC 09582−1137 is included in EXOTIME as a southern hemisphere target.

The discovery of pulsations in PG 1325+101 by Silvotti et al. (2002b) was also taken advantage of by characterizing the star asteroseismologically (Charpinet et al., 2006; Silvotti et al., 2006). This target is also extensively being observed within EXOTIME.

Schuh et al. (2010) present a portion of the observations currently available, describe the treatment of the data and display the first, still relatively short, $O - C$ diagrams for the EXOTIME targets HS 0444+0458 and HS 0702+6034. Not surprisingly, these illustrate the need for further observations and interpretation. The analysis resorts to tools provided by Montgomery & O'Donoghue (1999) and Lenz & Breger (2005).

6 Sub-stellar companions of evolved stars

6.1 Post-main-sequence evolution

For single stars with initial masses below $\approx 2.2\,M_\odot$ on the main sequence, the canonical steps in the evolution are: hydrogen shell burning following core hydrogen exhaustion, core mass increase (due to the shell burning) and core contraction with increasing electron degeneracy simultaneous to radius swelling on the first giant branch (RGB), ended by onset of a core helium flash, which brings the star onto the horizontal branch. Their masses now lie in the range roughly $0.5\,M_\odot < M < 1.0\,M_\odot$, with a canonical core mass of $0.46\,M_\odot$ and variable envelope masses with or without hydrogen shell burning that determine the location on the horizontal branch. The hottest, bluest objects are the least massive and have thin envelopes that put the most extreme of these objects with inert hydrogen shells on the EHB. This sequence is terminated at the point where the helium main sequence coincides with the horizontal branch helium core mass.

After core helium exhaustion the complex asymptotic giant branch evolution with H and He shell burning, thermal pulses, convection and dredge-up processes, and extreme mass loss unfolds. The subsequent post-AGB phase sees the creation of a planetary nebula from the envelope material lost and the emergence of the stellar remnant on its way to the white dwarf cooling track.

Figure 4: The subdwarf B (sdB) star V391 Pegasi oscillates in short-period p modes and long-period g modes, making it one of the known hybrid pulsators among sdBs. As a by-product of the effort to measure secular period changes in the p modes due to evolutionary effects on a time scale of almost a decade, the $O-C$ diagram has revealed an additional sinusoidal component attributed to a periodic shift in the light travel time caused by a planetary-mass companion around the sdB star in a 3.2 year orbit. In the above artistic impression, the V391 Pegasi system is shown at an earlier evolutionary stage in one of the proposed scenarios where, roughly 10^8 years ago, the star, at maximum red giant expansion, almost engulfed the planet. Image courtesy of HELAS, the European Helio- and Asteroseismology Network, funded by the European Union under Framework Programme 6; Mark Garlick, artist.

As the subdwarf B stars are in between the RGB and AGB giant expansion phases, or more precisely, will not undergo AGB evolution, their observation allows to separate the effects of the first giant branch on planets from those of the asymptotic giant branch.

6.2 The fate of planets around evolving stars

The starting point of discussions on planets around evolved stars is often that relic planets around white dwarfs are expected simply due to the considerable number of planets found around main sequence stars. The question then is at what initial masses and orbital separations the planets actually can survive both giant expansion phases of their host star without being engulfed, disrupted, evaporated or ejected. Debes & Sigurdsson (2002) consider the effects of mass loss, planet-planet interactions, and orbit stability and conclude that while inner planets will perish, far-out small bodies and distant planets have the potential to create a new dust disk. Such as disk could then pollute the central white dwarf's atmosphere to create a DAZ, and be observable via its IR excess. – While the search for planets around single white dwarfs has not been successful yet, the existence of dusk disks has indeed been established observationally.

In addition to orbital effects, Villaver & Livio (2007) consider the thermal conditions during the planetary nebula phase and establish new regions in the orbit and mass parameter space where planets survive all the way to the white dwarf stage of their host stars. Focusing more on the orbital evolution, Haghighipour (2008) investigates the issue of survival for planets in binary systems. A consequence in all of these studies is that finding planets at parameter combinations that correspond to previously "forbidden regions" immediately requires that second generation planet formation scenarios must have been at work. This can in principle include migration and significant accretion of previously existing planets in newly formed disks.

Of course, these investigations are predominantly concerned with the more violent effects during the AGB phase. However, once a planet has made it safely into the orbit of a subdwarf B host star, it can be expected to continue to exist without too much hassle when the host evolves to its final white dwarf stage.

A first scenario that explained the case of V391 Pegasi b basically in the context of single star evolution assumes that the planet is an old first generation planet which survived a common envelope phase. The situation where the star at maximum red giant expansion almost engulfed the planet is depicted in Fig. 4. Assuming this scenario, the existence of V391 Pegasi b proves observationally that gas-giant planets can survive the first red giant expansion in orbits similar to that of our Earth.

Silvotti (2008) puts forward alternative scenarios. In a second variant, V391 Pegasi b would have been an outer planet, never really under a strong direct influence of the expanding host star. The exceptional mass loss of the host star, turning it into a sdB, would instead have been triggered by another closer-in planet that got potentially destroyed in the process.

Scenario three explains V391 Pegasi b as a young second generation planet formed in a disk resulting from the merger of two He white dwarfs. The planet would then be a helium planet (compare also Livio et al. 2005).

In the case of the planets around HW Vir-like systems, the situation is somewhat easier as there is no dispute about the origin of the enhanced mass loss in a common envelope ejection phase. Besides the argument by Lee et al. (2009) that the planets were formed in a circumbinary disk, this leaves room for an interesting variety of possible planet formation scenarios, including a second generation scenario (cited from Heber 2009):

> "At birth, HW Vir's binary components must have been much further apart than they are today. During this [red giant] mass-loss episode the planets as well as the stellar companion may have gained mass. Rauch (2000) suggested that the low-mass companion in AA Dor may have grown from an initial planet by mass accretion during the CE-ejection phase. Could that have happened to the HW Vir system? As a matter of pure speculation, the star may have been born with three massive planets: the innermost launched a CE ejection, spiraled in, and accreted so much material as to turn into a low-mass star. [...] Another speculation concerns in situ formation; that is, could the planet have been formed during the CE phase?"

6.3 The fate of evolving stars with planets

As already implied above, it has recently been suggested that the presence of planets may have implications for the formation of sdB stars. This marks the renaissance of an older idea: Soker (1998) suggested that planets may constitute the *second parameter* influencing the irregular morphology of the (blue part of the) horizontal branch in globular clusters and elliptical galaxies. The physical processes discussed as candidates for the second parameter in horizontal branch morphology are

- age of the globular cluster,
- deep helium mixing and radiative levitation,
- large He abundance or fast rotation,
- stellar density in the cluster,
- planets enhancing mass loss on the red giant branch.

Quite similarly to this last suggestion (Soker, 1998), Soker (2010) points out that in addition to strengthening the blue component of the normal horizontal branch, enhanced mass loss due to planets could also be decisive in forming subdwarf B stars. The initial prediction – that the engulfed planet that helps to shed the envelope will in most cases be destroyed in the process – remains valid, so this hypothesis is challenging to test.

However, V391 Pegasi, while most probably not capable to have caused the enhanced mass loss of its host star itself, may well be regarded as a possible tracer for (former?) inner planets that could indeed have been able to cause the amount of mass loss required (compare the second scenario in the previous section). The scenario as a general explanation for the formation of a significant part of the single sdB star population may have a number of unsolved problems, mostly related to our poor understanding of the common envelope ejection mechanism. But, independently of the theory, EXOTIME and other programs now start to built up an empirical data base of low-mass companion statistics that further discussions can be based on. This should help to clarify the role of an – to date perhaps mostly still undiscovered – population of very low-mass companions to apparently single subdwarf B stars in the formation of these objects.

Should it one day become possible to determine the composition of the low-mass objects found, and should these turn out to be planets that were most likely formed in the helium disk of a He white dwarf merger event, the origin of the host stars will also have been cleared up unambiguously.

7 Summary

This article has attempted to highlight the added value of incorporating stellar pulsations in the comprehensive investigation of various stellar and planetary systems. The potential of links between asteroseismology and exoplanet science was first illustrated for solar-like oscillators. Accurate asteroseismic radii for (solar-like) host stars of transiting exoplanets translate into good radius determinations for the transit planets.

In the context of connections with coherent pulsations, as found in evolved stars, the focus was directed towards the investigation of stellar and planetary systems at late stages of stellar evolution. The asteroseismology-exoplanet-connection was explored in particular for post-RGB stars. The exotic class of subdwarf B stars, of which many show pulsations, was introduced as a group with several new planet discoveries. In discussing the fate of planetary systems based on known planets around extreme horizontal branch stars (e.g. V391 Pegasi), the question arises whether these are first or second generation planets. Data from programs such as EXOTIME, which make use of the timing method to find more such systems, can fill a gap with respect to the sensitivity to planets in wide orbits and applicability to evolved stars. With respect to the puzzle of subdwarf B formation, planets were proposed to (partly) be at the origin of subdwarf B stars, and also to play a role as the second parameter for the horizontal branch morphology in globular clusters and elliptical galaxies.

The aspects this article summarizes are all from the pre-Kepler era. It will be fascinating to see how our understanding in a great many areas evolves with the new data.

Acknowledgements

The author would like to thank the Astronomische Gesellschaft for the opportunity to present this work during the 2009 Ludwig Biermann Award Lecture. Warm thanks go to all of my collaborators and students who have been part of these and many other research projects, including K. Werner, S. Dreizler and his group, R. Silvotti, R. Lutz, R. Kruspe and B. Löptien. Out of the various sources for data, I want to single out the observers or operators at the Calar Alto, MONET, HET, and Keck telescopes (special thanks to A. Reiners and G. Basri for providing the V391 Pegasi spectrum), and also thank U. Heber and T. Rauch for providing grids of model spectra. Spectral energy distributions (SEDs) of the hydrogen NLTE models that were calculated with

TMAP, the Tübingen Model-Atmosphere Package, were retrieved via TheoSSA[4], a service provided by the German Astrophysical Virtual Observatory (GAVO[5]). The preparation of this article has benefited from up-to-date information on systems and WWW links of the Extra-Solar Planets Encyclopaedia maintained by J. Schneider (www.exoplanet.eu).

References

Backer, D.C.: 1993, in: J.A. Phillips, S.E. Thorsett, S.R. Kulkarni (eds.), *Planets Around Pulsars*, ASPC 36, p. 11

Backer, D.C., Foster, R.S., Sallmen, S.: 1993, Nature 365, 817

Bazot, M., Vauclair, S., Bouchy, F., Santos, N.C.: 2005, A&A 440, 615

Bennett, D.P., Anderson, J., Beaulieu, J.P., et al.: 2009, in: *Astronomy, Vol. 2010, astro2010: The Astronomy and Astrophysics Decadal Survey*, p. 18

Borucki, W.J., Koch, D., Jenkins, J., et al.: 2009, Sci 325, 709

Borucki, W.J., Koch, D., Basri, G., et al.: 2010, Sci 327, 977

Bouchy, F., Bazot, M., Santos, N.C., Vauclair, S., Sosnowska, D.: 2005, A&A 440, 609

Campbell, B., Walker, G.A.H., Yang, S.: 1988, ApJ331, 902

Catala, C., ESA Plato Science Study team: 2009, *PLATO assessment study report*, Tech. Rep. SRE-2009-4, ESA

Charpinet, S., Fontaine, G., Brassard, P., et al.: 1997, ApJ483, L123

Charpinet, S., Silvotti, R., Bonanno, A., et al.: 2006, A&A 459, 565

Charpinet, S., Fontaine, G., Brassard, P., Chayer, P.: 2008, Communications in Asteroseismology 157, 168

Christensen-Dalsgaard, J., Arentoft, T., Brown, T.M., et al.: 2008, Journal of Physics Conference Series 118, 012039

Christensen-Dalsgaard, J., Kjeldsen, H., Brown, T.M., et al.: 2010, ApJL 713, 164

Costa, J.E.S., Kepler, S.O.: 2008, A&A 489, 1225

Costa, J.E.S., Kepler, S.O., Winget, D.E.: 1999, ApJ522, 973

Debes, J.H., Sigurdsson, S.: 2002, ApJ572, 556

Döllinger, M.P., Hatzes, A.P., Pasquini, L., et al.: 2007, A&A 472, 649

Dorman, B., Rood, R.T., O'Connell, R.W.: 1993, ApJ419, 596

Dreizler, S., Schuh, S.L., Deetjen, J.L., Edelmann, H., Heber, U.: 2002, A&A 386, 249

Fontaine, G., Brassard, P., Charpinet, S., et al.: 2003, ApJ597, 518

Fontaine, G., Brassard, P., Charpinet, S., et al.: 2008, in: U. Heber, C.S. Jeffery, R. Napiwotzki (eds.), *Hot Subdwarf Stars and Related Objects*, ASPC 392, p. 231

---

[4] http://vo.ari.uni-heidelberg.de/ssatr-0.01/TrSpectra.jsp?
[5] http://www.g-vo.org

Francœur, M., Fontaine, G., Green, E.M., et al.: 2010, Ap&SS, in press

Frink, S., Mitchell, D.S., Quirrenbach, A., et al.: 2002, ApJ576, 478

Geier, S., Karl, C., Edelmann, H., Heber, U., Napiwotzki, R.: 2008, Mem. Soc. Astron. Ital. 79, 608

Geier, S., Edelmann, H., Heber, U., Morales-Rueda, L.: 2009a, ApJ702, L96

Geier, S., Heber, U., Edelmann, H., et al.: 2009b, Journal of Physics Conference Series 172, 012008

Geier, S., Heber, U., Tillich, A., et al.: 2010, Ap&SS84, in press

Gilliland, R.L., Brown, T.M., Christensen-Dalsgaard, J., et al.: 2010, PASP122, 131

Green, E.M., Fontaine, G., Reed, M.D., et al.: 2003, ApJ583, L31

Haghighipour, N.: 2008, in: J. Mason (ed.), *Exoplanets: Detection, Formation, Properties, Habitability*, p. 223

Han, Z., Podsiadlowski, P., Maxted, P.F.L., Marsh, T.R., Ivanova, N.: 2002, MNRAS336, 449

Han, Z., Podsiadlowski, P., Maxted, P.F.L., Marsh, T.R.: 2003, MNRAS341, 669

Hatzes, A.P., Cochran, W.D., Endl, M., et al.: 2003, ApJ599, 1383

Hatzes, A.P., Guenther, E.W., Endl, M., et al.: 2005, A&A 437, 743

Hatzes, A.P., Cochran, W.D., Endl, M., et al.: 2006, A&A 457, 335

Heber, U.: 2009, ARA&A47, 211

Israelian, G., Delgado Mena, E., Santos, N.C., et al.: 2009, Nature 462, 189

Jeffery, C.S., Saio, H.: 2006, MNRAS372, L48

Jeffery, C.S., Saio, H.: 2007, MNRAS378, 379

Kalas, P., Graham, J.R., Chiang, E., et al.: 2008, Sci 322, 1345

Kepler, S.O., Winget, D.E., Nather, R.E., et al.: 1991, ApJ378, L45

Kepler, S.O., Mukadam, A., Winget, D.E., et al.: 2000, ApJ534, L185

Kepler, S.O., Costa, J.E.S., Castanheira, B.G., et al.: 2005, ApJ634, 1311

Kilkenny, D.: 2010, Ap&SS82, in press

Kilkenny, D., Koen, C., O'Donoghue, D., Stobie, R.S.: 1997, MNRAS285, 640

Kilkenny, D., Stobie, R.S., O'Donoghue, D., et al.: 2006, MNRAS367, 1603

Kilkenny, D., Fontaine, G., Green, E.M., Schuh, S.: 2010, IBVS 5927, 1

Kruspe, R., Schuh, S., Silvotti, R., Traulsen, I.: 2008, Communications in Asteroseismology 157, 325

Lagrange, A., Gratadour, D., Chauvin, G., et al.: 2009, A&A 493, L21

Latham, D.W., Stefanik, R.P., Mazeh, T., Mayor, M., Burki, G.: 1989, Nature 339, 38

Lee, J.W., Kim, S.-L., Kim, C.-H., et al.: 2009, AJ137, 3181

Lenz, P., Breger, M.: 2005, Communications in Asteroseismology 146, 53

Livio, M., Pringle, J.E., Wood, K.: 2005, ApJ632, L37

Lutz, R., Schuh, S., Silvotti, R., et al.: 2008a, in: U. Heber, C.S. Jeffery, R. Napiwotzki (eds.), *Hot Subdwarf Stars and Related Objects*, ASPC 392, p. 339

Lutz, R., Schuh, S., Silvotti, R., Kruspe, R., Dreizler, S.: 2008b, Communications in Asteroseismology 157, 185

Lutz, R., Schuh, S., Silvotti, R., et al.: 2009a, A&A 496, 469

Lutz, R., Schuh, S., Silvotti, R., Kruspe, R., Dreizler, S.: 2009b, Communications in Asteroseismology 159, 94

Marcy, G.W., Butler, R.P.: 1995, BAAS 27, 1379

Marois, C., Macintosh, B., Barman, T., et al.: 2008, Sci 322, 1348

Mayor, M., Queloz, D.: 1995, Nature 378, 355

Montgomery, M.H., O'Donoghue, D.: 1999, Delta Scuti Star Newsletter 13, 28

Mullally, F., Winget, D.E., De Gennaro, S., et al.: 2008, ApJ676, 573

Neuhäuser, R., Mugrauer, M., Fukagawa, M., Torres, G., Schmidt, T.: 2007, A&A 462, 777

Østensen, R.H.: 2009, Communications in Asteroseismology 159, 75

Østensen, R., Heber, U., Silvotti, R., et al.: 2001a, A&A 378, 466

Østensen, R., Solheim, J.-E., Heber, U., et al.: 2001b, A&A 368, 175

Perryman, M.A.C.: 2000, Reports on Progress in Physics 63, 1209

Podsiadlowski, P., Han, Z., Lynas-Gray, A.E., Brown, D.: 2008, in: U. Heber, C.S. Jeffery, R. Napiwotzki (eds.), *Hot Subdwarf Stars and Related Objects*, ASPC 392, p. 15

Provencal, J.L.: 1997, in: D. Soderblom (ed.), *Planets Beyond the Solar System and the Next Generation of Space Missions*, ASPC 119, p. 123

Qian, S., Dai, Z., Liao, W., et al.: 2009a, ApJ706, L96

Qian, S., Liu, L., Zhu, L.: 2009b, PASA 26, 7

Qian, S., Zhu, L., Boonrucksar, S., Xiang, F., He, J.: 2009c, PASJ61, 333

Qian, S.-B., Zhu, L.-Y., Zola, S., et al.: 2009d, ApJ695, L163

Qian, S., Liao, W., Zhu, L., Dai, Z.: 2010a, ApJ708, L66

Qian, S., Liao, W., Zhu, L., et al.: 2010b, MNRAS401, L34

Qian, S., Zhu, L., Liu, L., et al.: 2010c, Ap&SS50, in press

Randall, S.K., van Grootel, V., Fontaine, G., Charpinet, S., Brassard, P.: 2009, A&A 507, 911

Reed, M.D., Terndrup, D.M., Zhou, A.-Y., et al.: 2007, MNRAS378, 1049

Richer, H.B., Ibata, R., Fahlman, G.G., Huber, M.: 2003, ApJ597, L45

Schuh, S.: 2008, Journal of Physics Conference Series 118, 012015

Schuh, S., Huber, J., Green, E.M., et al.: 2005, in: D. Koester, S. Moehler (eds.), *White Dwarfs*, ASPC 334, p. 530

Schuh, S., Huber, J., Dreizler, S., et al.: 2006, A&A 445, L31

Schuh, S., Kruspe, R., Lutz, R., Silvotti, R.: 2009, Communications in Asteroseismology 159, 91

Schuh, S., Silvotti, R., Lutz, R., et al.: 2010, Ap&SS130, in press

Silvotti, R.: 2008, in: U. Heber, C.S. Jeffery, R. Napiwotzki (eds.), *Hot Subdwarf Stars and Related Objects*, ASPC 392, p. 215

Silvotti, R., Janulis, R., Schuh, S.L., et al.: 2002a, A&A 389, 180

Silvotti, R., Østensen, R., Heber, U., et al.: 2002b, A&A 383, 239

Silvotti, R., Bonanno, A., Bernabei, S., et al.: 2006, A&A 459, 557

Silvotti, R., Schuh, S., Janulis, R., et al.: 2007, Nature 449, 189

Soker, N.: 1998, AJ116, 1308

Soker, N.: 2010, Ap&SS, in press

Soriano, M., Vauclair, S.: 2009, astro-ph/0903.5475

Soriano, M., Vauclair, S., Vauclair, G., Laymand, M.: 2007, A&A 471, 885

Thorsett, S.E., Arzoumanian, Z., Taylor, J.H.: 1993, ApJ412, L33

Thorsett, S.E., Arzoumanian, Z., Camilo, F., Lyne, A.G.: 1999, ApJ523, 763

Tillich, A., Heber, U., O'Toole, S.J., Østensen, R., Schuh, S.: 2007, A&A 473, 219

Vauclair, S.: 2008, astro-ph/0809.0249

Vauclair, S., Castro, M., Charpinet, S., et al.: 2006, in: M. Fridlund, A. Baglin, J. Lochard, L. Conroy (eds.), *The CoRoT Mission Pre-Launch Status – Stellar Seismology and Planet Finding*, ESA SP-1306, p. 77

Villaver, E., Livio, M.: 2007, ApJ661, 1192

Wolszczan, A.: 1994, Sci 264, 538

Wolszczan, A., Frail, D.A.: 1992, Nature 355, 145

Ludwig Biermann Award Lecture

Stellar archaeology: Exploring the Universe with metal-poor stars

Anna Frebel

Harvard-Smithsonian Center for Astrophysics
60 Garden St., MS-20
Cambridge, MA 02138, USA
afrebel@cfa.harvard.edu

Abstract

The abundance patterns of the most metal-poor stars in the Galactic halo and small dwarf galaxies provide us with a wealth of information about the early Universe. In particular, these old survivors allow us to study the nature of the first stars and supernovae, the relevant nucleosynthesis processes responsible for the formation and evolution of the elements, early star- and galaxy formation processes, as well as the assembly process of the stellar halo from dwarf galaxies a long time ago. This review presents the current state of the field of "stellar archaeology" – the diverse use of metal-poor stars to explore the high-redshift Universe and its constituents. In particular, the conditions for early star formation are discussed, how these ultimately led to a chemical evolution, and what the role of the most iron-poor stars is for learning about Population III supernovae yields. Rapid neutron-capture signatures found in metal-poor stars can be used to obtain stellar ages, but also to constrain this complex nucleosynthesis process with observational measurements. Moreover, chemical abundances of extremely metal-poor stars in different types of dwarf galaxies can be used to infer details on the formation scenario of the halo and the role of dwarf galaxies as Galactic building blocks. I conclude with an outlook as to where this field may be heading within the next decade. A table of ~ 1000 metal-poor stars and their abundances as collected from the literature is provided in electronic format.

1 Introduction

As Carl Sagan once remarked, *If you wish to make an apple pie from scratch, you must first create the Universe.* An apple contains at least 16 different elements[1], and the human body is even more complex, having at least trace amounts of nearly 30

---

[1] http://www.food-allergens.de/symposium-vol1(3)/data/apple/apple-composition.htm

elements[2], all owing to a 14-billion year long manufacturing process called cosmic chemical evolution. Thus, the basis of chemically complex and challenging undertakings such as cooking and baking, not to mention the nature of life, will ultimately be gained through an understanding of the formation of the elements that comprise organic material. It is thus important to examine how the constituents of an apple, and by extension the stuff of life and the visible Universe were created: baryonic matter in the form of elements heavier than primordial hydrogen and helium. In this review I will describe how the chemical abundances observed in the most metal-poor stars are employed to unravel a variety of details about the young Universe, such as early star formation, nucleosynthesis in stars and supernovae (SNe), and the formation process(es) of the Galactic halo. This concept is often called "stellar archaeology" and is frequently used to address a number of important, outstanding questions:

- What is the nature of Population III stars? Are the yields of the first SNe different from today's? Can we find the signatures of theorized pair-instability SNe?

- What drove early star formation? How/where did the first low-mass stars and the first galaxies form?

- What are the main nucleosynthesis processes and sites that are responsible for forming the elements from the Big Bang until today?

- How did chemical evolution proceed? How do stellar chemistry and halo kinematics correlate? How can we use abundances to learn about the halo formation process?

- Was the old halo built from accreted satellites? Can we identify accreted dwarf galaxies in the halo? Did the first stars form in dwarf galaxies?

Even though some of these question appear to refer to completely unrelated topics, metal-poor stars do provide us with a powerful tool to study a very broad range of astrophysical issues ranging from nuclear astrophysics to early galaxy formation. Metal-poor stars represent the local equivalent to the high-redshift Universe, and thus provide a unique tool to address a wide range of near and far-field cosmological topics.

Each section of this review discusses one of those topics and can be read independently, although it is advisable to also peruse Sect. 2. It sets the overall stage by introducing the first stars, early low-mass star formation and whether stellar archaeology is a valid concept to study the high-redshift Universe. Section 3 then discusses the most iron-poor stars and their connection to early SNe, whereas in Sect. 4, the observed signatures of neutron-capture nucleosynthesis are presented. In particular, the r-process and nucleo-chronometry are discussed. Section 5 deals with the metal-poor stars recently found in small dwarf galaxies and what we can learn from their chemical abundances about the early formation process of the Galactic halo. Conclusions and an outlook at given in Sect. 6.

---

[2] http://chemistry.about.com/cs/howthingswork/f/blbodyelements.htm

Since there exist a large range of metal-poor stars in terms of their metallicities and chemical signatures, Beers & Christlieb (2005) suggested a classification scheme. In this review I will make extensive use of their term "extremely metal-poor stars", referring to stars with [Fe/H] < -3.0. "Ultra metal-poor" then refers to [Fe/H] < -4.0, and "hyper metal-poor" to [Fe/H] < -5.0. This already shows that the main metallicity indicator used to determine any stellar metallicity is the iron abundance, [Fe/H], which is defined as [A/B]$= \log_{10}(N_A/N_B)_\star - \log_{10}(N_A/N_B)_\odot$ for the number N of atoms of elements A and B, and $\odot$ refers to the Sun. With few exceptions, [Fe/H] traces the overall metallicity of the objects fairly well.

2 The early Universe

2.1 The first stars

According to cosmological simulations that are based on the Λ cold dark matter model of hierarchical structure growth in the Universe, the first stars formed in small minihalos some few hundred million years after the Big Bang. Due to the lack of cooling agents in the primordial gas, significant fragmentation was largely suppressed so that these first objects were very massive (of the order to $\sim 100\,M_\odot$; e.g. Bromm & Larson 2004 and references therein). This is in contrast to low-mass stars dominating today's mass function. These objects are referred to as Population III (Pop III) as they formed from metal-free gas.

These objects soon exploded as SNe to either collapse into black holes (progenitor masses of $25 < M/M_\odot < 140$ and $M/M_\odot > 260$) or to die as energetic pair-instability SNe (PISN, $140 < M/M_\odot < 260$; Heger & Woosley 2002). During their deaths, these objects provided vast amounts of ionizing radiation (and some of the first metals in the case of the PISNe) that changed the conditions of the surrounding material for subsequent star formation even in neighboring minihalos. Hence the second generation of stars might have been less massive ($M_\star \sim 10\,M_\odot$). Partially ionized gas supports the formation of the H_2, and then the HD molecule which in turn facilitates more effective cooling than what is possible in neutral gas. Also, any metals or dust grains left behind from PISNe would have similar cooling effects. This may then have led to the first more regular metal-producing SNe, although not all higher mass SNe must necessarily end in black hole formation. Umeda & Nomoto (2003) suggested that some $25\,M_\odot$ stars undergo only a partial fallback, so that some of the newly created metals get ejected into the surrounding gas. By that time, most likely enough metals were present to ensure sufficient gas fragmentation to allow for low-mass ($<1\,M_\odot$) star formation. Those stars that formed from any metal-enriched material are referred to as Population II (Pop II) stars. More metal-rich stars like the Sun that formed in a much more metal-rich Universe are called Population I.

The concept of stellar archaeology entails that the most extreme versions of Pop II stars preserve in their surface composition the individual SN yields of previous Pop III stars. Studying those "chemical fingerprints" in the oldest, most metal-poor stars can thus reveal a great deal about the first nucleosynthesis events in the Universe. Indeed, several metal-poor star abundance patterns have been fitted with calculated Pop III SN yields. Moreover, one may also seek to find a PISN signature (a pronounced odd-even abundance signature) in metal-poor stars. This has, however, not yet occurred.

2.2 Early low-mass star formation and the connection to carbon-enhanced metal-poor stars

Early Pop II stars began to form from the enriched material left behind by the first stars. The actual formation process of these initial low-mass ($M \leq 0.8$ M$_\odot$) Pop II stars (i.e. the most metal-poor stars) that live longer than a Hubble time, is not well-understood so far. Ideas for the required cooling processes necessary to induce sufficient fragmentation of the near-primordial gas include cooling through metal enrichment ("critical metallicity") or dust, cooling based on enhanced molecule formation due to ionization of the gas, as well as more complex effects such as turbulence and magnetic fields (Bromm et al., 2009).

Fine-structure line cooling through neutral carbon and singly-ionized oxygen has been suggested as a main cooling agent facilitating low-mass star formation (Bromm & Loeb, 2003). These elements were likely produced in vast quantities in Pop III objects (e.g. Meynet et al. 2006a). Gas fragmentation is then induced once a critical metallicity of the interstellar medium (ISM) is achieved. The existence of such a critical metallicity can be probed with large numbers of *carbon and oxygen-poor* metal-poor stars. Frebel et al. (2007b) developed an "observer-friendly" description of the critical metallicity that incorporates the observed C and/or O stellar abundances; $D_{\mathrm{trans}} = \log(10^{[C/H]} + 0.3 \times 10^{[O/H]}) \geq -3.5$. Any low-mass stars still observable today then has to have C and/or O abundances above the threshold of $D_{\mathrm{trans}} = -3.5$ (see Fig. 1 in Frebel et al. 2007b). At metallicities of [Fe/H] $\gtrsim -3.5$, most stars have C and/or O abundances that are above the threshold since they follow the solar C and O abundances simply scaled down to their respective Fe values. Naturally, this metallicity range is not suitable for directly probing the first low-mass stars. Below [Fe/H] ~ -3.5, however, the observed C and/or O levels must be higher than the Fe-scaled solar abundances to be above the critical metallicity. Indeed, none of the known lowest-metallicity stars has a D_{trans} below the critical value, consistent with this cooling theory. Some stars, however, have values very close to $D_{\mathrm{trans}} = -3.5$. HE 0557–4840, at [Fe/H] $= -4.75$ (Norris et al., 2007), falls just onto the critical limit (Bessell 2009, priv. comm.). A star in the ultra-faint dwarf galaxy Boötes I has $D_{\mathrm{trans}} = -3.2$ (at [Fe/H] $= -3.7$; and assuming that its oxygen abundance is twice that of carbon). An interesting case is also the most metal-poor star in the classical dwarf galaxy Sculptor, which has an upper limit of carbon of [C/H] < -3.6 at [Fe/H] $= -3.8$ (Frebel et al., 2010b). Despite some still required up-correction of carbon to account for atmospheric carbon-depletion of this cool giant, the star could potentially posses a sub-critical D_{trans} value.

Overall, more such "borderline" examples are crucial to test for the existence of a critical metallicity. If fine-structure line cooling were the dominant process for low-mass star formation, two important consequences would follow: (1) Future stars to be found with [Fe/H] $\lesssim -4.0$ are predicted to have these significant C and/or O overabundances with respect to Fe. (2) The so-far unexplained large fraction of

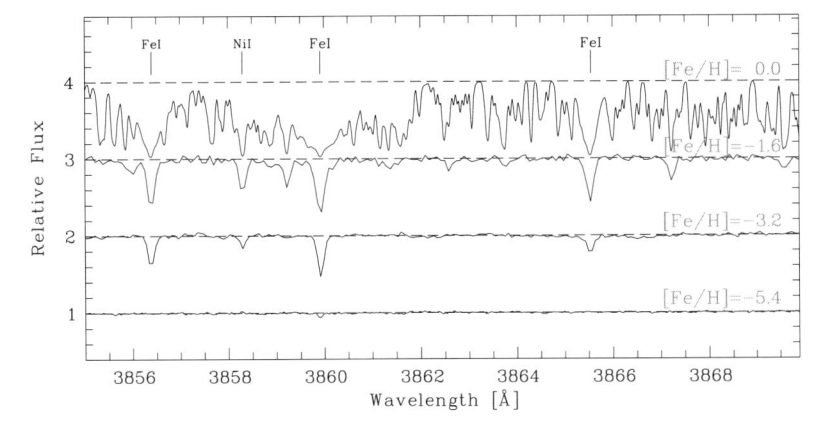

Figure 1: Spectral comparison of stars in the main-sequence turn-off region with different metallicities. Several absorption lines are marked. The variations in line strength reflect the different metallicities. From top to bottom: Sun with [Fe/H] = 0.0, G66-30 with [Fe/H] = −1.6 (Norris et al., 1997c), G64-12 [Fe/H] = −3.2 (Frebel et al., 2005), and HE1327–2326 with [Fe/H] = −5.4 (Frebel et al., 2005).

metal-poor objects that have large overabundances of carbon with respect to iron ([C/Fe] > 1.0) may reflect an important physical cause. About 20% of metal-poor stars with Fe/H $\lesssim -2.5$ exhibit this behavior (e.g. Beers & Christlieb 2005). Moreover, at the lowest metallicities, this fraction is even higher. All three stars with [Fe/H] < −4.0 are extremely C-rich, well in line with the prediction of the line cooling theory.

It should also be mentioned that cooling through dust grains might have been responsible for the transition from Pop III to Pop II star formation. Dust created during the first SNe explosions or mass loss during the evolution of Pop III stars may induce fragmentation processes (e.g., Schneider et al. 2006) that lead to the formation of subsolar-mass stars. The critical metallicity in this scenario is a few orders of magnitude below that of C and O line cooling. If some metal-poor stars are found to be below $D_{\mathrm{trans}} = -3.5$, they may still be consistent with the critical value set by dust cooling.

2.3 Validating stellar archaeology

Stellar archaeology is based on long-lived low-mass metal-poor main-sequence and giant stars whose chemical abundances are thought to reflect the composition of the ISM during their formation. A vital assumption is that the stellar surface compositions have not been significantly altered by any internal mixing processes given that these stars are fairly unevolved despite their old age. But are there other means by which the surface composition could be modified? Accretion of interstellar matter while a star orbits in the Galaxy for $\sim 10\,\mathrm{Gyr}$ has long been suggested as a mechanism to affect the observed abundance patterns. Iben (1983) calculated a basic "pollution limit" of [Fe/H] = −5.7 based on Bondi-Hoyle accretion. He predicted that no stars with Fe abundances below this value could be found since they would have accreted too much enriched material.

Assuming that stars with such low-metallicities exist (for example low-mass Pop III stars if the IMF was Salpeter-like, and not top-heavy), significant amounts of interstellar accretion could masquerade the primordial abundances of those putative low-mass Pop III stars. Analogously, stars with very low abundances, say [Fe/H] < −5.0, could principally be affected also. Frebel et al. (2009) carried out a kinematic analysis of a sample of metal-poor stars to assess their potential accretion histories over the past 10 Gyr in a Milky Way-like potential. The amount of accreted Fe was calculated based on the total accreted material over 10 Gyr. The overall chemical evolution with time was taken into account assuming the ISM to have scaled solar abundances.

The stellar abundances were found to be little affected by accretion. The calculated, "accreted abundances" were often lower than the observed measurements by several orders of magnitude. Generally, this confirms that accretion does not significantly alter the observed abundance patterns, even in an extreme case in which a star moves once through a very large, dense cloud. The concept of stellar archaeology is thus viable. Nevertheless, since there is a large accretion dependency on the space velocity it becomes obvious that kinematic information is vital for the identification of the lowest-metallicity stars in the Milky Way and the interpretation of their abundances.

2.4 The metallicity distribution function

Large numbers of Galactic metal-poor stars found in objective-prism surveys in both hemispheres have provided great insight into the history and evolution of our Galaxy (e.g., Beers & Christlieb 2005). However, there are still only very few stars known ($\lesssim 20$) with metallicities below [Fe/H] < −3.5. Recently, Schörck et al. (2009) presented a new metallicity distribution function (MDF) for halo stars that is corrected for various selection effects and other biases. The number of known metal-poor stars declines significantly with decreasing metallicity (below [Fe/H] < −2.0) as illustrated in their Figs. 10 and 18. The new bias-corrected MDF also shows how rare metal-poor stars really are, but also that past targeted ("biased") searches for metal-poor stars have been extremely successful at identifying these objects. The biggest achievements in terms of the most iron-deficient stars was the push to a significantly lower stellar metallicity [Fe/H] almost a decade ago: From a long-standing [Fe/H] = −4.0 (CD −38° 245; Bessell & Norris 1984) to [Fe/H] = −5.2 (HE 0107 5240; Christlieb et al. 2002[3]), and more recently, down to [Fe/H] = −5.4 (HE 1327−2326; Frebel et al. 2005). Overall, only three stars are known with iron abundances of [Fe/H] < −4.0. The recently discovered star HE 0557−4840 (Norris et al., 2007) with [Fe/H] < −4.8 bridges the gap between [Fe/H] = −4.0 and the two hyper Fe-poor objects. Objects in the very tail of the MDF provide a unique

---

[3] Applying the same non-LTE correction to the Fe I abundances of HE 0107−5240 and HE 1327−2326 leads to a final abundance of [Fe/H] = −5.2 for HE 0107−5240.

observational window onto the time very shortly after the Big Bang. They provide key insights into the very beginning of Galactic chemical evolution.

To illustrate the progression of metallicity from metal-rich to the most metal-poor stars, Fig. 1 shows spectra around the strongest optical Fe line at 3860 Å of the Sun and four other metal-poor main-sequence stars. The number of atomic absorption lines detectable in the spectra decreases with increasing metal-deficiency. In HE 1327−2326, only the intrinsically strongest metal lines remain observable, and these are extremely weak. If a main-sequence star with even lower Fe value was discovered, no Fe lines would be measurable anymore. In the case of a giant, the lines would be somewhat stronger due to its cooler temperature and thus allow for the discovery of a [Fe/H] $\lesssim -6$ object.

3 Studying the early Universe with metal-poor stars

3.1 Searching for the most metal-poor stars

Over the past two decades, the quest to find the most metal-poor stars to study the chemical evolution of the Galaxy led to a significant number of stars with metallicities down to [Fe/H] ~ -4.0 (see Beers & Christlieb 2005 for a more detailed review). Those stars were initially selected as candidates from a large survey, such as the HK survey (Beers et al., 1992) and the Hamburg/ESO survey (Wisotzki et al., 1996). A large survey is required to provide numerous low-resolution spectra to search for weak-lined stellar candidates. Those spectra have to cover the strong Ca II K line at 3933 Å because the strength of this line indicates the metallicity of the star, and can be measured in low-quality spectra. If this line is sufficiently weak as a function of the star's estimated effective temperature, an object is selected as a candidate metal-poor star. For all candidates, medium-resolution spectra ($R \sim 2000$) are required to more accurately determine the Ca II K line strength for a more robust estimate for the Fe abundance. This line is still the best indicator for the overall metallicity [Fe/H] of a metal-poor star in such spectra. In the Sloan Digital Sky Survey and LAMOST survey, the survey spectra themselves are already of medium-resolution allowing for a quicker and more direct search for metal-poor stars. To confirm the metallicity, and to measure elemental abundances from their respective absorption lines besides that of iron, high-resolution optical spectroscopy is required. Only then the various elements become accessible for studying the chemical evolution of the Galaxy. Those elements include carbon, magnesium, calcium, titanium, nickel, strontium, and barium, and trace different enrichment mechanisms, events and timescales. Abundance ratios [X/Fe] as a function of [Fe/H] can then be derived for the lighter elements ($Z < 30$) and neutron-capture elements ($Z > 38$). The final number of elements hereby depends on the type of metal-poor star, the wavelength coverage of the data, and the data quality itself.

3.2 Chemical evolution of the Galaxy

Generally, there are several main groups of elements observed in metal-poor stars, with each group having a common, main production mechanism; (1) α-elements (e.g. Mg, Ca, Ti) are produced through α-capture during various burning stages of late stellar evolution, before and during SN explosions. These yields appear very robust with respect to parameters such as mass and explosion energy; (2) Fe-peak elements ($23 < Z < 30$) are synthesized in a host of different nucleosynthesis processes before and during SN explosions such as radioactive decay of heavier nuclei or direct synthesis in explosive burning stages, neutron-capture onto lower-mass Fe-peak elements during helium and later burning stages and alpha-rich freeze-out processes. Their yields also depend on the explosion energy; (3) light and heavy neutron-capture elements ($Z > 38$) are either produced in the slow (s-) process occurring in thermally pulsing AGB stars (and then transferred to binary companions or deposited into the ISM through stellar winds) or in the rapid (r-) process most likely occurring in core-collapse SN explosions. For more details on SN nucleosynthesis see e.g., Woosley & Weaver (1995).

The α-element abundances in metal-poor halo stars with [Fe/H] ~ -1.5 are enhanced by ~ 0.4 dex with respect to Fe (see Fig. 2). This reflects a typical core-collapse SN signature because at later times (in chemical space at about [Fe/H] ~ -1.5) the onset of SN Ia provides a significant contribution to the overall Galactic Fe inventory. As a consequence, the [α/Fe] ratio decreases down to the solar value at [Fe/H] ~ 0.0. The general uniformity of light element abundance trends down to [Fe/H] ~ -4.0 led to the conclusion that the ISM must have been already well-mixed at very early times (Cayrel et al., 2004). Otherwise it would be hard to understand why so many of the most metal-poor stars have almost identical abundance patterns. However, despite the well-defined abundance trends, some stars, particularly those in the lowest metallicity regime show significant deviations. Some stars have been found to be very α-poor (e.g. Ivans et al. 2003) and others are strongly overabundant in Mg and Si (e.g., Aoki et al. 2002b; Frebel et al. 2005).

Among the Fe-peak elements, many have subsolar abundance trends at low metallicity (e.g. [Cr,Mn/Fe]) which become solar-like as the metallicity increases (see Fig. 3). It is not clear whether these large underabundances are of cosmic origin or have to be attributed to modelling effects such as that of non-LTE (Bergemann & Gehren 2008; Sobeck et al. 2007). Trends of other elements are somewhat overabundant at low metallicity (Co) or relatively unchanged throughout (Sc, Ni). All elements with $Z < 30$ hereby have relatively tight abundance trends.

How can all those signatures be understood? The observed abundances of the most metal-poor stars with "classical" halo signatures have successfully been reproduced with Pop III SN yields. Tominaga et al. (2007) model the averaged abundance pattern of four non-carbon-enriched stars with $-4.2 < $ [Fe/H] < -3.5 with the elemental yields of a massive, energetic (~ 30–$50\,\mathrm{M}_\odot$) Pop III hypernova. The abundances can also be fitted with integrated yields of Pop III SNe Heger & Woosley (2008). Special types of SNe or unusual nucleosynthesis yields can then be considered for stars with chemically peculiar abundance patterns. It is, however, often difficult to explain the entire abundance pattern. Additional metal-poor stars as well

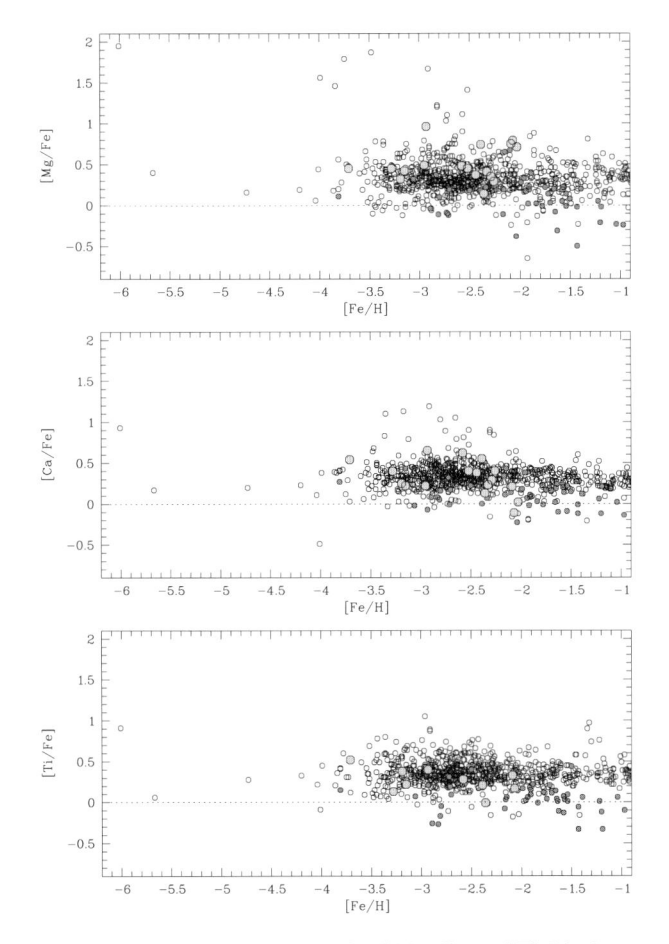

Figure 2: Light element abundance trends of Mg, Ca, and Ti. Black open circles represent halo stars, red fille circles are stars in the classical dwarf galaxies, and green filled circles show stars in the ultra-faint dwarf galaxies. Both the x- and y-axis have the same scale for easy comparisons. Only stars with high-resolution abundance analyses are shown. The abundance data can be found in an electronic table. The scatter in the data likely reflects systematic differences between literature studies. Assuming this, systematic uncertainties in abundance analyses may be around ~ 0.3 dex.

as a better understanding of the explosion mechanism and the impact of the initial conditions on SNe yields are required to arrive at a more solid picture of exactly how metal-poor stars reflect early SNe yields.

On the contrary, the abundances of the neutron-capture elements in metal-poor stars are "all over the place". Strontium has an extremely large scatter (~ 3 dex). This indicates that different nucleosynthetic processes must have contributed to its Galactic inventory, or that neutron-capture yields are very environmentally-sensitive. Ba has even more scatter at [Fe/H] ~ -3.0 (see Fig. 4). Other heavier neutron-capture elements, such as Eu, have somewhat less scatter but this may be due to their generally weak and usually more difficult to detect absorption lines. What is apparent, though, is that at the lowest metallicities, core-collapse SNe must have dominated the chemical evolution (below [Fe/H] = -3.0). Hence, the r-process is responsible for the neutron-capture elements at this early time. The s-process contribution occurred at somewhat later times, driven by the evolutionary timescales of stars with $\sim$2–8 M$_\odot$ to become AGB stars.

To illustrate the extent of the chemical evolution of the Galaxy we have collected abundance data of metal-poor stars from the literature (Figs. 2, 3, and 4). All abundances [X/Fe] have been recalculated with the latest solar abundances of Asplund et al. (2009). In the many cases where a star has been studied more than once, the study with the most elements (i.e. likely with the highest quality data) was usually picked. No other processing of the literature data has been done. Hence, systematic differences between different studies remain. The tables containing all the abundances as well as abundance plots are available at the CDS via http://cdsarc.u-strasbg.fr/cgi-bin/qcat?J/AN/331/474[4]. References are assigned to each set of abundances in the table. They are also alphabetically listed in the Appendix. For many stars, a key tag has been assigned to each star corresponding to its chemical properties. The keys are listed on the website as well. A similar collection of data can be found in the SAGA database (Suda et al., 2008) which is independent of the current collection.

3.3 The most iron-poor stars

The first star with a record-low iron abundance was found in 2001. The faint ($V = 15.2$) red giant HE 0107$-$5240 has [Fe/H] = -5.2 (Christlieb et al., 2002). In 2004, the bright ($V = 13.5$) subgiant HE 1327$-$2326 was discovered (Aoki et al., 2006b; Frebel et al., 2005). HE 1327$-$2327 has an even lower iron abundance of [Fe/H] = -5.4. This value corresponds to $\sim 1/250\,000$ of the solar iron abundance. Interestingly, the entire mass of iron in HE 1327$-$2326 is actually 100 times less than that in the Earth's core. At the same time the star is of course of the order of a million times more massive than the Earth. A third star with [Fe/H] = -4.75 (Norris et al., 2007) was found in 2006. The metallicity of the giant HE 0557$-$4840 is in between the two <-5.0 stars and the next most metal-poor stars are [Fe/H] ~ -4.0. Hence it sits right in the previously claimed "metallicity gap" (between [Fe/H] ~ -4.0 and [Fe/H] ~ -5.0; e.g. Shigeyama et al. 2003) showing that the scarcity of stars below [Fe/H] = -4.0 is not a physical cause but a merely an observational incompleteness. All three objects were found in the Hamburg/ESO survey making it the so far most successful database for metal-poor stars.

The most striking features in both <-5.0 stars are the extremely large overabundances of CNO elements. HE 0557$-$4840 partly shares this signature by also having a fairly large [C/Fe] ratio. Other elemental ratios [X/Fe] are somewhat enhanced in HE 1327$-$2327 with respect to the stars with $-4.0 <$ [Fe/H] < -2.5, but less so for the two giants. No neutron-capture element is detected in HE 0107$-$5240

---

[4]also at http://www.cfa.harvard.edu/$\sim$afrebel/abundances/abund.html

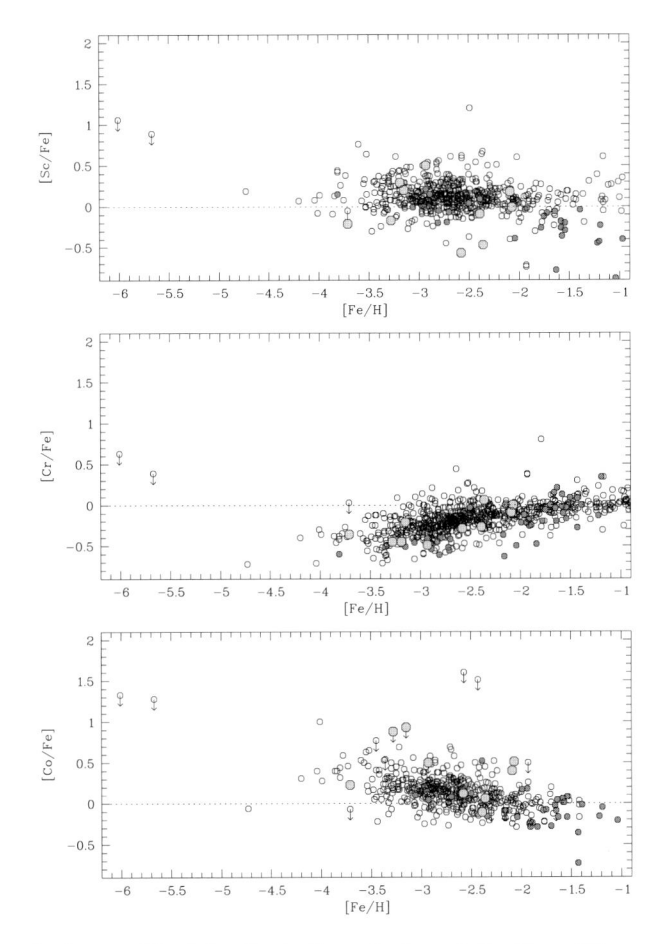

Figure 3: Same as Fig. 2, but for the Fe-peak elements Sc, Cr, and Co.

or HE 0557−4840, whereas, unexpectedly, Sr is observed in HE 1327−2326. The Sr may originate from the neutrino-induced νp-process operating in SN explosions (Fröhlich et al., 2006). Despite expectations, Li could not be detected in the relatively unevolved subgiant HE 1327−2326. The upper limit is $\log \epsilon(\mathrm{Li}) < 1.6$, where $\log \epsilon(A) = \log_{10}(N_A/N_H) + 12$. This is surprising, given that the primordial Li abundance is often inferred from similarly unevolved metal-poor stars (Ryan et al., 1999). Furthermore, the upper limit found from HE 1327−2326, however, strongly contradicts the WMAP value ($\log \epsilon(\mathrm{Li}) = 2.6$) from the baryon-to-photon ratio (Spergel et al., 2007). This may indicates that the star formed from extremely Li-poor material.

HE 0107−5240 and HE 1327−2326 immediately became benchmark objects to constrain various theoretical studies of the early Universe, such as the formation of the first stars (e.g., Yoshida et al. 2006), the chemical evolution of the early ISM (e.g., Karlsson & Gustafsson 2005) or calculations of Pop III SN yields. Their highly individual abundance patterns have been successfully reproduced by several differ-

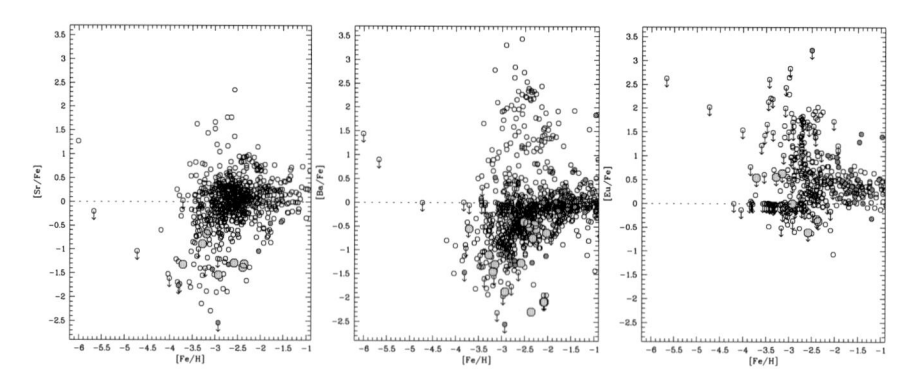

Figure 4: Same as Fig. 2, but for the neutron-capture elements Sr, Ba, and Eu. For these elements, the scatter is much beyond systemtic differences between individual studies and thus indicates a cosmic origin.

ent SNe scenarios. This makes HE 0107−5240 and HE 1327−2326 early, extreme Pop II stars that possibly display the "fingerprint" of just one Pop III SN. Umeda & Nomoto (2003) first matched the yields of a faint 25 $M_\odot$ SN that underwent a mixing and fallback process to the observed abundances of HE 0107−5240. To achieve a simultaneous enrichment of a lot of C and only little Fe, large parts of the Fe-rich SN ejecta have to fall back onto the newly created black hole. Using yields from a SN with similar explosion energy and mass cut, Iwamoto et al. (2005) then reproduced the abundance pattern of HE 1327−2326 also. Trying to fit the observed stellar abundances, Heger & Woosley (2008) are employing an entire grid of Pop III SN yields to search for the best match to the data. A similar progenitor mass range as the Umeda & Nomoto (2003) 25 $M_\odot$ was found to be the best match to have provided the elemental abundances to the ISM from which these Pop II stars formed. Meynet et al. (2006b) explored the influence of stellar rotation on elemental yields of 60 $M_\odot$ near-zero-metallicity SNe. The stellar mass loss rate of rotating massive Pop III stars qualitatively reproduces the CNO abundances observed in HE 1327−2326 and other carbon-rich metal-poor stars.

Limongi et al. (2003) were able to reproduce the abundances of HE 0107−5240 through pollution of the birth cloud by at least two SNe. Suda et al. (2004) proposed that the abundances of HE 0107−5240 would originate from a mass transfer of CNO elements from a postulated companion, and from accretion of heavy elements from the ISM. However, neither HE 0107−5240 nor HE 1327−2326 show radial velocity variations that would indicate binarity (R. Lunnan in preparation).

4 Neutron-capture nucleosynthesis observed in metal-poor stars

All elements heavier than the Fe-peak are created through neutron-capture processes in stars during stellar evolution (via AGB nucleosynthesis; slow s-process) and SN explosions (via the rapid r-process). We here summarize the most important constraints on the various neutron-capture processes provided by different types of metal-poor stars. A short review of r-process nucleosynthesis as observed in metal-poor stars can also be found in Frebel (2009). For a detailed review on how neutron-capture processes drove the chemical evolution of the Galaxy, we refer the interested reader to Sneden et al. (2008).

4.1 The r-process signature

The r-process occurs when a seed nucleus (e.g., a Fe or C nucleus) is under intensive neutron bombardment. In particular, when the neutron-capture rate exceeds that of the β-decay, large numbers of neutrons can be captured quickly to build up even the heaviest nuclei in the periodic table ($38 < Z < 92$). The astrophysical site(s) that can accommodate the required extreme environment for this process have not yet been clearly identified. Neutron-driven winds emerging from a proto-neutron-star which formed after a SN explosion (perhaps with a progenitor of 8–10 $M_\odot$) seem to be promising locations (Qian & Wasserburg, 2003). Neutron-star mergers have also been considered, but their long evolutionary timescales make them unsuitable as the primary r-process site in the early Galaxy (Argast et al., 2000). Not knowing the site, and hence, the initial conditions, makes it difficult to study the r-process theoretically. Furthermore, nuclear physics experiments on heavy, exotic r-process nuclei are technically out of reach. This leaves few experimental constraints on the production of the heaviest elements in the Universe.

Fortunately, about 5 % of metal-poor stars with [Fe/H] < -2.5 contain a strong enhancement of neutron-capture element[5] associated with the r-process (Beers & Christlieb, 2005). In those stars, we can observe the majority of elements in the periodic table: the usual light, α- and iron-peak, *and* the full set of optically available light ($38 < Z < 56$) and heavy ($56 < Z < 92$) neutron-capture elements. These neutron-capture elements were not produced in the observed metal-poor stars themselves, but in a previous-generation SN explosion. The so-called r-process metal-poor stars formed from material that was chemically pre-enriched by this SN. We are thus able to study the "chemical fingerprint" of individual r-process events that likely occurred shortly before the formation of the observed star. Hence, the r-process stars fortuitously bring together astrophysics and nuclear physics by acting as a "cosmic laboratory" for both fields of study, and provide crucial experimental data on heavy-element production.

The picture that has been emerging from these observation is simple and yet astounding. The abundance patterns of the neutron-capture elements particularly the heaviest ones with $Z > 56$, agree extremely well with each other in all r-process stars (Barklem et al., 2005; Frebel et al., 2007a; Hill et al., 2002; Ivans et al., 2006; Johnson & Bolte, 2001; Sneden et al., 1996) and with the scaled solar r-process pattern (e.g. Burris et al. 2000). Overall, this behavior suggests that the r-process is universal and leads to the same elemental ratios in the early Universe as at much

---

[5] Stars with [r/Fe] > 1.0; r represents the average abundance of elements from the r-process.

later times, when the Sun was born[6]. The universality offers a unique constraint on any theoretical modeling of the r-process as the end-product is "known" through the r-process stars. It also enables calculating elemental ratios involving radioactive elements such as thorium and uranium that can be compared with observed measurements of the left-over radioactive stellar material. This, in turn, makes possible nucleochronometric age dating of the oldest stars.

Among the lighter neutron-capture elements there are abundance deviations between the scaled solar r-process pattern and to some extent among the individual stars. Since it is not clear if these lighter are produced by a single r-process, additional new process were invoked in order to explain the entire spread of observed neutron-capture abundances in this mass range (e.g., Aoki et al. 2005; Travaglio et al. 2004; Wanajo & Ishimaru 2005). Suggestions include a "weak" r-process acting as a "failed" r-process only producing elements with $Z < 56$. An example of this process may the r-poor star HD122563 (Honda et al., 2006, 2007) showing a comparable, exclusive light neutron-capture element enhancement. Unraveling the exact details about the different r-processes should shed new light on the astrophysical site(s) of the r-process.

A "clean" r-process signature can only be observed in those stars that formed before the onset of the s-process ([Fe/H] ~ -2.6; Simmerer et al. 2004) and at times when the galactic inventory of lighter elements was still low so that spectral contamination with metal lines is minimal. Only the most metal-poor stars can thus be examined for the r-process signature. The first strongly enhanced r-process star was discovered more than a decade ago in the HK survey, CS 22892-052 (Sneden et al., 1996). A second star was found a few years thereafter, CS 31082-001 (Cayrel et al., 2001). Recognizing the need to find more such stars to enable detailed studies of the r-process, Christlieb et al. (2004) initiated a large campaign to observe metal-poor candidate stars from the main Hamburg/ESO survey. It led to the discovery of several strongly r-process-enhanced stars (Barklem et al., 2005; Hayek et al., 2009) and dozens of mildly enriched ones. Two more such stars were found more recently elsewhere (Frebel et al., 2007a; Lai et al., 2008).

The metallicity of all the strongly r-enriched stars is coincidentally [Fe/H] ~ -3.0 or just above it. Regarding their light-elements, no unusual abundances are found, and their halo-typical enhancements in α-elements suggest a core-collapse SN the source responsible for their overall elemental signature, including the neutron-capture elements. This is in line with SNe being the early contributors to galactic chemical evolution whereas AGB enrichment sets in only at somewhat later times due to their evolutionary timescales and the presence of seed nuclei (i.e., from Fe-peak elements created in previous SNe generations).

4.2 Nucleo-chronometry of the oldest stars

The long-lived radioactive isotopes ^{232}Th (half-life of 14 Gyr) and ^{238}U (4.5 Gyr) are suitable for measurements of cosmic timescales. They also have transitions in the optical range so that, in principle, Th and U abundances can be measured in stellar spec-

---

[6]The Sun's r-process pattern can be obtained by subtracting the calculated s-process component from the measured total abundance pattern.

tra of r-enriched stars. However, suitable stars for these challenging measurements are difficult to find: Cool metal-poor red giants that exhibit very strong overabundances of r-process elements. Their carbon abundances should be low (i.e. subsolar) since CH features blend with many important neutron-capture lines (e.g. U, Pb), rendering them unmeasurable in carbon-rich r-process stars such as CS 22892-052. Moreover, the stars need to be bright (preferably $V < 13$) so that high-resolution data with very high S/N can be collected in reasonable observing times. Many neutron-capture features are very weak and often partially blended and thus require exceptionally high data quality. The two most important examples are the extremely weak U line at 3859 Å and the even weaker Pb line at 4057 Å. These two lines are the strongest optical transitions of these element. Subject to the availability of a very suitable star, a successful U measurement requires a $R > 60\,000$ spectrum with a S/N of at least 350 per pixel at 3900 Å. A Pb measurement may be attempted in a spectrum with S/N ~ 500 at 4000 Å.

Fortunately, Th and several other stable r-process elements (most notably Eu) can be detected in lower S/N data, and have led to a number of stellar age measurements. Such ages can be derived from a ratio of a radioactive element to a stable r-process nuclei (i.e. Th/r, U/r, U/Th; with r being stable elements such as Eu, Os, and Ir), and comparing them with calculations of their initial r-process productions (e.g. Schatz et al. 2002). Since the giant CS 22892–052 is very C-rich, U could not be measured but its Th/Eu ratio yielded an age of 14 Gyr (Sneden et al., 2003). Johnson & Bolte (2001) measured similar ages of five mildly r-enriched stars, and other studies have produced ages of another few (Barklem et al., 2005; Cowan et al., 2002; Hayek et al., 2009; Lai et al., 2008; Westin et al., 2000).

Compared to Th/Eu, the U/Th ratio is more robust to uncertainties in the theoretically derived production ratio because Th and U haves similar atomic masses (for which uncertainties would largely cancel out; e.g., Kratz et al. 2004; Wanajo et al. 2002). Hence, stars displaying Th *and* U are the more desired r-process stars. For a similar reason, stable elements of the 3rd r-process peak ($76 \leq Z \leq 78$) are best used in combination with the actinides (Th, $Z = 90$ and U, $Z = 92$). The first U detection was made in the giant CS 31082-001 (Cayrel et al., 2001) giving an age of 14 Gyr as derived from the U/Th ratio. Other chronometer ratios in this star, such as Th/Eu, however, yielded *negative* ages. This is due to unusually high Th and U abundances combined with an underabundance in Pb relative to the majority of r-process enriched metal-poor stars. By now, there are four known r-process stars (Hayek et al., 2009; Hill et al., 2002; Honda et al., 2004; Lai et al., 2008) with such high Th/Eu ratios ($\sim 20\%$ of r-process stars). Since only the elements heavier than the 3rd r-process peak are (equally) affected (Roederer et al., 2009), the U/Th ratio still gives a reasonable age in CS31082-01. This behavior was termed an "actinide-boost" (Schatz et al., 2002), but it has become clear that these stars likely have a different origin (Kratz et al., 2007) than "normal" r-process stars. It also supports the conjecture that there are multiple r-process sites (Hill et al., 2002).

Only one star has so far been discovered for which measurements of Th and U provide stellar age determinations from more than just one chronometer ratio. The giant HE 1523−0901 ($V = 11.1$) was found in a sample of bright metal-poor stars (Frebel et al., 2006) from the Hamburg/ESO Survey. It has the strongest enhance-

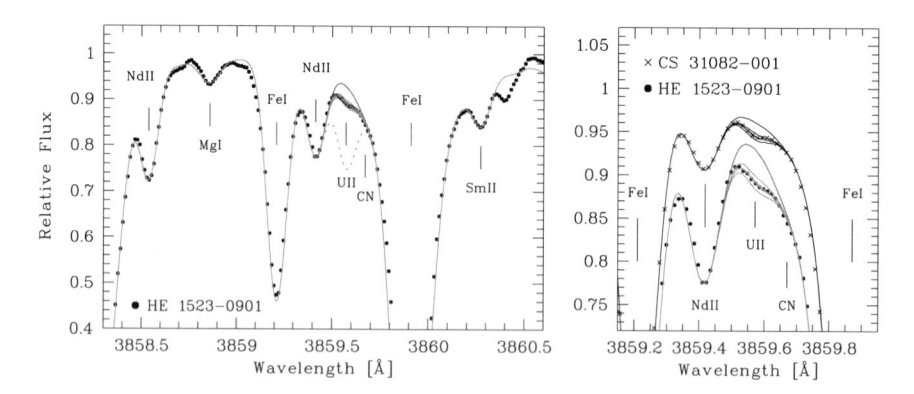

Figure 5: Spectral region around the U II line in HE 1523−0901 (filled dots) and CS 31082-001 (crosses; *right panel* only). Overplotted are synthetic spectra with different U abundances. The dotted line in the left panel corresponds to a scaled solar r-process U abundance present in the star if no U had decayed. Figure taken from Frebel et al. (2007a).

ment in r-process elements so far, [r/Fe] = 1.8 (Frebel et al., 2007a), and among the measured neutron-capture elements are Os, Ir, Th, and U. In fact, the U detection in this star is currently the most reliable one of only *three* stars with such measurements (in CS31082-01, and a somewhat uncertain detection in BD +17° 3248; Cowan et al. 2002). Figure 5 shows the spectral region around the U line from which the U abundance was deduced. The averaged stellar age of ~ 13 Gyr (Frebel et al., 2007a) is based on seven chronometers Th/r, U/r and U/Th involving combinations of Eu, Os, Ir, Th and U.

Unfortunately, realistic uncertainties for any such age measurements range from ~ 2 to ~ 5 Gyr (see Schatz et al. 2002 and Frebel et al. 2007a for a discussion). Nevertheless, the stellar ages of old stars naturally provide an important independent lower limit to the age of the Galaxy, and hence, the Universe. The currently accepted value for the age of the Universe is 13.7 Gyr (Spergel et al., 2007), making these r-process stars some of the oldest known objects. This is in line with the low metallicities of the stars that already indicates a very early formation time.

In the absence of an age-metallicity relation for field halo stars, the nucleo-chronometric ages thus demonstrate that metal-deficient stars are indeed quite ancient, and well suited for studying the early Universe. By extension, this furthermore suggests that metal-poor stars with similarly low Fe abundances but no excess in r-process elements should also be old. Moreover, the commonly made assumption about the low mass (0.6 to 0.8 $M_\odot$) of these long-lived objects is justified as well.

4.3 A stellar triumvirate: Th, U and Pb

Since different r-process models and the associated parameters usually yield different r-process abundance distributions, particularly in the heavy mass range, self-consistency constraints are very valuable. Explaining the stellar abundance triumvirate of Th, U *and* Pb can provide such constraints. These three elements are intimately coupled not only with each other but also to the conditions (and potentially also the environment) of the r-process. Pb is the β- plus α-decay end-product of all decay chains in the mass region between Pb and the onset of dominant spontaneous fission above Th and U. It is also built up from the decay of Th and U isotopes. Together with the Th and U measurements, a known Pb abundance provides important constraints also on those poorly understood decay channels. They are of critical importance for the successful modeling of the r-process which, in turn, could provide improved initial production ratios necessary for stellar age dating.

A new spectrum of HE 1523−0901 with S/N ~ 500 was recently obtained with VLT/UVES. Preliminary results indicate a detection of this extremely difficult line measurement with an abundance of $-0.4 < \log \epsilon(\text{Pb}) < -0.3$ (Frebel et al. 2010, in preparation). To learn about the different contributions to the production of lead, decay contributions of ^{238}U into ^{206}Pb, ^{232}U into ^{208}Pb, and ^{235}U into ^{207}Pb can be calculated (whereby the last one is based on a theoretically derived ratio of ^{235}U/^{238}U). The total abundance of these three decays channels amounts to $\log \epsilon(\text{Pb}) = -0.72$ which leaves "room" for the direct and indirect decay channels that likely produce the main portion of the Pb in the star. Using r-process model calculations, predictions were derived for the total Pb to be found in HE 1523−0901. A site-independent model of the classical "waiting-point" approximation yielded $\epsilon(\text{Pb}) = -0.346$ (Frebel & Kratz, 2009) for a decay time of 13 Gyr which is in agreement with the preliminary abundances. At the same time, this prediction is less in agreement with the measured value of $\log \epsilon(\text{Pb}) = -0.55$ in CS31082-001 (Plez et al., 2004). This low Pb values is difficult to understand theoretically (despite the actinide boost and high Th and U), and is currently subject to much debate.

4.4 The s-process signatures

Neutron-capture elements are also produced in the interiors of low and intermediate-mass asymptotic giant branch (AGB) stars through the s-process and later dredged up to the surface. Unlike the r-process, the s-process is not universal because two different sites seem to host s-process nucleosynthesis. The "main" component of the s-process occurs in the helium shells of thermally pulsing lower mass AGB stars and is believed to account for elements with $Z \geq 40$ (e.g., Arlandini et al. 1999; Karakas et al. 2002). The so-called "weak" component occurs in the He- and C-burning cores of more massive stars, and preferentially produces elements around $Z \sim 40$.

The s-process leads to a different characteristic abundance pattern of neutron-capture elements than the r-process. The signature is also observed in some metal-poor stars, and their abundances follow that of the scaled solar s-process. The s-process metal-poor stars received their s-enriched material during a mass transfer event across a binary system from a former more massive companion that went through its AGB phase (e.g., Aoki et al. 2001; Bisterzo et al. 2009; Norris et al. 1997a). During this period, large amounts of carbon are also transferred to the companion producing a typical carbon-rich, s-rich metal-poor stellar signature. Some metal-poor stars display signatures from both r- and s-process enhancement in their abundance pattern (e.g., Barbuy et al. 2005; Cohen et al. 2003; Jonsell et al. 2006; Roederer et al. 2008). Several different scenarios have been invoked to explain the combination of the two neutron-capture processes originating at two very different astrophysical sites. However, no unambiguous explanation has yet been found.

5 Tracing the formation of the Galactic halo with metal-poor stars

Simulations of the hierarchical assembly of galaxies within the cold dark matter framework (Diemand et al. 2007; Springel et al. 2008) show that the Milky Way halo was successively built up from small dark matter substructures, often referred to as galactic building blocks, as long ago suggested by Searle & Zinn (1978). However, these simulations only include dark matter, and it remains unclear to what extent small dark halos contain luminous matter in the form of stars and gas. This question is particularly important with respect to the so-called "missing-satellite" problem which reflects the mismatch of the number of observed dwarf galaxies surrounding the Milky with the predicted number of substructures for a Milky Way-like halo. Studying the onset of star formation and associated chemical evolution in dwarf galaxies thus provides some of the currently missing information for our understanding of how the observed properties of small satellites relate to the dark matter substructures that build up larger galaxies.

5.1 Chemical evolution of dwarf galaxies and their connection to the stellar halo

The connection between the surviving dwarfs and those that dissolved to form the halo is best addressed by examining in detail the stellar chemical compositions of present-day dwarf galaxies. Establishing detailed chemical histories of these different systems can provide constraints on their dominant chemical enrichment events, as well as the formation process of the Milky Way. Specifically, detailed knowledge of the most metal-poor (hence, oldest) stars in a given system allow insight into the earliest phases of star formation before the effects of internal chemical evolution were imprinted in stars born later with higher metallicity (for further details on this topic we refer the reader to the recent reviews by Tolstoy et al. 2009 and Koch 2009).

If the old, metal-poor halo was assembled from dwarf galaxies, the metallicities of stars in dwarf galaxies must reach values as low as (or lower) than what is currently found in the Galactic halo. Moreover, the abundance ratio of those low-metallicity stars must be roughly equal to those of equivalent stars in the halo. Assuming that the currently observable dwarf galaxies are the survivors of such an early accretion process, they provide an opportunity to examine their stellar chemistry in search for such an accretion process. Until recently, the common wisdom was, however, that the "classical" dwarf galaxies in the Local Group (e.g., Carinae, Sextans, Sculptor, and Fornax), would not contain any stars below [Fe/H] $= -3$ (e.g., Shetrone et al. 2003) despite the fact that many halo stars exist with such low Fe values, and even some with metallicities as low as [Fe/H] ~ -5.0. Furthermore, the higher-metallicity stars were found to have abundance ratios different from comparable halo stars. Most strikingly, the α-element abundances are not enhanced, indicating different enrichment mechanisms and timescales in these systems. This sparked a debate about the viability of the Searle & Zinn (1978) paradigm, and in trying to explain the origin of the metal-poor stellar halo. However, it has now been shown that this claim stems merely from biases in earlier search techniques (Cohen & Huang, 2009; Kirby et al., 2009; Starkenburg et al., 2010). With improved methods for identifying the lowest-metallicity objects (e.g., Kirby et al. 2008), extremely metal-poor stars with [Fe/H] < -3 have already been identified in several dwarf galaxies (Frebel et al., 2010a; Geha et al., 2009; Kirby et al., 2008; Norris et al., 2008).

Ultra-faint dwarf Galaxies   Kirby et al. (2008) extended the metallicity-luminosity relationship to the ultra-faint ($L < 10^5$ $L_\odot$) dwarfs and showed that the lowest luminosity dwarfs have indeed the lowest average metallicities, with several individual stars having [Fe/H] < -3. The combined MDF of all these systems goes down to [Fe/H] ~ -3.3, and the shape appears similar to that of the Milky Way halo. The ultra-faint dwarf galaxies have large internal metallicity spreads, confirming earlier results at lower spectral resolution (e.g., Simon & Geha 2007; Norris et al. 2008). They span more than 1 dex in [Fe/H] in some of them. Such spreads indicate early star formation in multiple proto-dwarf galaxies that later merged, extended star formation histories, or incomplete mixing in the early ISM, or all of the above.

The three brightest stars in each of Ursa Major II (UMa II) and Coma Berenices (ComBer) and two stars in Hercules are the first stars in the ultra-faint dwarf galaxies to have been observed with high-resolution spectroscopy. Two of them (in UMa II) are also the first known extremely metal-poor stars which are not members of the halo field population. Details on the observations and analysis techniques are given in Frebel et al. (2010a) and Koch et al. (2008). For the UMa II and ComBer stars, chemical abundances and upper limits of up to 26 elements were determined for each star. The abundance results demonstrate that the evolution of the elements in the ultra-faint dwarfs is very similar to that of the Milky Way, and likely dominated by massive stars. The α-elements in these two ultra-faint dwarf stars are overabundant, showing the halo-typical core-collapse SNe signature. This is the first evidence that the abundance patterns of light elements ($Z < 30$) in the ultra-faint dwarfs are remarkably similar to the Milky Way halo. The agreement suggests that the metal-poor end of the MW halo population could have been built up from destroyed dwarf galaxies. Similar abundance results were also found by other studies (Feltzing et al. 2009; Norris et al. 2010; Simon et al. 2010) but also indications that the chemical evolution in these small systems may have been very inhomogeneous.

The neutron-capture abundances are extremely low in the ultra-faint dwarf stars. The observed Sr and Ba values are up to two orders of magnitude below the abundances found in typical MW halo stars with similar Fe content. However, a very large spread (up to 3 dex) in these elements is found among halo stars themselves. The large depletions in the dwarf galaxy stars are thus not inconsistent with the halo data since similarly low values are found in numerous objects. Interestingly though, the low neutron-capture abundances may represent a typical signature of stars in dwarf galaxies. Similarly low values are also found in Hercules (Koch et al., 2008) and Draco (Fulbright et al., 2004) despite their sometimes relatively high Fe values of [Fe/H] ~ -2.0 (in Hercules).

"Classical" dSphs By applying the new search techniques also to the more luminous dwarf galaxy Sculptor, the first extremely metal-poor star in a classical dwarf was recently discovered (in a sample of 380 stars; Kirby et al. 2009). The metallicity of [Fe/H] ~ -3.8 was confirmed from a high-resolution spectrum taken with Magellan/MIKE (Frebel et al., 2010b). Only nine stars in the halo have even lower Fe abundances than this object. This remarkable finding suggests that more such low-metallicity stars could soon be identified in the more luminous systems (see also Starkenburg et al. 2010). The new star also shows that a metallicity spread of ~ 3 dex is present in Sculptor. The chemical abundances obtained from the high-resolution spectrum reveal a similar picture to what has been found in the ultra-faint dwarf stars. The Sculptor star, at [Fe/H] ~ -3.8, also has a remarkably similar chemical make-up compared to that of the Milky Way halo at the lowest metallicities. This is in contrast to what is found at higher metallicities in these brighter dwarfs which have lower α-abundances than comparable halo stars (e.g., Geisler et al. 2005; Shetrone et al. 2003). There is increasing evidence, though that a transition of halo-typical abundance ratios may take place around a metallicity of [Fe/H] $= -3.0$ (Aoki et al., 2009; Cohen & Huang, 2009).

5.2 The origin of the metal-poor tail of the halo

These new observational results are broadly consistent with the predictions of currently favored cosmological models (e.g. Robertson et al. 2005; Johnston et al. 2008). The majority of the mass presumably in the inner part of the stellar halo (at [Fe/H] ~ -1.2 to -1.6) was formed in much larger systems such as the Magellanic Clouds. A scenario where the ultra-faint dwarf galaxies contributed some individual metal-poor stars that are now found primarily in the outer Galactic halo (although not exclusively) is supported. However, these systems may not have been sufficiently numerous to account for the entire metal-poor end of the Fe metallicity distribution of the Milky Way halo. Since the classical dSphs have more stellar mass and have been shown to also contain at least some of the most metal-poor stars (Frebel et al., 2010b; Kirby et al., 2009), they could have been a major source of the lowest-metallicity halo stars. Additional observations of more extremely metal-poor stars in the various dwarf galaxies are highly desirable in the quest to determine individual MDFs for each of these galaxies, and how those would compare with each other and the Milky Way.

6 Outlook: what is possible with stellar archaeology?

Old metal-poor stars have long been used to learn about the conditions of the early Universe. This includes the origin and evolution of the chemical elements, the relevant nucleosynthesis processes and sites and the overall chemical and dynamical history of the Galaxy. By extension, metal-poor stars provide constraints on the nature of the first stars and their initial mass function, the chemical yields of first/early SNe, as well as early star and galaxy formation processes including the formation of the galactic halo. Finding more of the most metal-poor stars (e.g., stars with [Fe/H] < −5.0) would enormously help to address all of these topics in more detail. However, these stars are extremely rare (Schörck et al., 2009) and difficult to find. The most promising way forward is to survey larger volumes further out in the Galactic halo.

But how feasible is it to identify stars with even lower metallicities? Frebel et al. (2009) calculated the minimum observable Fe and Mg abundances in the Galaxy by combining the critical metallicity of $[C/H]_{min} = -3.5$ (the criterion for the formation of the first low-mass stars by Bromm & Loeb 2003) with the maximum carbon-to-iron ratio found in any metal-poor star. The resulting minimum Fe value is $[Fe/H]_{min} = -7.3$. Analogously, the minimum Mg value is $[Mg/H]_{min} = -5.5$. If $[C/H]_{min}$ was lower, e.g., $[C/H]_{min} = -4.5$, as suggested by recent dust cooling computations, the minimum observable Fe and Mg abundances would accordingly be lower. Spectrum synthesis calculations suggest these low abundance levels are indeed measurable from each of the strongest Fe and Mg lines in suitably cool metal-deficient giants.

Future surveys such as Skymapper and LAMOST will provide an abundance of new metal-poor candidates as well as new faint dwarf galaxies. By accessing such stars in the outer Galactic halo and dwarf galaxies we will be able to gain a more complete census of the chemical and dynamical history of our own Galaxy. Also, the lowest metallicity stars are expected to be in the outer halo (e.g., Carollo et al. 2007). Their corresponding kinematic signatures may prevent them from accreting too much enriched material from the ISM during their lives so that their surface composition would not be altered (i.e. increased, Frebel et al. 2009). Hence, selecting for the most metal-poor candidates will increasingly rely on our ability to combine chemical abundance analyzes with kinematic information. Future missions such as GAIA will provide accurate proper motions for many object that have no kinematic information available, including for most of the currently known metal-poor giants.

However, many, if not most, of these future metal-poor candidates will be too faint to be followed up with the high-resolution spectroscopy necessary for detailed abundance analyzes. This is already an issue for many current candidates leaving the outer halo a so far largely unexplored territory: The limit for high-resolution work is $V \sim 19$ mag, and one night's observing with 6–10 m telescopes is required for the minimum useful signal-to-noise (S/N) ratio of such a spectrum. With the light-collecting power of the next generation of optical telescopes, such as the Giant Magellan Telescope, the Thirty Meter Telescope or the European ELT, and if they are equipped with high-resolution spectrographs, it would be possible to not only reach out into the outer halo in search of the most metal-poor stars, but also provide spectra with very high-S/N ratio of somewhat brighter stars. For example, r-process enhanced stars which provide crucial empirical constraints on the nature of this nucleosynthesis process require exquisite data quality, e.g. for uranium and lead measurements that are currently only possible for the very brightest stars.

It appears that the hunt for the metal-deficient stars in dwarf galaxies may have just begun since these dwarfs host a large fraction of low-metallicity stars, perhaps even much higher than what has so far been inferred for the Galactic halo (Schörck et al., 2009). The detailed abundance patterns of the stars in UMa II, ComBer, Leo IV and Sculptor are strikingly similar to that of the Milky Way stellar halo, thus renewing the support for dwarf galaxies as the building blocks of the halo. Future discoveries of additional faint dwarf galaxies will enable the identification of many more metal-poor stars in new, low-luminosity systems. But also the brighter dSphs have to be revisited for their metal-poor content (Kirby et al., 2009). More stars at the lowest metallicities are clearly desired to better quantify the emerging chemical signatures and to solidify our understanding of the early Galaxy assembly process. Together with advances in the theoretical understanding of early star and galaxy formation and SNe yields, a more complete picture of the evolution of the Milky Way Galaxy and other systems may soon be obtained. Only in this way can the hierarchical merging paradigm for the formation of the Milky Way be put on firm observational ground.

Acknowledgements

I am very grateful to the Astronomische Gesellschaft and its selection committee for awarding me the 2009 Ludwig-Biermann Award. Lars Hernquist and Ian Roederer have given useful comments to the manuscript, and John Norris provided the spectrum of G66-30 shown in Fig. 1. I warmly thank my many wonderful collaborators who have always inspired me, and make working in this field a great pleasure: Wako Aoki, Martin Asplund, Tim Beers, Volker Bromm, Norbert Christlieb, Karl-Ludwig Kratz, Evan Kirby, John Norris, Ian Roederer, Josh Simon, Chris Sneden and many others. This work has been supported by a Clay Postdoctoral Fellowship administered by the Smithsonian Astrophysical Observatory. This research has made use of NASA's Astrophysics Data System. This research has made use of the SIMBAD database, operated at CDS, Strasbourg, France.

References

Aoki, W., Norris, J.E., Ryan, S.G., Beers, T.C., Ando, H.: 2000, ApJ 536, L97

Aoki, W., Ryan, S.G., Norris, J.E., et al.: 2001, ApJ 561, 346

Aoki, W., Ando, H., Honda, S., Iye, M., et al.: 2002a, PASJ 54, 427

Aoki, W., Norris, J.E., Ryan, S.G., Beers, T.C., Ando, H.: 2002b, ApJ 576, L141

Aoki, W., Norris, J.E., Ryan, S.G., Beers, T.C., Ando, H.: 2002c, ApJ 567, 1166

Aoki, W., Ryan, S.G., Norris, J.E., Beers, T.C., Ando, H., Tsangarides, S.: 2002d, ApJ 580, 1149

Aoki, W., Honda, S., Beers, T.C., et al.: 2005, ApJ 632, 611

Aoki, W., Bisterzo, S., Gallino, R., Beers, T.C., Norris, J.E., Ryan, S.G., Tsangarides, S.: 2006a, ApJ 650, L127

Aoki, W., Frebel, A., Christlieb, N., et al.: 2006b, ApJ 639, 897

Aoki, W., Beers, T.C., Christlieb, N., Norris, J.E., Ryan, S.G., Tsangarides, S.: 2007a, ApJ 655, 492

Aoki, W., Honda, S., Beers, T.C., et al.: 2007b, ApJ 660, 747

Aoki, W., Honda, S., Sadakane, K., & Arimoto, N. 2007c, PASJ, 59, L15

Aoki, W., Beers, T.C., Sivarani, T., et al.: 2008, ApJ 678, 1351

Aoki, W., Arimoto, N., Sadakane, K., et al.: 2009, A&A 502, 569

Argast, D., Samland, M., Gerhard, O.E., Thielemann, F.-K.: 2000, A&A 356, 873

Arlandini, C., Käppeler, F., Wisshak, K., Gallino, R., Lugaro, M., Busso, M., Straniero, O.: 1999, ApJ 525, 886

Arnone, E., Ryan, S.G., Argast, D., Norris, J.E., Beers, T.C.: 2005, A&A 430, 507

Asplund, M., Grevesse, N., Sauval, A.J., Scott, P.: 2009, ARA&A 47, 481

Barbuy, B., Spite, M., Spite, F., Hill, V., Cayrel, R., Plez, B., Petitjean, P.: 2005, A&A 429, 1031

Barklem, P.S., Christlieb, N., Beers, T.C., et al.: 2005, A&A 439, 129

Beers, T.C., Christlieb, N.: 2005, ARA&A 43, 531

Beers, T.C., Preston, G.W., Shectman, S.A.: 1992, AJ 103, 1987

Bergemann, M., Gehren, T.: 2008, A&A 492, 823

Bessell, M.S., Norris, J.: 1984, ApJ 285, 622

Bisterzo, S., Gallino, R., Straniero, O., Aoki, W.: 2009, PASA 26, 314

Bonifacio, P., Spite, M., Cayrel, R., et al.: 2009, A&A 501, 519

Bromm, V., Larson, R.B.: 2004, ARA&A 42, 79

Bromm, V., Loeb, A.: 2003, Nature 425, 812

Bromm, V., Yoshida, N., Hernquist, L., McKee, C.F.: 2009, Nature 459, 49

Burris, D.L., Pilachowski, C.A., Armandroff, T.E., Sneden, C., Cowan, J.J., Roe, H.: 2000, ApJ 544, 302

Carollo, D., Beers, T.C., Lee, Y.S., et al.: 2007, Nature 450, 1020

Carretta, E., Gratton, R., Cohen, J.G., Beers, T.C., Christlieb, N.: 2002, AJ 124, 481

Cayrel, R., Hill, V., Beers, T., et al.: 2001, Nature 409, 691

Cayrel, R., Depagne, E., Spite, M., et al.: 2004, A&A 416, 1117

Christlieb, N., Bessell, M.S., Beers, T.C., et al.: 2002, Nature 419, 904

Christlieb, N., Beers, T.C., Barklem, P.S., et al.: 2004, A&A 428, 1027

Cohen, J.G., Huang, W.: 2009, ApJ 701, 1053

Cohen, J.G., Christlieb, N., Qian, Y.-Z., Wasserburg, G.J.: 2003, ApJ 588, 1082

Cohen, J.G., Christlieb, N., McWilliam, A., et al.: 2004, ApJ 612, 1107

Cohen, J.G., McWilliam, A., Christlieb, N., et al.: 2007, ApJ 659, L161

Cohen, J.G., Christlieb, N., McWilliam, A., et al.: 2008, ApJ 672, 320

Cohen, J.G., McWilliam, A., Shectman, S., et al.: 2006, AJ 132, 137

Collet, R., Asplund, M., Trampedach, R.: 2006, ApJ 644, L121

Cowan, J.J., Sneden, C., Burles, S., et al.: 2002, ApJ 572, 861

Diemand, J., Kuhlen, M., Madau, P.: 2007, ApJ 667, 859

Feltzing, S., Eriksson, K., Kleyna, J., Wilkinson, M.I.: 2009, A&A 508, L1

François, P., Depagne, E., Hill, V., et al.: 2007, A&A 476, 935

Frebel, A.: 2009, in PoS, Vol. in press, Nuclei in the Cosmos

Frebel, A., Kratz, K.: 2009, in: E.E. Mamajek, D.R. Soderblom, R.F.G. Wyse (eds.), *The Ages of Stars*, IAU Symp. 258, p. 449

Frebel, A., Aoki, W., Christlieb, N., et al.: 2005, Nature 434, 871

Frebel, A., Christlieb, N., Norris, J.E., et al.: 2006, ApJ 652, 1585

Frebel, A., Christlieb, N., Norris, J.E., Thom, C., Beers, T.C., Rhee, J.: 2007a, ApJ 660, L117

Frebel, A., Johnson, J.L., Bromm, V.: 2007b, MNRAS 380, L40

Frebel, A., Norris, J.E., Aoki, W., et al.: 2007c, ApJ 658, 534

Frebel, A., Collet, R., Eriksson, K., Christlieb, N., Aoki, W.: 2008, ApJ 684, 588

Frebel, A., Johnson, J.L., Bromm, V.: 2009, MNRAS 392, L50

Frebel, A., Simon, J.D., Geha, M., Willman, B.: 2010a, ApJ 708, 560

Frebel, A., Kirby, E., Simon, J.D.: 2010b, Nature 464, 72

Fröhlich, C., Martínez-Pinedo, G., Liebendörfer, M., et al.: 2006, Phys. Rev. Lett. 96, 142502

Fulbright, J.P.: 2000, AJ 120, 1841

Fulbright, J.P., Rich, R.M., Castro, S.: 2004, ApJ 612, 447

Geha, M., Willman, B., Simon, J.D., Strigari, L.E., Kirby, E.N., Law, D.R., Strader, J.: 2009, ApJ 692, 1464

Geisler, D., Smith, V.V., Wallerstein, G., Gonzalez, G., Charbonnel, C.: 2005, AJ 129, 1428

Hayek, W., Wiesendahl, U., Christlieb, N., et al.: 2009, A&A 504, 511

Heger, A., Woosley, S.E.: 2002, ApJ 567, 532

Heger, A., Woosley, S.E.: 2008, astro-ph/0803.3161

Hill, V., Plez, B., Cayrel, R., et al.: 2002, A&A 387, 560

Honda, S., Aoki, W., Kajino, T., et al.: 2004, ApJ 607, 474

Honda, S., Aoki, W., Ishimaru, Y., Wanajo, S., Ryan, S.G.: 2006, ApJ 643, 1180

Honda, S., Aoki, W., Ishimaru, Y., Wanajo, S.: 2007, ApJ 666, 1189

Iben, I.: 1983, Mem. Soc. Astron. Ital. 54, 321

Ito, H., Aoki, W., Honda, S., Beers, T.C.: 2009, ApJ 698, L37

Ivans, I.I., Sneden, C., James, C.R., et al.: 2003, ApJ 592, 906

Ivans, I.I., Sneden, C., Gallino, R., Cowan, J.J., Preston, G.W.: 2005, ApJ 627, L145

Ivans, I.I., Simmerer, J., Sneden, C., Lawler, J.E., Cowan, J.J., Gallino, R., Bisterzo, S.: 2006, ApJ 645, 613

Iwamoto, N., Umeda, H., Tominaga, N., Nomoto, K., Maeda, K.: 2005, Sci 309, 451

Johnson, J.A.: 2002, ApJS 139, 219

Johnson, J.A., Bolte, M.: 2001, ApJ 554, 888

Johnson, J.A., Bolte, M.: 2002a, ApJ 579, L87

Johnson, J.A., Bolte, M.: 2002b, ApJ 579, 616

Johnson, J.A., Bolte, M.: 2004, ApJ 605, 462

Johnston, K.V., Bullock, J.S., Sharma, S., Font, A., Robertson, B.E., Leitner, S.N.: 2008, ApJ 689, 936

Jonsell, K., Edvardsson, B., Gustafsson, B., Magain, P., Nissen, P.E., Asplund, M.: 2005, A&A 440, 321

Jonsell, K., Barklem, P.S., Gustafsson, B., Christlieb, N., Hill, V., Beers, T.C., Holmberg, J.: 2006, A&A 451, 651

Karakas, A.I., Lattanzio, J.C., Pols, O.R.: 2002, PASA 19, 515

Karlsson, T., Gustafsson, B.: 2005, A&A 436, 879

Kirby, E. N., Guhathakurta, P., Bolte, M., Sneden, C., & Geha, M. C. 2009, ApJ, 705, 328

Kirby, E.N., Simon, J.D., Geha, M., Guhathakurta, P., Frebel, A.: 2008, ApJ 685, L43

Koch, A.: 2009, AN 330, 675

Koch, A., McWilliam, A., Grebel, E.K., Zucker, D.B., Belokurov, V.: 2008, ApJ 688, L13

Kratz, K.-L., Farouqi, K., Pfeiffer, B., Truran, J.W., Sneden, C., Cowan, J.J.: 2007, ApJ 662, 39

Kratz, K.-L., Pfeiffer, B., Cowan, J.J., Sneden, C.: 2004, New A Rev. 48, 105

Lai, D.K., Johnson, J.A., Bolte, M., Lucatello, S.: 2007, ApJ 667, 1185

Lai, D.K., Bolte, M., Johnson, J.A., Lucatello, S., Heger, A., Woosley, S.E.: 2008, ApJ 681, 1524

Lai, D.K., Rockosi, C.M., Bolte, M., et al.: 2009, ApJ 697, L63

Limongi, M., Chieffi, A., Bonifacio, P.: 2003, ApJ 594, L123

Lucatello, S., Gratton, R., Cohen, J.G., Beers, T.C., Christlieb, N., Carretta, E., Ramírez, S.: 2003, AJ 125, 875

Masseron, T., van Eck, S., Famaey, B., et al.: 2006, A&A 455, 1059

McWilliam, A.: 1998, AJ 115, 1640

McWilliam, A., Preston, G.W., Sneden, C., Searle, L.: 1995, AJ 109, 2757

Meynet, G., Ekström, S., Maeder, A.: 2006a, A&A 447, 623

Meynet, G., Ekström, S., Maeder, A.: 2006b, A&A 447, 623

Norris, J.E., Beers, T.C., Ryan, S.G.: 2000, ApJ 540, 456

Norris, J.E., Christlieb, N., Korn, A.J., et al.: 2007, ApJ 670, 774

Norris, J.E., Gilmore, G., Wyse, R.F.G., Wilkinson, M.I., Belokurov, V., Wyn Evans, N., Zucker, D.B.: 2008, ApJ 689, L113

Norris, J.E., Ryan, S.G., Beers, T.C.: 1997a, ApJ 488, 350

Norris, J.E., Ryan, S.G., Beers, T.C.: 1997b, ApJ 489, L169

Norris, J.E., Ryan, S.G., Beers, T.C., Deliyannis, C.P.: 1997c, ApJ 485, 370

Norris, J. E., Ryan, S. G., & Beers, T. C. 2001, ApJ, 561, 1034

Norris, J.E., Ryan, S.G., Beers, T.C., Aoki, W., Ando, H.: 2002, ApJ 569, L107

Norris, J.E., Yong, D., Gilmore, G., Wyse, R.F.G.: 2010, ApJ 711, 350

Plez, B., Hill, V., Cayrel, R., et al.: 2004, A&A 428, L9

Preston, G.W., Sneden, C.: 2000, AJ 120, 1014

Preston, G.W., Sneden, C.: 2001, AJ 122, 1545

Preston, G.W., Sneden, C., Thompson, I.B., Shectman, S.A., Burley, G.S.: 2006, AJ 132, 85

Qian, Y.-Z., Wasserburg, G.J.: 2003, ApJ 588, 1099

Robertson, B., Bullock, J.S., Font, A.S., Johnston, K.V., Hernquist, L.: 2005, ApJ 632, 872

Roederer, I.U., Frebel, A., Shetrone, M.D., et al.: 2008, ApJ 679, 1549

Roederer, I.U., Kratz, K., Frebel, A., Christlieb, N., Pfeiffer, B., Cowan, J.J., Sneden, C.: 2009, ApJ 698, 1963

Ryan, S.G., Norris, J.E.: 1991, AJ 101, 1835

Ryan, S.G., Norris, J.E., Beers, T.C.: 1996, ApJ 471, 254

Ryan, S.G., Norris, J.E., Beers, T.C.: 1999, ApJ 523, 654

Schatz, H., Toenjes, R., Pfeiffer, B., Beers, T.C., Cowan, J.J., Hill, V., Kratz, K.-L.: 2002, ApJ 579, 626

Schneider, R., Omukai, K., Inoue, A.K., Ferrara, A.: 2006, MNRAS 369, 1437

Schörck, T., Christlieb, N., Cohen, J.G., et al.: 2009, A&A 507, 817

Searle, L., Zinn, R.: 1978, ApJ 225, 357

Shetrone, M., Venn, K.A., Tolstoy, E., Primas, F., Hill, V., Kaufer, A.: 2003, AJ 125, 684

Shetrone, M.D., Côté, P., Sargent, W.L.W.: 2001, ApJ 548, 592

Shigeyama, T., Tsujimoto, T., Yoshii, Y.: 2003, ApJ 586, L57

Simmerer, J., Sneden, C., Cowan, J.J., Collier, J., Woolf, V.M., Lawler, J.E.: 2004, ApJ 617, 1091

Simon, J.D., Geha, M.: 2007, ApJ 670, 313

Simon, J.D., Frebel, A., McWilliam, A., Kirby, E.N., Thompson, I.B.: 2010, astro-ph/1001.3137

Sivarani, T., Bonifacio, P., Molaro, P., et al.: 2004, A&A 413, 1073

Sivarani, T., Beers, T.C., Bonifacio, P., et al.: 2006, A&A 459, 125

Sneden, C., McWilliam, A., Preston, G.W., Cowan, J.J., Burris, D.L., Amorsky, B.J.: 1996, ApJ 467, 819

Sneden, C., Cowan, J.J., Lawler, J.E., et al.: 2003, ApJ 591, 936

Sneden, C., Cowan, J.J., Gallino, R.: 2008, ARA&A 46, 241

Sobeck, J.S., Lawler, J.E., Sneden, C.: 2007, ApJ 667, 1267

Spergel, D.N., Bean, R., Doré, O., et al.: 2007, ApJS 170, 377

Spite, M., Depagne, E., Nordström, B., Hill, V., Cayrel, R., Spite, F., Beers, T.C.: 2000, A&A 360, 1077

Spite, M., Cayrel, R., Plez, B., et al.: 2005, A&A 430, 655

Springel, V., Wang, J., Vogelsberger, M., et al.: 2008, MNRAS 391, 1685

Starkenburg, E., Hill, V., Tolstoy, E., et al.: 2010, astro-ph/1002.2963

Suda, T., Aikawa, M., Machida, M.N., Fujimoto, M.Y., Iben, I.J.: 2004, ApJ 611, 476

Suda, T., Katsuta, Y., Yamada, S., et al.: 2008, PASJ 60, 1159

Tolstoy, E., Hill, V., Tosi, M.: 2009, astro-ph/0904.4505

Tominaga, N., Umeda, H., Nomoto, K.: 2007, ApJ 660, 516

Travaglio, C., Gallino, R., Arnone, E., Cowan, J., Jordan, F., Sneden, C.: 2004, ApJ 601, 864

Umeda, H., Nomoto, K.: 2003, Nature 422, 871

Wanajo, S., Ishimaru, Y.: 2005, in: V. Hill, P. François, F. Primas (eds.), *From Lithium to Uranium: Elemental Tracers of Early Cosmic Evolution*, IAU Symp. 228, p. 435

Wanajo, S., Itoh, N., Ishimaru, Y., Nozawa, S., Beers, T.C.: 2002, ApJ 577, 853

Westin, J., Sneden, C., Gustafsson, B., Cowan, J.J.: 2000, ApJ 530, 783

Wisotzki, L., Köhler, T., Groote, D., Reimers, D.: 1996, A&AS 115, 227

Woosley, S.E., Weaver, T.A.: 1995, ApJS 101, 181

Yoshida, N., Omukai, K., Hernquist, L., Abel, T.: 2006, ApJ 652, 6

Zacs, L., Nissen, P.E., Schuster, W.J.: 1998, A&A 337, 216

A References for the collection of the literature data

Aoki et al. (2000), Aoki et al. (2001), Aoki et al. (2002d), Aoki et al. (2002a), Aoki et al. (2002c), Aoki et al. (2002b), Aoki et al. (2005), Aoki et al. (2006a), Aoki et al. (2007a), Aoki et al. (2007c), Aoki et al. (2007b), Aoki et al. (2008), Aoki et al. (2009), Aoki et al. (2006b), Arnone et al. (2005), Barklem et al. (2005), Barbuy et al. (2005), Bonifacio et al. (2009), Burris et al. (2000), Carretta et al. (2002), Christlieb et al. (2004), Cohen et al. (2003), Cohen et al. (2004), Cohen et al. (2006), Cohen et al. (2007), Cohen et al. (2008), Cohen & Huang (2009), Collet et al. (2006), Cowan et al. (2002), François et al. (2007), Frebel et al. (2007c), Frebel et al. (2008), Frebel et al. (2010a), Frebel et al. (2010b), Fulbright (2000), Fulbright et al. (2004), Feltzing et al. (2009), Geisler et al. (2005), Hayek et al. (2009), Hill et al. (2002), Plez et al. (2004), Honda et al. (2004), Honda et al. (2006), Honda et al. (2007), Ito et al. (2009), Ivans et al. (2003), Ivans et al. (2005), Ivans et al. (2006), Johnson & Bolte (2002a), Johnson & Bolte (2002b), Johnson (2002), Johnson & Bolte (2001), Johnson & Bolte (2004), Aoki et al. (2006a), Jonsell et al. (2005), Jonsell et al. (2006), Koch et al. (2008), Lai et al. (2007), Lai et al. (2008), Lai et al. (2009), Lucatello et al. (2003), Masseron et al. (2006), McWilliam (1998); McWilliam et al. (1995), Norris et al. (2000), Norris et al. (2001), Norris et al. (2002), Norris et al. (2007), Norris et al. (2008), Norris et al. (1997b), Norris et al. (1997a), Norris et al. (1997c), Preston & Sneden (2001), Preston & Sneden (2000), Preston et al. (2006), Roederer et al. (2008), Ryan & Norris (1991), Ryan et al. (1996), Shetrone et al. (2001), Shetrone et al. (2003), Simon et al. (2010), Sivarani et al. (2004), Sivarani et al. (2006), Sneden et al. (2003), Spite et al. (2005), Spite et al. (2000), Westin et al. (2000), Cayrel et al. (2004), Zacs et al. (1998).

Quantitative solar spectroscopy

Klaus Wilhelm

Max-Planck-Institut für Sonnensystemforschung
Max-Planck-Str. 2, 37191 Katlenburg-Lindau, Germany
wilhelm@mps.mpg.de

Abstract

Quantitative solar spectroscopy must be based on calibrated instrumentation. The basic requirement of a calibration, i.e., a comparison between the instrument under test and a primary laboratory standard through appropriate procedures, will be briefly reviewed, and the application to modern space instruments will be illustrated. Quantitative measurements of spectral radiances with high spectral and spatial resolutions as well as spectral irradiances yield detailed information on temperatures, electron densities, bulk and turbulent motions, element abundances of plasma structures in various regions of the solar atmosphere – from the photosphere to the outer corona and the solar wind. The particular requirements for helioseismology and magnetic-field observations will not be covered in any depth in this review. Calibration by a laboratory standard is necessary, but not sufficient, because an adequate radiometric stability can only be achieved together with a stringent cleanliness concept that rules out a contamination of the optical system and the detectors as much as possible. In addition, there is a need for calibration monitoring through inter-calibration and other means.

1 Introduction

The Sun as central star of our planetary system is of paramount importance not only for the conditions on the Earth, other planets, moons, and many smaller bodies gravitationally bound to the Sun, but also for providing the opportunity to study a main sequence star in detail with high spatial resolution. The impact of the solar electromagnetic radiation on the Earth is mainly controlled by the total solar irradiance (TSI) and its spectral distribution – the solar spectral irradiance (SSI) – as a function of the wavelength, λ (or, alternatively, the frequency, ν). The interaction of this radiation with the atmosphere and the solid Earth (including all forms of life) can only be understood with quantitative information on the input parameters based on calibrated instrumentation. The basic concepts of radiometric and spectral calibrations and their application to solar observations will therefore be outlined in Sect. 2. Space instruments will, in general, be taken as examples, because they can observe the Sun undisturbed by the Earth's atmosphere. However, the principles apply to all measurements after appropriate correction procedures. The crucial point is that the observations are compared to laboratory-based standards, thereby ensuring the homogeneity of various measurements and providing a baseline for short-term and long-term investigations of any solar variability (cf. Quinn & Fröhlich 1999; Lean 2000; Willson & Mordvinov 2003).

2 Calibration concepts

The need for quantitative measurements in many human activities has been recognized since early history. Nevertheless, only during the French Revolution a serious attempt was made to devise a coherent system of standards for calibrated measurements. The first historic standards of platinum represented the metre as unit of length and the kilogram as unit of mass in 1799. The system evolved into the International System of Units (SI: Le système international d'unités) (BIPM 2006; see also NIST 2008), which is the basis of the national calibration legislation in most countries, and is used by scientists worldwide. In Table 1, some derived units of physical quantities are compiled that are relevant in the context of spectroradiometry. In addition, four (out of seven) base units are listed in Table 2.

Table 1: Some derived SI units used in spectroradiometry.

Quantity	Symbol[a]	Unit
Radiant energy[b]	Q	J
Radiant flux, power[c]	Φ	W
Spectral flux	Φ_λ	$W\,nm^{-1}$
Irradiance	E	$W\,m^{-2}$
Spectral irradiance	E_λ	$W\,m^{-2}\,nm^{-1}$
Radiance[d]	L	$W\,m^{-2}\,sr^{-1}$
Spectral radiance	L_λ	$W\,m^{-2}\,sr^{-1}\,nm^{-1}$
Radiant intensity	I	$W\,sr^{-1}$
Spectral intensity	I_λ	$W\,sr^{-1}\,nm^{-1}$
Doppler velocity	v_D	$m\,s^{-1}$

[a] Recommendations only.
[b] The unit name of energy is joule (1 J = 1 kg m² s⁻²),
[c] that of radiant flux is watt (1 W = 1 J s⁻¹),
[d] and for solid angle steradian (1 sr = 1 m m⁻²).

Table 2: Relevant base units of SI.

Quantity	Symbol	Unit Name	Unit	Year and Current Definition
Mass	m	kilogram	kg	1889; international prototype
Thermodyn. temperature	T	kelvin	K	1968; triple point of water
Length	l	metre	m	1983; speed of light in vacuum
Time	t	second	s	1997; ^{133}Cs at 0 K

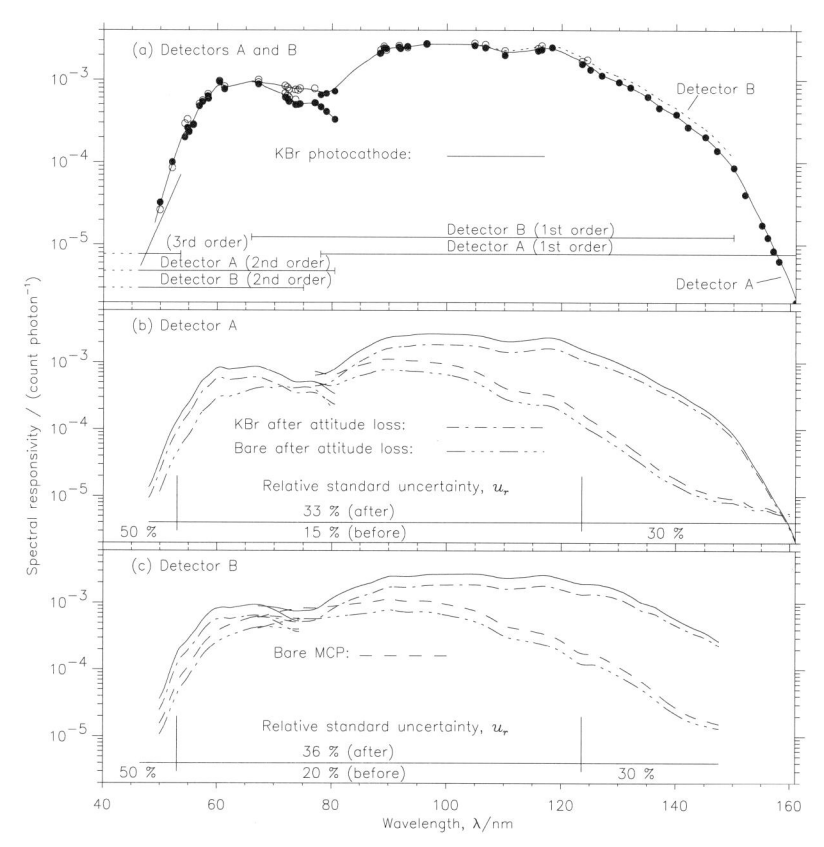

Figure 1: Spectral responsivities of the SUMER spectrometer (with two detectors, A and B), and the corresponding relative uncertainties. (*a*) First-order, second-order, and third-order responsivities for both detectors (interpolated between calibration points); (*b*) Independent detector A assessment; and (*c*) results for detector B. The relative uncertainties in the central wavelength range (indicated by vertical bars at 53 nm and 124 nm) are derived from the laboratory calibration and are smaller than those of the in-flight calibration extensions obtained from line-ratio and stellar observations. The changes caused by the attitude loss of SOHO in the year 1998 have been determined by calibration tracking and inter-calibration activities (Wilhelm et al. 2002).

The calibration of a detector instrument can be performed by establishing its response to the action of a certain primary standard. A detector might have a dark output without any input. This offset has to be taken into account appropriately. For radiometric measurements, synchrotron radiation constitutes a source standard, because the spectral radiant flux emitted can be calculated from the parameters of the electron or positron storage ring (Schwinger 1949; cf. Hollandt et al. 2002; Vest et al. 2006). Another source standard is black-body emission, for which the spectral flux follows from the Planck law. Electric substitution radiometers and double-ionization chambers, on the other hand, are detector standards that can be used to characterize a radiation source.

In most cases, secondary standards have to be engaged as transfer standards between primary standards and the instrumentation to be calibrated, because the operational requirements of the primary standard and those of the test specimen are often not compatible. It is, for instance, impractical to calibrate a vacuum-ultraviolet (VUV) spectrometer designed for operation on a spacecraft directly at a synchrotron facility. Cleanliness (cf. Sect. 2.3) requirements and schedule conflicts would cause extreme complications.

2.1 Radiometric calibration

As examples, the calibration procedures for the VUV imaging spectrometers CDS[1] and SUMER[2] on SOHO[3] as well as the EIS[4] instrument on the Japanese spacecraft Hinode will be summarized: transfer standards equipped with hollow-cathode plasma-discharge lamps were calibrated with BESSY I at the PTB laboratory[5]. This was done by a comparison of the radiation characteristics of both standards with the help of a VUV monochromator. The polarization of the synchrotron beam required special precautions. The calibration of the transfer standards were carried out before (and after) they were employed to characterize the spectral response of the instruments at the various integration facilities. The plasma discharge was operated with constant current and a stabilized voltage drop during the calibration runs ensuring a reproducibility of the radiant flux in certain spectral lines within ± 2.5 % over several hours, and ± 5 % after a change of the filling gas (Hollandt et al. 1996; Wilhelm et al. 2000; Lang et al. 2002, 2006).

For space observations, it is not only essential to trace the laboratory calibration of the instrument at the end of its development phase to primary standards, but also to avoid any subsequent deterioration as much as possible, and, in addition, to track the calibration status through all phases of the mission, i.e., transport, spacecraft integration and tests, launch, commissioning as well as operations. The procedures include in-flight calibration with calibration lamps, inter-calibration between instruments, degradation monitoring using redundant instrumentation, etc. (cf. articles in Pauluhn et al. 2002). Balloon, sounding-rocket and STS[6] payloads with prospects of recovery might offer the opportunity of a re-calibration after the missions.

In Fig. 1, the responsivities of SUMER are shown as a typical result of such calibration activities. Note that the radiant energy is measured here as the number of photons with energy $h\nu = hc_0/\lambda$, where h is Planck's constant and c_0 the speed of light in vacuum. This convention is often adopted in radiometry.

---

[1] Coronal Diagnostic Spectrometer (Harrison et al. 1995)
[2] Solar Ultraviolet Measurements of Emitted Radiation (Wilhelm et al. 1995)
[3] Solar and Heliospheric Observatory of ESA and NASA
[4] Extreme-ultraviolet Imaging Spectrometer (Culhane et al. 2007)
[5] Berlin electron storage ring for synchrotron radiation, Physikalisch-Technische Bundesanstalt
[6] Space Transportation System

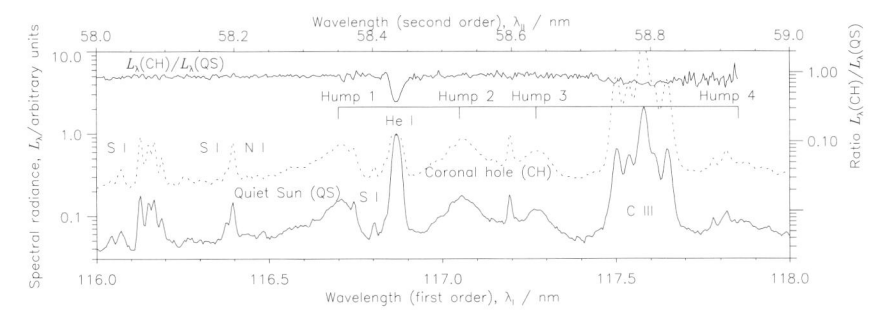

Figure 2: Radiance spectra of the Sun in the wavelength ranges from 116 nm to 118 nm in the first order covering four spectral humps, and from 58 nm to 59 nm in the second order with the superimposed He I line at 58.43 nm. The quiet-Sun spectrum observed near the centre of the disk on 28 April 1996 is plotted as solid line, and a polar coronal-hole spectrum from 5 June 1996 as dotted line. The radiances of both spectra are normalized to the maxima of the helium line.

Of particular importance is the evaluation of the uncertainty level relevant for a certain calibration[7]. In the vacuum-ultraviolet (VUV) range, the relative uncertainties of the spectral radiances are, with the present technologies, rather large (cf. Sect. 5), but can be substantially reduced for TSI and some SSI measurements (see Sect. 6). In view of the importance for the terrestrial climate and its evolution, very good accuracies are, indeed, mandatory in this field of research, and further improvements are required.

2.2 Wavelength calibration

The wavelengths of many spectral lines from atoms or ions can be measured in the laboratory with very good accuracies. Line lists are available (see, e.g., Kelly 1987; Ralchenko 2005) that allow us to compare the wavelengths of solar spectral lines to those of known lines. In the laboratory and for instruments with spectral calibration lamps, this directly leads to wavelength determinations[8].

In many cases, however, solar lines of atoms or singly-ionized species have to be used for wavelength calibrations, and it is important to eliminate the shifts introduced by gas or plasma motions in the solar atmosphere through the Doppler effect

$$\Delta\lambda = \lambda - \lambda_0 = \lambda_0 \frac{v_D}{c_0}, \qquad (1)$$

written in non-relativistic form with λ_0 the rest wavelength of a spectral line. A limiting velocity uncertainty of approximately 100 m s^{-1} has been deduced by Allende Prieto & García López (1998) and Dravins (2008) for photospheric lines. The solar

---

[7] For a detailed discussion of measurement uncertainties see Bernhard & Seckmeyer (1999). In this review standard uncertainties are quoted, unless other coverage factors are explicitly given.

[8] In order to get true laboratory wavelengths, corrections are required to account for shifts caused by the gravitational potential of the Sun (cf. Eq. 2) and any relative motion of the detector with respect to the emitter.

rotation speed of ± 2 km s^{-1} at the equatorial limb positions is a suitable test signal for Doppler-shift measurements.

The gravitational red shift (Einstein 1911) of photospheric lines is

$$\frac{\Delta\lambda}{\lambda_0} = \frac{1}{c_0^2}\frac{G M_\odot}{R_\odot} \approx 2 \times 10^{-6}, \tag{2}$$

where G, $M_\odot$, $R_\odot$ are the gravitation constant, the mass and the radius of the Sun, respectively, affects all wavelengths. So, when a comparison between solar wavelengths of known lines and the line under test provides a wavelength determination, the shift is nearly constant for lines in a narrow spectral range and is taken into account, if laboratory wavelengths of the calibration lines are used. The gravitational shift was confirmed in laboratory experiments through the Mössbauer effect by Pound & Rebka (1959), and was subsequently observed in the solar spectrum by Blamont & Roddier (1961) and others (cf. Brault 1962; Snider 1972), but Freundlich (1914) presented, based on measurements of Evershed, indications of such a shift much earlier.

2.3 Cleanliness requirements

The radiometric calibration depends on the stability of the spectroscopic responsivities, which, for space missions, can be severely affected by any obstruction of the apertures or a contamination of the optical elements and detectors. Solar ultraviolet (UV) radiation leads to photo-activated polymerization of contaminating hydrocarbons and, as a result, to a permanent degradation of the system. A stringent cleanliness programme is therefore of great importance to ensure particulate and molecular cleanliness of the instruments and the spacecraft. Detailed accounts of the very successful SOHO cleanliness programme are available in the literature (e.g., Thomas 2002; Schühle 2003).

3 The solar electromagnetic spectrum

The solar electromagnetic radiation extends over 14 decades from the γ-ray range to radio waves[9]; for an overview see Wilhelm (2009) and the references therein. The extreme ranges do not significantly contribute to the quiet-Sun energy output, which is concentrated between the X-ray range and the infrared from $\lambda \approx 12.3$ pm ($h\nu = 100$ keV) to $\lambda \approx 1$ mm ($h\nu = 1.2$ meV). The full range can only be observed from space.

An important task is to measure the wavelengths of the solar absorption and emission lines. Line shifts and broadenings caused by the Doppler effect allow – together with spatial resolution – dynamical studies of selected phenomena. Quantitative radiance measurements of many spectral lines are important for solar plasma diagnostics and provide information on plasma densities, temperatures, and element abundances. The morphologies of the photosphere, chromosphere, transition region, and corona can also be studied as well as their interactions.

---

[9]The 10.7 cm radio emission of the Sun is an indicator for the solar activity level.

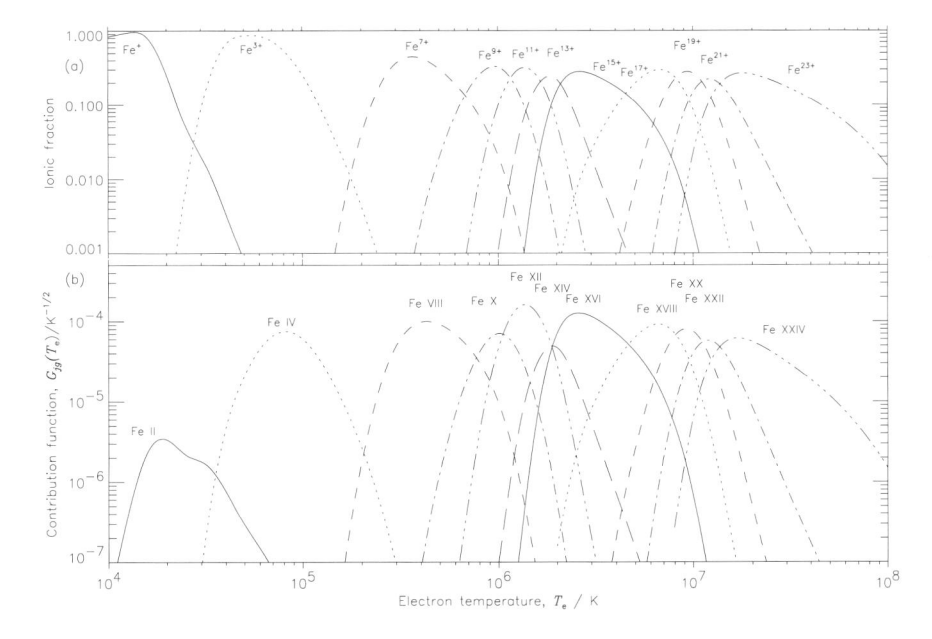

Figure 3: (*a*) Fractions of some of the iron ions Fe^+ to Fe^{23+} as a function of the electron temperature, T_e, in a plasma in ionization equilibrium (after Mazzotta et al. 1998). (*b*) Examples of contribution functions of emission lines (most of them in the VUV range) in the Fe II to Fe XXIV spectra.

4 Solar wavelengths

Accurate wavelength measurements of solar lines[10] are essential for an identification of the lines and a determination of the emitting molecules, atoms or ions. A very specific identification is illustrated in Fig. 2: very broad features, called humps in the figure, had been observed in the VUV spectrum near 117 nm (Wilhelm et al. 2005). Avrett et al. (2006) identified them as auto-ionization lines of sulfur broadened (as a consequence of Heisenberg's uncertainty principle) by the extremely short lifetimes of the energy levels.

On a relative basis, line shifts and broadenings have to be measured in a wide range with high precision in order to investigate dynamical processes of the Sun. Very low speeds of less than a metre per second are of interest in helioseismology studies (cf., e.g., Kosovichev et al. 1997). They can be observed, for instance, by the SOHO MDI[11] instrument in the photospheric Ni I line at 676.8 nm with the help of filters and tunable interferometers. The Dopplergrams obtained provide information

---

[10] By means of Fourier transform spectrometers (FTS) (Brault 1996) or, in general, with grating instruments (cf. Lemaire 2009).
[11] Michelson Doppler Imager (Scherrer et al. 1995)

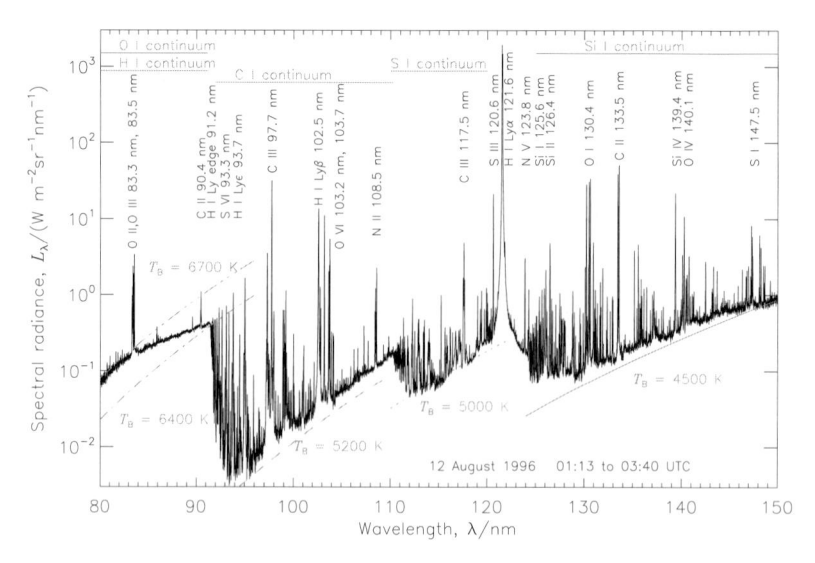

Figure 4: Spectral radiance of the quiet Sun in the VUV range from a region near the centre of the disk. The spectral radiances of some continua expected for certain brightness temperatures, T_B, are also shown. The relative uncertainty below 124 nm is 15 % and above this limit 30 %.

on helioseismic oscillation modes, and – via polarization measurements – on the LOS component of the magnetic field[12].

On the other hand, Doppler shifts indicating bulk motions of the plasma with line-of-sight (LOS) velocities of more than six hundred kilometres per second occur during flare events (Innes et al. 2001). Line broadenings imply thermal and/or turbulent motions of the emitting atoms or ions. A quantitative separation of thermal and non-thermal motions, ξ, can, in general, be accomplished, if measurements of many lines are available from different emitting species with a wide range of atomic masses, m_i, and charge states, and assumptions about the non-thermal processes involved, using the relation

$$c_0 \frac{\Delta \lambda}{\lambda_0} = \sqrt{\frac{2 k_B T_i}{m_i} + \xi^2}, \qquad (3)$$

with T_i the ion temperature and k_B the Boltzmann constant (cf. Mariska 1992; Tu et al. 1998).

In some cases, wavelength measurements of lines from high-ionization stages can more accurately be made by observing them in the solar atmosphere rather than in the laboratory, for instance, the line at 77.0428 nm ± 0.3 pm of the Ne VIII spectrum (Dammasch et al. 1999) and the Mg X line at 62.4968 nm ± 0.7 pm (Peter & Judge 1999). Other examples are the lines in the Ne IX, Na X, Mg XI, and Si XIII spectra of the helium-like transitions of the ions Ne^{8+}, Na^{9+}, Mg^{10+}, and Si^{12+}.

---
[12] In this context, the so-called second solar spectrum – scattered polarized radiation at the limb – can also be mentioned (cf. Gandorfer 2001).

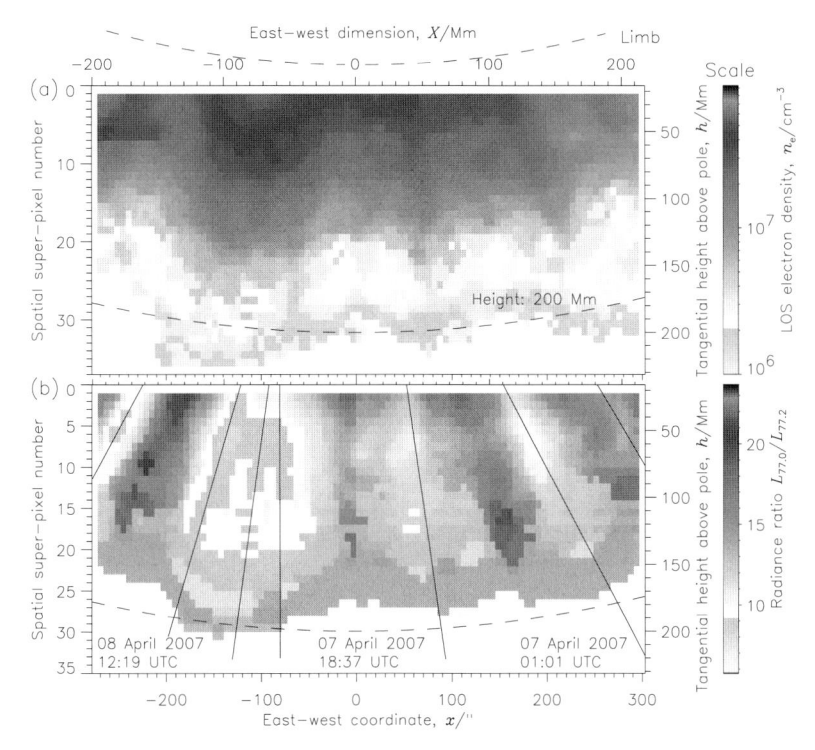

Figure 5: Observations in the southern coronal hole of the Sun. (*a*) The density-sensitive line pair (Si VIII 144.0 nm, 144.6 nm) yields the electron density, n_e, averaged along the LOS (beyond a height of ≈ 200 Mm noise effects disturb the measurements). (*b*) The contribution functions of Ne VIII at 77.0 nm and Mg VIII at 77.2 nm overlap considerably. Variations of the radiance ratio, $L_{77.0}/L_{77.2}$ can, therefore be interpreted as changes of the element abundances caused by the first ionization potential (FIP) effect, neon and magnesium being high and low FIP elements, respectively. Very clear signatures of small n_{Ne}/n_{Mg} ratios in the denser plumes (their central projections are indicated in the lower panel) contrast with inter-plume regions (Curdt et al. 2008).

The wavelength of the sodium ion transition $1s2s\,{}^3S_1 - 1s2p\,{}^3P_2$ could be measured at (111.1770 ± 0.0017) nm for the first time (Curdt et al. 2000). These helium-like lines are seen in solar flare spectra and have formation temperatures between 4 MK and 10 MK.

A rough estimate of the electron temperature, T_e, of an emitting plasma follows from the formation temperature, T_F, of a spectral line. This is the temperature where the contribution function

$$G_{jg}(T_e) = \frac{n_g}{n_X}\frac{1}{\sqrt{T_e}}\exp\left(\frac{-\Delta\epsilon_{gj}}{k_B T_e}\right) \tag{4}$$

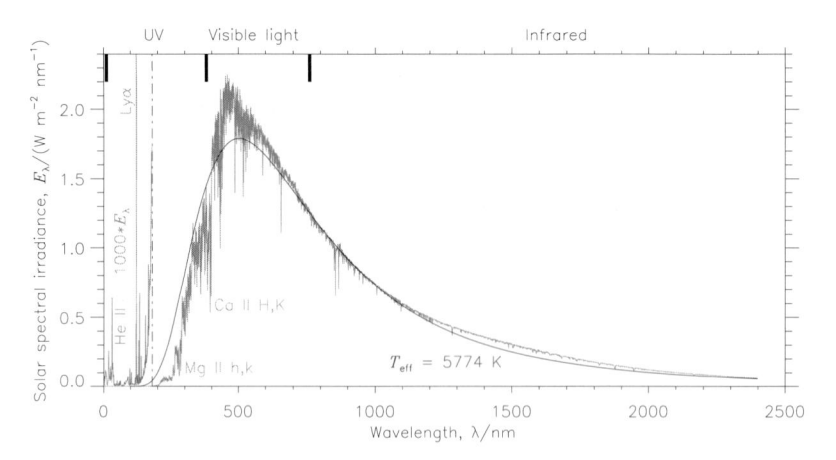

Figure 6: Solar spectral irradiance (SSI), E_λ, of the quiet Sun. Below $\lambda = 180$ nm, the values are displayed as 1000 E_λ. The spectrum is here dominated by emission lines of atoms and ions. The H I Ly α line is off the scale. At longer wavelengths, the thermal radiation of the photosphere is modulated by the Fraunhofer absorption lines. The smooth curve shows the calculated radiation of the Sun as a "black body" with $T_{\text{eff}} = 5774$ K and a radius $R_\odot = 696.35$ Mm. The relative uncertainties are 3 % to 4 % in the green range and $\approx$30 % in the red one.

has its maximum, with the ionic fraction, n_g/n_X (n_g: number density of the ions in the ground state, n_X: number density of element X, cf. Mazzotta et al. 1998). The transition energy between levels g and j is $\Delta \epsilon_{gj}$ (cf. Wilhelm et al. 2004).

Famous examples are the red and green coronal lines Fe X at 637.4 nm and Fe XIV at 530.3 nm of forbidden transitions in highly ionized iron, which led Grotrian (1939) and Edlén (1943) to conclude that the outer layers of the solar atmosphere are much hotter than the photosphere. To illustrate this point, some of the ionic fractions of iron and the corresponding contribution functions are shown in Fig. 3.

5 Spectral radiance

The spectral radiance, $L_\lambda(\vartheta, t)$, is defined by the relation

$$dQ(\lambda, \vartheta, t) = L_\lambda(\vartheta, t) \cos \vartheta \, dS \, d\omega \, dt \, d\lambda, \tag{5}$$

where dQ is the differential radiant energy emitted into the solid angle $d\omega$ from $dS \cos\vartheta$, the projected area normal to the direction of $d\omega$ during the time interval $(t, t + dt)$ and in the wavelength interval $(\lambda, \lambda + d\lambda)$. An average value of the spectral radiance, $\overline{L_\lambda}$, over certain solid angle, time, and wavelength intervals can be obtained from a measurement of the energy

$$\Delta Q = \overline{L_\lambda} \, \Delta\Omega \, \Delta t \, \Delta\lambda \, A \tag{6}$$

through an aperture area, A, of a spectrometer (cf. Wilhelm 2002). If the wavelength interval $\Delta\lambda$ covers the profile of a spectral line at λ, $L_{\text{line}} = (\overline{L_\lambda} - L_{\text{back}})\Delta\lambda$ represents – after a suitable background subtraction – its line radiance.

The solar spectral radiance in the wavelength range from 330 nm to 1099 nm can reliably be determined from high-altitude ground stations. This was done in the pioneering work of Labs & Neckel (1962, 1967) from the Jungfraujoch in the Swiss Alps during 1961 and 1965; later improved with the help of Kitt Peak FTS data (Neckel & Labs 1984; see also Kurucz 2005).

Radiance measurements outside the optical window of the terrestrial atmosphere must be made from space platforms. Figure 4 depicts the VUV radiance spectrum of a quiet-Sun region with many emission lines and some continua in the wavelength range from 80 nm to 150 nm. A spectral atlas from 67.0 nm to 148.6 nm covering a quiet-Sun region, a coronal hole, and a sunspot has been compiled by Curdt et al. (2001). In this range, the Lyman series and continuum of H I are of particular interest for radiative transfer calculations. Calibrated high-resolution spectral observations of the high members of the series and the adjacent continuum have been presented by Warren et al. (1998).

At longer wavelengths, the emission lines fade out and absorption features – the Fraunhofer lines – become prominent (cf. Fig. 6) as a consequence of the thermal structure of the solar atmosphere. Such spectra provide important information about the composition and other properties of the Sun. For five transition-region and coronal lines as well as for the H I Lyman continuum at 88 nm, Schühle et al. (2000) found increases (up to a factor of two) of the line radiances in quiet-Sun areas in the time period between the solar minimum in 1996 and mid-1999.

Schemes have been developed to improve the temperature measurements with the help of line radiance ratios of temperature-sensitive line pairs. Applying the same method to line-ratio measurements of density-sensitive pairs provides a determination of the electron densities in the solar atmosphere (cf. Wilhelm et al. 2004, 2007). Electron-density and element-abundance diagnostics are indicated in Fig. 5 for coronal plumes in a polar coronal hole.

Although the lines in Fig. 4 appear to be rather narrow, they can be resolved into profiles by modern spectrometers (see Sect. 4), and, in addition, their Doppler shifts with respect to other lines can be determined. High-temperature lines, such as the Mg IX lines, are shifted to shorter wavelengths relative to cooler lines, signifying outflows in coronal holes as the source of the fast solar wind, whereas transition-region lines, on average, are shifted to longer wavelengths, i.e., they indicate downflows.

6 Spectral irradiance

Integration of the spectral radiance over the full solar disk and the low corona yields the spectral irradiance of the Sun, although, in general, the irradiance is measured with full-disk instruments without spatial resolution. Such integrations were, however, performed using full-disk radiance observations of the N v line at 123.8 nm and the C IV line at 154.8 nm obtained with SUMER in 1996, in oder to compare the PTB-calibrated instrument with the Solar Stellar Irradiance Comparison Experiment

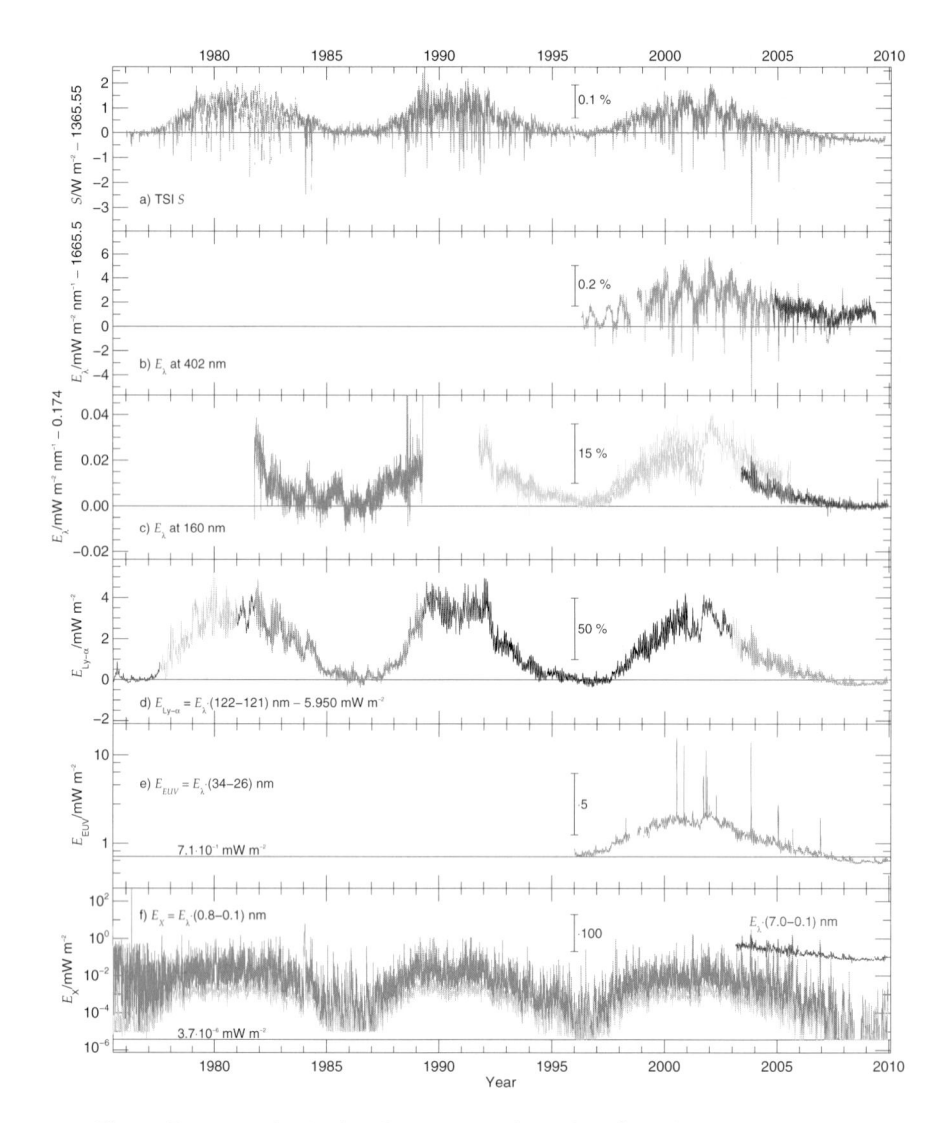

Figure 7: (a) Total solar irradiance composite (TSI), S, during solar cycles 21 to 23. The expanded uncertainty of TSI composites is estimated to be $U = 0.35$ W m^{-2} with a coverage factor $k = 2$ by Dewitte et al. (2004). The mean enhancements during solar maximum periods stem from hot faculae in active regions interrupted by cool sunspot groups. The bars in this panel and the following ones indicate relative variations based on the mean values at solar minimum. (b) SSI results in visible light for cycle 23. (c) Time series of SSI in the VUV. (d) Irradiance composite in the range from 121 nm to 122 nm, including the H I Lyα line. (e) Extreme-ultraviolet (EUV) irradiance data for solar cycle 23. The prominent increases correspond to strong flare activity. (f) The X-ray irradiance (daily mean values: green; extrema: red), and measurements integrated over a wider range. The relative variations between solar minimum and maximum increase from Panel a to Panel f by approximately a factor of 100 000 (Data compilation: C. Fröhlich).

on the Upper Atmosphere Research Satellite (SOLSTICE/UARS) (Rottman et al. 1993; Woods et al. 1993) radiometrically calibrated at the Synchrotron Ultraviolet Radiation Facility (SURF-II) at NIST. Agreement within a factor of 1.14 was found for the N V line and approximately 1.10 for the C IV line (Wilhelm et al. 1999).

In the VUV range, between about 10 nm and 290 nm, measurements can only be performed from above the atmosphere of the Earth with space probes, satellites, or rockets. These platforms are moving with respect to the source (the Sun). The necessary corrections both for radiance and irradiance measurements are treated in Wilhelm & Fröhlich (2010). Reference spectra of the SSI at 1 ua over a wide range are derived from space measurements made during distinct time periods – for instance, that shown in Fig. 6 from data of Thuillier et al. (2004) for a period in 1992. More recent reference spectra during Carrington Rotation 2068 in 2008 (Whole Heliosphere Interval) have been obtained by Woods et al. (2009). Such spectra are used for many applications, e.g., in weather and climate models.

The solar H I Lyα line at $\lambda_0 = 121.5664$ nm exhibits a pronounced self-reversal. Observations from Earth-orbiting spacecraft show, in addition, a central absorption feature caused by hydrogen atoms in the geocorona (cf. Lemaire et al. 1998). Since the interaction of the strong Lyα line with the cool hydrogen in the solar system is mainly controlled by the spectral (photon) irradiance at the line centre, $E^{\text{ph}}_{\lambda_0}$, it is important to establish a relation between this central value and the total line irradiance, $E^{\text{ph}}_{\text{Ly}\alpha}$, which can be routinely monitored by instruments with low spectral resolution. Measurements of the Lyα line profile obtained with SUMER/SOHO (positioned at the Lagrange point L1 outside the geocorona) led to the relation

$$\frac{E^{\text{ph}}_{\lambda_0}}{10^{16}\,\text{s}^{-1}\,\text{m}^{-2}\,\text{nm}^{-1}} = 0.64 \left(\frac{E^{\text{ph}}_{\text{Ly}\alpha}}{10^{15}\,\text{s}^{-1}\,\text{m}^{-2}} \right)^{1.21} \pm 0.08, \tag{7}$$

where the uncertainty gives the standard deviation of the measurements during the solar cycle 23 and the derived relation (Emerich et al. 2005). This relation updates an earlier one by Vidal-Madjar (1975).

Soft and hard X-ray emissions of the Sun as well as γ-rays and radio waves are highly variable. Figure 7 demonstrates – with the help of a few time series – the solar irradiance variability during the solar cycles 21 to 23 in wavelength ranges from the visible to X-rays. The data were obtained with the following space experiments using filter radiometers or spectrometers for the wavelength selection: in the visible light by the Sun Photometer/Variability of solar Irradiance and Gravity Oscillations (SPM/VIRGO) (Fröhlich & Wehrli 2006) and the Spectral Irradiance Monitor/Solar Radiation and Climate Experiment (SIM/SORCE) (Rottman et al. 2005); in the VUV by the Solar Mesospheric Explorer (SME) (Rottman 1988), by SOLSTICE/UARS (Rottman 2000), the Solar Ultraviolet Spectral Irradiance Monitor (SUSIM/UARS) (Floyd et al. 1998), and SOLSTICE/SORCE (Rottman et al. 2006); from 121 nm to 122 nm by Hinteregger et al. (1981), Woods et al. (2000), and Woods & Eparvier (2006); in the EUV by the Solar EUV Monitor (SEM/SOHO) (Judge et al. 2002); the X-ray irradiance by the X-ray Sensor/Geostationary Operational Environmental Satellite (XRS/GOES) (Kahler & Kreplin 1991) and the XUV Photometer System (XPS/SORCE) (Woods & Rottman 2005).

For comparison, the daily averages of the TSI are plotted in Panel (a) as a composite from many measurements (cf. Fröhlich 2006, 2009). The Total Irradiance Monitor (TIM/SORCE) measured 1360.84 W m^{-2} during the present minimum conditions (Woods et al. 2009), and thus ≈ 5 W m^{-2} less than shown in the composite. The different locations of the aperture stops in the radiometers and the infrared radiation emitted by a hot stop might cause the discrepancy (Fröhlich 2010).

The relative variability of the TSI is of the order of $\pm 0.1\,\%$. The corresponding variations in the visible and near-UV ranges are below 1 % during an 11-year cycle. It is interesting that only the TSI is showing a minimum at the beginning of 2008 with a relative decrease of almost 0.02 % with respect to the other minima. This is about a fifth to a quarter of the mean solar cycle variation (Fröhlich 2009b).

7 Conclusions

Solar spectroscopy provides wavelengths of emission lines with certain accuracies (which in some cases can be better than laboratory values). Wavelength determinations are essential for line identifications. Together with atomic physics data, approximate electron temperatures can be deduced. Line shifts and broadenings caused by the Doppler effect of moving emitters yield bulk and turbulent speeds as well as ion temperatures.

Accurate spectral irradiance and radiance measurements can only be obtained with calibrated instrumentation. Spectral irradiance measurements traceable to primary laboratory standards characterize the current electromagnetic energy output of the Sun. Radiance observations with high spatial resolution in various regions of the solar atmosphere provide information on the prevailing plasma conditions. Line ratios from different line radiances allow refined studies of electron temperatures, electron densities, and element abundances in various solar regions. Cleanliness requirements are essential for minimizing any degradation of the calibration status of the instrumentation, and inter-calibration schemes must be employed for monitoring its stability.

Acknowledgements

I thank Martin C.E. Huber and Udo Schühle for helpful comments on the manuscript.

References

Allende Prieto, C., García López, R.J.: 1998, A&AS 131, 431

Avrett, E.H., Kurucz, R.L., Loeser, R.: 2006, A&A 452, 651

Bernhard, G., Seckmeyer, G.: 1999, JGR 104, 14 321

Blamont, J.E., Roddier, F.: 1961, Phys. Rev. Lett. 7, 437

Brault, J.W.: 1962, PhD Thesis, Princeton University

Brault, J.W.: 1996, Appl. Opt. 35, 2891

Bureau International des Poids et Mesures (BIPM): 2006, *Le système international d'unités (SI)*, 8e édition, Sèvres, France

Culhane, J.L., Harra, L.K., James, A.M., et al.: 2007, Sol. Phys. 243, 19

Curdt, W., Landi, E., Wilhelm, K., Feldman, U.: 2000, Phys. Rev. A 62, 022502-1

Curdt, W., Brekke, P., Feldman, U., Wilhelm, K., Dwivedi, B.N., Schühle, U., Lemaire, P.: 2001, A&A 375, 591

Curdt, W., Wilhelm, K., Feng, L., Kamio, S.: 2008, A&A 481, L61

Dammasch, I.E., Wilhelm, K., Curdt, W., Hassler, D.M.: 1999, A&A 346, 285

Dewitte, S., Crommelynck, D., Mekaoui, S., Joukoff, A.: 2004, Sol. Phys. 224, 209

Dravins, D.: 2008, A&A 492, 199

Edlén, B.: 1943, ZAp 22, 30

Einstein, A.: 1911, AnP 340, 898

Emerich, C., Lemaire, P., Vial, J.-C., Curdt, W., Schühle, U., Wilhelm, K.: 2005, Icarus 178, 429

Floyd, L., Reiser, P., Crane, P., Herring, L.C., Prinz, D.K., Brueckner, G.E.: 1998, Sol. Phys. 177, 79

Freundlich, E.: 1914, AN 198, 265

Fröhlich, C.: 2006, Space Sci. Rev. 125, 53

Fröhlich, C.: 2009, A&A 501, L27

Fröhlich, C.: 2010, ISSIR 9, in press

Fröhlich, C., Wehrli, C.: 2006, *Comparison of the WRC-85 Solar Spectral Irradiance with RSSV1 and the SPM of VIRGO/SOHO*, SORCE Science Meeting, Washington, USA

Gandorfer, A.M.: 2001, in: M. Sigwarth (ed.), *Advanced Solar Polarimetry – Theory, Observation, and Instrumentation*, ASPC 236, p. 109

Grotrian, W.: 1939, NW 27, 214 (name misspelled as "Grotian" in publication)

Harrison, R.A., Sawyer, E.C., Carter, M.K., et al.: 1995, Sol. Phys. 162, 233

Hinteregger, H.E., Fukui, K., Gilson, B.R.: 1981, Geophys. Res. Lett. 8, 1147

Hollandt, J., Schühle, U., Paustian, W., Curdt, W., Kühne, M., Wende, B., Wilhelm, K.: 1996, Appl. Opt. 35, 5125

Hollandt, J., Kühne, M., Huber, M.C.E., Wende, B.: 2002, ISSIR 2, 51

Innes, D.E., Curdt, W., Schwenn, R., et al.: 2001, ApJ 549, L249

Judge, D.L., Ogawa, H.S., McMullin, D.R., Gangopadhyay, P., Pap, J.M.: 2002, AdSpR 29, 1963

Kahler, S.W., Kreplin, R.W.: 1991, Sol. Phys. 133, 371

Kelly, R.L.: 1987, JPCRD 16, 1

Kosovichev, A.G., Schou, J., Scherrer, P.H., et al.: 1997, Sol. Phys. 170, 43

Kurucz, R.L.: 2005, Mem. Soc. Astron. Ital. Suppl. 8, 189

Labs, D., Neckel, H.: 1962, ZAp 55, 269

Labs, D., Neckel, H.: 1967, ZAp 65, 133

Lang, J., Thompson, W.T., Pike, C.D., Kent, B.J., Foley, C.R.: 2002, ISSIR 2, 105

Lang, J., Kent, B.J., Paustian, W., et al.: 2006, Appl. Opt. 45, 8689

Lean, J.: 2000, Geophys. Res. Lett. 27, 2425

Lemaire, P.: 2009, ISSIR 9, in press

Lemaire, P., Emerich, C., Curdt, W., Schühle, U., Wilhelm, K.: 1998, A&A 334, 1095

Mariska, J.T.: 1992, *The Solar Transition Region*, Cambridge University Press, Cambridge

Mazzotta, P., Mazzitelli, G., Colafrancesco, S., Vittorio, N.: 1998, A&AS 133, 403

Neckel, H., Labs, D.: 1984, Sol. Phys. 90, 205 (see Erratum: 1984, Sol. Phys. 92, 391)

National Institute of Standards and Technology (NIST): 2008, *Guide for the Use of the International System of Units (SI)*, NIST SP 811, Gaithersburg, USA

Pauluhn, A., Huber, M.C.E., von Steiger, R.: 2002, ISSIR 2, V

Peter, H., Judge, P.G.: 1999, ApJ 522, 1148

Pound, R.V., Rebka, G.A.: 1959, Phys. Rev. Lett. 3, 439

Quinn, T.J., Fröhlich, C.: 1999, Nat 401, 841

Ralchenko, Y.: 2005, Mem. Soc. Astron. Ital. Suppl. 8, 96

Rottman, G.J.: 1988, AdSpR 8, 53

Rottman, G.: 2000, Space Sci. Rev. 94, 83

Rottman, G.J., Woods, T.N., Sparn, T.P.: 1993, J. Geophys. Res. 98, 10667

Rottman, G., Harder, J., Fontenla, J., Woods, T., White, O.R., Lawrence, G.M.: 2005, Sol. Phys. 230, 205

Rottman, G.J., Woods, T.N., McClintock, W.: 2006, AdSpR 37, 201

Scherrer, P.H., Bogart, R.S., Bush, R.I., et al.: 1995, Sol. Phys. 162, 129

Schühle, U.: 2003, in: S.L. Keil, S.V. Avakyan (eds.), *Innovative Telescopes and Instrumentation for Solar Astrophysics*, SPIE 4853, p. 88

Schühle, U., Wilhelm, K., Hollandt, J., Lemaire, P., Pauluhn, A.: 2000, A&A 354, L71

Schwinger, J.: 1949, Phys. Rev. 75, 1912

Snider, J.L.: 1972, Phys. Rev. Lett. 28, 853

Thomas, R.: 2002 ISSIR 2, 91

Thuillier, G., Floyd, L., Woods, T.N., Cebula, R., Hilsenrath, E., Hersé, M., Labs, D.: 2004, in: J.M. Pap et al. (eds.), *Solar Variability and its Effects on Climate*, GMS 141, p. 171

Tu, C.-Y., Marsch, E., Wilhelm, K., Curdt, W.: 1998, ApJ 503, 475

Vest, R.E., Barad, Y., Furst, M.L., Grantham, S., Tarrio, C., Shaw, P.-S.: 2006, AdSpR 37, 283

Vidal-Madjar, A.: 1975, Sol. Phys. 40, 69

Warren, H.P., Mariska, J.T., Wilhelm, K.: 1998, ApJS 119, 105

Wilhelm, K.: 2002, ISSIR 2, 37

Wilhelm, K.: 2009, in: J. Trümper (ed.), *Landolt-Börnstein*, New Ser. VI 4B, p. 10

Wilhelm, K., Fröhlich, C.: 2010, ISSIR 9, in press

Wilhelm, K., Curdt, W., Marsch, E., et al.: 1995, Sol. Phys. 162, 189

Wilhelm, K., Woods, T.N., Schühle, U., Curdt, W., Lemaire, P., Rottman, G.J.: 1999, A&A 352, 321

Wilhelm, K., Schühle, U., Curdt, W., Dammasch, I.E., Hollandt, J., Lemaire, P., Huber, M.C.E.: 2000, Metro 37, 393

Wilhelm, K., Schühle, U., Curdt, W., et al.: 2002, ISSIR 2, 145

Wilhelm, K., Dwivedi, B.N., Marsch, E., Feldman, U.: 2004, Space Sci. Rev. 111, 415

Wilhelm, K., Schühle, U., Curdt, W., et al.: 2005, A&A 439, 701

Wilhelm, K., Marsch, E., Dwivedi, B.N., Feldman, U.: 2007, Space Sci. Rev. 133, 103

Willson, R.C., Mordvinov, A.V.: 2003, Geophys. Res. Lett. 30, 1199

Woods, T.N., Rottman, G.J., Ucker, G.J.: 1993, J. Geophys. Res. 98, 10679

Woods, T.N., Tobiska, W.K., Rottman, G.J., Worden, J.R.: 2000, J. Geophys. Res. 105, 27 195

Woods, T.N., Rottman, G.: 2005, Sol. Phys. 230, 375

Woods, T.N., Eparvier, F.G.: 2006, AdSpR 37, 219

Woods, T.N., Chamberlin, P.C., Harder, J.W., et al.: 2009, Geophys. Res. Lett. 36, L01101

Metallicity and kinematical clues to the formation of the Local Group

Rosemary Wyse

Johns Hopkins University
Department of Physics & Astronomy
Baltimore, MD 21218, USA
wyse@pha.jhu.edu

Abstract

The kinematics and elemental abundances of resolved stars in the nearby Universe can be used to infer conditions at high redshift, trace how galaxies evolve and constrain the nature of dark matter. This approach is complementary to direct study of systems at high redshift, but I will show that analysis of individual stars allows one to break degeneracies, such as between star formation rate and stellar Initial Mass Function, that complicate the analysis of unresolved, distant galaxies.

1 Introduction

These are exciting times to study local galaxies, due to a convergence of capabilities:

– Large observational surveys of individual stars in Local Group galaxies are feasible, using wide-field imagers and multi-object spectroscopy, complemented by space-based imaging and spectroscopy, with in the near-future the astrometric satellite Gaia providing 6-dimensional phase-space information. Hence, imaging surveys need matched spectroscopic surveys for full physics.

– High-redshift surveys are now starting to quantify the stellar populations and morphologies of galaxies at high look-back times, equal to the ages of old stars nearby.

– Large, high-resolution simulations of structure formation are allowing predictions of Galaxy formation in a cosmological context.

I will focus on the first of these capabilities, which provides the data that enables the fast-growing field of Galactic Archaeology. In this approach, analysis of the properties of low-mass old stars in our own Galaxy, and in nearby galaxies, allows us to do cosmology, locally. There are copious numbers of stars nearby that have ages greater than ~ 10 Gyr: these formed at look-back times corresponding to redshifts >2, and for a subset, perhaps as early as the epoch of reionization. The 'fossil record' is comprised of the conserved quantities that reveal conditions at the time the star formed, such as the chemical abundances in the interstellar medium. This is a complementary approach to direct study of galaxies at high redshift; the two approaches to understanding galaxy evolution analyse on the one hand snapshots of different galaxies at different times (redshifts) and on the other hand the evolution of one galaxy. A significant advantage of Galactic Archaeology, studying resolved stars, is that one can derive *separately* the stellar metallicity distributions and age distributions, and constrain variations in the massive-star initial mass function (IMF) from elemental ratios. One can then break degeneracies that hamper the interpretation of photometry and even spectra of the integrated light of galaxies, including the well-known age–metallicity case and star-formation-rate–IMF. Analysis of the motions of stars within a given galaxy can also provide the mass profile of the underlying potential, going from kinematics to dynamics.

The key questions that can be addressed, and which I will touch on in this review, include

- How do galaxies form?

    Star formation histories,

    mass assembly histories,

    link to black hole growth,

    the physics of 'feedback'.

- What is the nature of dark matter?

    Determines potential well shape,

    determines merger histories.

2 Testing the ΛCDM paradigm of structure formation

The standard model of structure formation is that of ΛCDM, where most of the gravitating material in the Universe is in the form of cold dark matter, and most of the energy is in a dark component that behaves as a constant density Cosmological Constant. This has proven to be an excellent description of the Universe on large scales, probed by the Cosmic Microwave background (e.g. Komatsu et al. 2009), and the large-scale power-spectrum of the distribution of galaxies (e.g. Percival et al. 2010). Gravity dominates on these scales, so that hot and cold dark matter fit equally well, and the physics of the constituents of dark matter is expected to be manifest only on smaller scales, such as sparse galaxy groups and individual galaxies (e.g. Ostriker & Steinhardt 2003). These are indeed the scales on which cracks are appearing in the ΛCDM armour (e.g. Wyse 2001; Peebles & Nusser 2010).

Structure formation in the ΛCDM paradigm is hierarchical, with smaller scales forming first, due to the shape of the power spectrum of primordial density fluctuations. This results in galaxy formation and evolution being largely driven by mergers. The outcome of a merger between two galaxies (i.e. dark matter haloes with associated baryonic material bound within them) depends in large part to the relative fractions of dissipationless material (dark matter, stars) and dissipational material (gas – ignoring dissipational dark matter models) (e.g. Tinsley & Larson 1979; Zurek, Quinn & Salmon 1988; Governato et al. 2009), the assumed physical conditions and equation of state of the dissipational material. The mass ratio and relative densities of the merging systems are also extremely important. Simulations that are dark-matter-only have the simplest physics, and – following on from the pioneering work of Moore et al. (1999) and of Klypin et al. (1999) – the highest resolution realizations for a Milky Way galaxy analogue show significant merging at all redshifts, and persistent substructure on all mass scales throughout the final, zero redshift, galaxy (e.g. Stadel et al. 2009). Substructure makes its presence felt through its gravitational interactions and assimilation through mergers, which can lead to heating of initially cold stellar components (e.g. Quinn & Goodman 1986; Kazantzidis et al. 2009). Thin disks are very susceptible to the destructive effects of mergers (of roughly equal mass), and in the ΛCDM scenario the disks we see today largely formed after merging ceases to be active, which is typically a redshift $z \sim 1$. The possibility that gas accretion into dark halos is dominated by cold streams, rather than the steady accretion of gas initially shock-heated to the halo virial temperature (White & Rees 1978), leads to the creation of fat, clumpy, turbulent disks of gas and stars at higher redshifts, $z \sim 2$ with active star formation (Ceverino, Dekel & Bournaud 2010), evolving into quiescent spheroid-dominated systems at lower redshift. The initial formation at $z \sim 2$ of thin disks – that are later destroyed through merging – is seen in high-resolution multi-phase hydrodynamic simulations within the ΛCDM framework (e.g. Scannapieco et al. 2009). The relationship of these predicted early thick and thin disks to the star-forming galaxies observed at redshifts ~ 2 (e.g. Genzel 2009) is as yet unclear, but intriguing.

A thin stellar disk is likely to be significantly affected by even very unequal mass-ratio mergers (minor mergers), with a significant fraction of the pre-merger orbital energy being absorbed into the internal degrees of freedom of the disk, resulting in heating, and creation of a thick stellar disk. Gas of course can radiate energy and cool, but the stellar component of the disk cannot. Adiabatic compression of the heated stellar disk,by gas settling later into the midplane,results in both a decrease of scale height of the stellar component and an increase of its velocity dispersion (e.g. Toth & Ostriker 1990; Elmgreen & Elmgreen 2006). Some of the orbital angular momentum will also be absorbed, in general resulting in a tilt of the disk plane. The relative density profiles and the details of the satellite's orbital parameters also are important determinants of the final outcome, in terms of the stellar disk structure (and of the fate of the satellite).

The robustness, masses and orbital parameters of substructure in ΛCDM, and the typical merging history of large galaxies, leads to the expectation that thick disks should be extremely common, and that the heating of thin disks should continue to late times. A vivid illustration of the results, using a pure N-body code and a cosmologically motivated retinue of satellite galaxies impinging on a stellar thin disk, assumed in place at redshift $z = 1$, are shown in Kazantzidis et al. (2009). As noted in several papers, in the concordance ΛCDM cosmology, a typical dark halo of the mass of that inferred for the Milky Way ($\sim 10^{12}$ M$_\odot$) will have suffered significant merging with relatively massive satellites (up to 10^{11} M$_\odot$) since a redshift of unity (e.g. Stewart et al. 2008; Fakhouri, Ma & Boylan-Kolchin, 2010). While the presence of gas may prevent the *destruction* of an existing thin disk, the *heating* is unavoidable, leaving an imprint in the stellar velocity dispersions (which as noted above, only increase with subsequent adiabatic growth of an embedded gas-rich thin disk).

Material is stripped from satellites as they merge, depending to first order on the relative density compared to that of the larger galaxy interior to the satellite orbit. The less dense, outer parts of satellites are thus stripped more easily, at larger distances from the central regions of the large galaxy. The typical orbits of merging substructure are elliptical, albeit with the eccentricity distribution peaking at close-to-parabolic orbits (e.g. Benson 2005; Khochfar & Burkert 2006). Debris from a given satellite that survives to the disk plane is broadly expected to be distributed in a thick torus, with radial extent indicating the location at which the satellite was disrupted, and some stars in the debris could have orbital parameters similar to the heated thin disk, now the thick disk (e.g. Huang & Carlberg 1997), and even to the surviving old thin disk (Abadi et al. 2003). The dark matter stripped from satellites during merging also builds up a dark-matter disk (Read et al. 2009), the mass of which is severely constrained, at least locally, by the vertical motions of stars, which are consistent with no significant disk dark matter (the 'K_Z' analysis and Oort limit, Kuijken & Gilmore 1989, 1991; Holmberg & Flynn 2004; Bienaymé et al. 2006). Torques during a minor merger will drive gas and stars inwards, to contribute to the inner disk and bulge (e.g. Mihos & Hernquist 1996).

In ΛCDM, the stellar halo of disk galaxies is postulated to be formed from disrupted satellite galaxies, with structure in coordinate space persisting for many dynamical times, even to the present times in the outer regions, where timescales are longest. Graphic illustration of the predicted structure, should the stellar halo consist entirely of accreted satellites, is given in Johnston et al. (2008; note that this is not a fully self-consistent model). Some fraction of the stars now in the inner halo/bulge may be disrupted disks from earlier stages of sub-halo merging, giving a 'dual halo' (e.g. Zolotov et al. 2009).

2.1 The Milky Way and M31 as templates

As noted above, the stellar halo, bulge, thick disk and even some part of (old?) thin disk of a typical large disk galaxy is predicted, in ΛCDM, to be created through the effects of mergers. One should see signatures of these origins in the stellar populations, since some memory of the initial conditions when a star is formed is retained. The Local Group provides an ideal test-bed of such predictions. Observational effects include

- coordinate space structure,

- kinematic (sub)structure,

- chemical abundance signatures,

- distinct age distributions,

- properties of surviving satellite galaxies.

Kinematic phase-space structure should survive longer than will coordinate-space structure, and chemical signatures of distinct populations are the most long-lived. Kinematics and chemical distributions are most robustly determined through spectroscopy.

Stars today probe conditions at a look-back time equal to their age; 10 Gyr corresponds to a redshift $z \sim 2$ for the concordance ΛCDM cosmology. Such stars, represented by main sequence, red giant stars and horizontal branch stars, are accessible in the Local Group with current capabilities. Only young, massive stars may be studied currently in galaxies beyond the Local Group; the wealth of information that can be derived from these was described by Rolf Kudritzki in his Schwarzschild lecture; an Extremely Large telescope is required to reach lower mass stars beyond the Local Group.

The Local Group consists of two large disk galaxies, M31 and the Milky Way galaxy, plus a low luminosity, very late-type disk galaxy, M33, a companion to M31. There are also retinues of gas-rich dwarf irregular (dIrr) and gas-poor dwarf spheroidal (dSph) satellite galaxies of each of the large disk galaxies, with many recent discoveries (discussed further below). We will discuss how photometric and (in particular) spectroscopic surveys of these galaxies reveal the important processes by which galaxies evolve.

3 Photometric surveys

3.1 The Milky Way galaxy

The first irrefutable evidence for ongoing merging between the Milky Way and its satellite galaxies was provided by the (serendipitous) discovery of the Sagittarius (Sgr) dSph galaxy, by Ibata, Gilmore & Irwin (1994). The coordinate-space overdensity of the core of this galaxy is illustrated in Fig. 1 of Wyse, Gilmore & Franx (1997). This galaxy was detected kinematically during a spectroscopic study of the bulge of the Milky Way (Ibata & Gilmore 1995). Member stars of this 'moving group' also have a redder colour distribution than the field bulge stars, a manifestation of their distinct age and metallicity distributions (determined by subsequent observations e.g. Layden & Sarajedini 2000; McWilliam & Smecker-Hane 2005; Sbordone et al. 2007; Siegel et al. 2007; Chou et al. 2010). The Sgr dSph contains no detectable gas, down to limits of $\sim$140 $M_\odot$ of atomic Hydrogen (Grcevich & Putman 2009 and refs. therein), and is clearly dark-matter dominated, based on the velocity dispersion of its member stars (e.g. Ibata et al. 1997). This lack of gas and dominance by dark matter are characteristic of the dSph satellite galaxies.

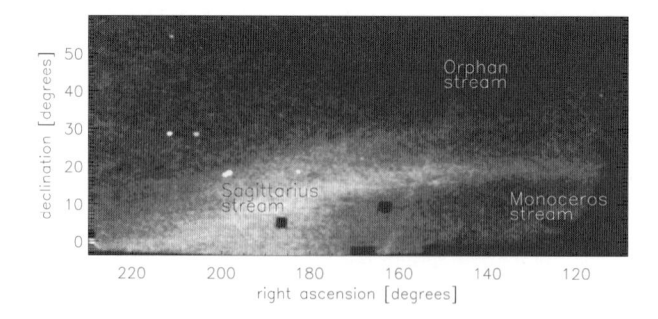

Figure 1: The 'Field of Streams' – and dots. This annotated version is courtesy of Gerry Gilmore and Vasily Belokurov (see Belokurov et al. 2006). The figure shows the distribution of main sequence turnoff stars on the sky, selected from the SDSS imaging data by the colour and magnitude cuts $19.0 < r < 22.0$ and $g - r < 0.4$, colour-coded in the figure by apparent magnitude, i.e. by distance from the Sun. Blue represents the nearest stars (at ~ 10 kpc), then green, then red (at ~ 30 kpc). The distribution of stars is clearly not uniform or smooth.

A new vitality was injected into the field by the advent of the Sloan Digital Sky Survey (SDSS) imaging data. The uniformly excellent photometry across a large fraction of the Northern sky revealed a very non-uniform distribution of resolved stars, dubbed the 'Field of Streams' (Belokurov et al. 2006; see Fig. 1 here).

The distribution of stars in the Northern sky shown in Fig. 1 is dominated by the two swathes of stars in the tidal streams from the Sgr dSph. Tidal debris from this system appears to be manifest not only in these two wraps (e.g. Fellhauer et al. 2006) but also as a significant part of the Virgo overdensity (e.g. Prior, Da Costa & Keller 2009). The outer stellar halo, at Galactocentric distances of greater than ~ 15 kpc and traced out to ~ 60 kpc, shows significant coordinate space structure (Bell et al. 2008; Watkins et al. 2009), the vast majority of which is associated with the Sgr dSph, and to a lesser extent with the Hercules-Aquila Cloud (Belokurov et al. 2007a), accounting for perhaps 10% of the total mass of the stellar halo ($M_{\text{star,halo}} \sim 10^9$ M$_\odot$).

The numerous 'dots' are also of extreme interest: these are gravitationally bound satellites of the Milky Way, ranging in luminosity from only a few hundred solar luminosities to millions of solar luminosities. Whether they are star clusters or satellite galaxies is only revealed by spectroscopic measurements of their internal kinematics, from radial velocities of member stars – galaxies, by definition, are held together by the gravity of dark matter haloes, while star clusters are baryon-dominated self-gravitating systems; this is discussed further below. Photometry alone provides distances – from the colour-magnitude diagram (CMD) of member stars – the scalelengths of the stellar distribution, and the total luminosity. These result in the size–absolute magnitude plot shown in Fig. 2, which includes known star clusters and dwarf galaxies (and is an annotated update of Fig. 1 in Gilmore et al. 2007).

The unoccupied region at the lower right is a selection effect – too few stars over too large an area – and the boundary approximately follows a line of constant V-band surface brightness equal to ~ 31 mag/arcsec^2 (see Belokurov et al. 2007b).

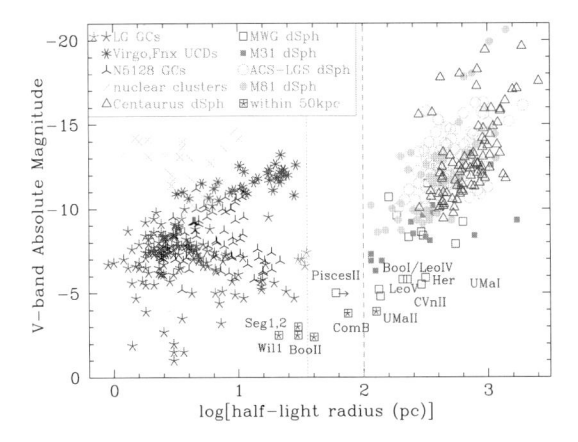

Figure 2: Distribution of star clusters and dwarf galaxies in the plane of stellar half-light radius vs. absolute V-band magnitude. The faint systems discovered in the SDSS imaging data that have been reported to be galaxies are identified by name. There are only three systems between the vertical red lines (dotted on the left, at a half-light radius of 36 pc, and dashed on the right, at a half-light radius of 100 pc).

There is a noticeable gap, emphasised in the Figure by the vertical red lines. As the SDSS imaging data have allowed lower and lower luminous systems to be discovered, they have largely been on either side of the gap (the newly discovered faint globular clusters are unlabelled). Indeed only three systems fall in the gap. Of these, two are closer than 50 kpc, within the realm of expected tidal effects – manifest in the Sgr streams – and one, Pisces II, is surrounded by a 'sea' of blue horizontal branch stars that were not included in the estimation of the half-light radius (Belokurov et al. 2010), and so will likely move out of the gap, to larger sizes, with deeper photometry. The nearby systems, within 50 kpc, are indicated in Fig. 2 and these tend to be rather elliptical and/or distorted (e.g. Martin, de Jong & Rix 2008), which may well indicate the effects of tides. Their present structure cannot be assumed to be the equilibrium structure at formation.

3.2 M31

This subsection should perhaps be labelled 'M31/M33', since deep imaging surveys are revealing hints of low surface brightness stellar features between the two, which, together with the wide-spread H I gas (Putman et al. 2009) are suggestive of past interactions. The most comprehensive wide-field survey to date is that of the PAndAS collaboration (McConnachie et al. 2009), building upon the pioneering work of Ibata et al. (2001) and Ferguson et al. (2002). Evolved stars in either of the hydrogen-shell burning phase (red giants) or core-helium burning phase (red clump and horizontal branch) are fairly straightforward to image at the distance of M31, at least in the lower surface-brightness outer disk and halo (avoiding source crowding and confusion). Similarly to the dominant stream from the Sgr dSph in the Milky Way, the structure in the stellar halo of M31 is dominated by the 'Giant Stream' (Ibata et al. 2004), together with significant other structure (e.g. Tanaka et al. 2010). Not only the Giant Stream but also smaller-scale structure can result from one accreted system (e.g. Fardal et al. 2008), but as in the Milky Way the situation is complex and many merged satellites may be involved (Font et al. 2008).

Again, many new satellites of M31 have been discovered by deep imaging surveys (e.g. Martin et al. 2009 and references therein); these are included in the satellite systems shown in Fig. 2; the M31 satellites also avoid the size gap. One might note that the earlier finding of significantly larger radii of M31 satellites compared to those of the Milky Way (McConnachie & Irwn 2006) is not held up by the fainter systems.

The colour of a red giant star of a given luminosity is more sensitive to chemical abundance than to age. Counts of red giants divided into two categories, red and blue, reveal significant chemical inhomogeneities across the outer disk and halo of M31 (Ferguson et al. 2002). The mean metallicity of the dominant 'halo' (really 'non-thin-disk') component of M31, derived from colours of the RGB stars, is significantly higher than that of the stellar halo of the Milky Way, being [Fe/H] ~ -0.6 dex (cf. Mould & Kristian 1986; Durrell, Harris & Pritchet 2004). This is equal to the mean metallicity of the Galactic thick disk at the solar neighbourhood, leading to speculation that these stars represent the thick disk of M31, and that the Galactic thick disk, rather than the halo, should be called 'Population II' to match the population in M31 that Baade resolved into stars (Wyse & Gilmore 1988). This high a value of mean chemical abundance, combined with the inferred old age from the CMD, implies that the stars formed within a fairly deep potential well, significantly more massive than a typical dwarf galaxy.

A large, high S/N spectroscopic survey is required to provide the kinematics and robust metallicities of both the field stars and satellite galaxies. This will allow testing of models for interactions between M31 and its satellites, including M33, and masses estimates for the surviving satellites. These have been initiated with 8m-class telescopes (e.g. Chapman et al. 2008; Letarte et al. 2009; Gilbert et al. 2009; Kalirai et al. 2010).

4 Spectroscopic surveys

While there exist photometric techniques, such as the colour of the RGB with an assumed age, or the UV excess of main sequence F/G stars (measured either by broadband filters or Strömgren photometry), to determine chemical abundances, and other photometric techniques for effective temperatures, spectroscopy remains the most robust and reliable means to determine stellar parameters, in particular the elemental abundances. Spectroscopy is also critical for determining kinematics for stars, particularly those far enough away that proper motions are too small to be measured (in the era prior to Gaia!)

There is a need for a variety of spectroscopic surveys at different spectral resolutions. Moderate resolution, here defined as a resolving power of a few thousand at optical wavelengths, provides spectra that are sufficient to determine line-of-sight motions to better than 10 km/s and metallicities to ~ 0.2 dex. With statistically large samples of stars these data can then be used to define the distribution functions of the major stellar populations under study, and investigate decompositions into different components. An individual star can then, in a probabilistic manner, be assigned to a given component. A significant amount of physics lies in the detailed shape and tails of the distribution functions, and these require large samples with well-defined selection functions.

High resolution, here defined as a resolving power of several tens of thousand at optical wavelengths, provides spectra that are sufficient to determine line-of-sight motions to much better than 1 km/s and individual elemental abundances to ~ 0.05 dex. This precision in the velocities allows the mapping of cold stellar substructure, such as tidal streams, and the internal kinematics of low luminosity satellite galaxies. Ideally one would have both high and moderate resolution spectra for stars drawn from the same parent sample, such as is possible with the RAVE spectroscopic survey (e.g. Steinmetz et al. 2006), as discussed below.

4.1 Masses and mass profiles of dwarf galaxies

Precise measurement, with echelle spectroscopy, of the line-of-sight velocities of member stars of dSph satellite galaxies was pioneered by Marc Aaronson, who derived a high mass-to-light ratio for the Draco dSph, based on velocities of three carbon stars. He obtained a *lower limit* to the central velocity dispersion of 6.5 km/s (Aaronson 1983), perfectly consistent with the modern value of 9 km/s (Walker et al. 2009). Mass-to-light ratios have been estimated for the systems in Fig. 2 leading to the conclusion that all systems to the right of the gap are embedded in dark matter haloes, i.e. are galaxies, while all equilibrium systems to the left of the gap are star clusters. This then provides the physical interpretation of the size distribution in terms of a minimum scale for systems dominated by dark matter (Gilmore et al. 2007). The scale in Fig. 2 is that of the light; given that baryons dissipate binding energy while gaseous, and become self-gravitating to form stars, this stellar half-light radius is expected to be a lower limit to the dark-matter scale-length. Such a minimum scale to dark matter of greater than ~ 100 pc is not expected if the dark matter is cold, and may be indicative of warm dark matter, such as sterile neutrinos (e.g. Kusenko 2009).

Derivation of mass profiles, as opposed to estimates of total mass, requires velocity information as a function of projected radius, not just for the central regions. The most straightforward analysis uses the Jeans equations, which relate the mass profile to the stellar light profile through the radial dependences of (line-of-sight) velocity dispersion and velocity dispersion anisotropy. However, this introduces a well-known mass–anisotropy degeneracy (Binney & Mamon 1982), which can be broken by use of the detailed line-of-sight velocity distribution, rather than its moments, such as the velocity dispersion (Gerhard 1993; Gerhard et al. 1998). This latter approach requires sample sizes of thousands of stars, well-distributed across the face of the galaxy, but with the central regions remaining critical, since it is here that the physics of the dark matter particle will be manifest most strongly. Such datasets are now being acquired (Gilmore et al. in preparation). The spherical Jeans' equation

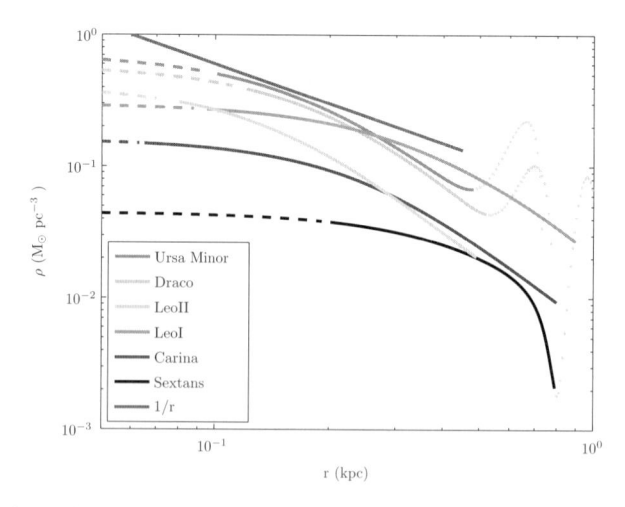

Figure 3: Derived mass density profiles from (isotropic) Jeans-equation analyses of the stellar velocity dispersion profiles of 6 dwarf spheroidal galaxies. Also shown is a r^{-1} profile, to which CDM-mass profiles are predicted to asymptote. The analysis is in each case reliable out to $r \sim 500$ pc. This figure is taken from Gilmore et al. (2007).

approach, with assumed isotropy, has the advantage of simplicity and when applied to a set of galaxies can identify any differences among them. The results of such an investigation, applied to six of the classical dSph satellites of the Milky Way is shown in Fig. 3, taken from Gilmore et al. (2007) – the derived mass density profile for each galaxy favours an inner core, rather than the cusped profile predicted for cold dark matter (Navarro, Frenk & White 1996). Further, all the profiles are very similar. This similarity is despite the fact that this set of galaxies covers a wide range in star-formation histories, implying that an appeal to feedback from star-formation to smooth out an inner cusp into a core would need to be tuned to each galaxy.

4.2 Elemental abundances: beyond metallicity

Different elements are created in stars of different masses and different evolutionary stages, on different timescales. Thus the pattern of elemental abundances seen in a stellar population reflects the star formation history and the stellar initial mass function (and the binary population) that provided the enrichment of the stars we observe now. A consequence is that old low-mass stars nearby provide insight into the stellar IMF at high redshift (look-back times at least equal to the age of the star).

Core-collapse supernovae, the death throes of massive, short-lived, stars, are the primary sources of the α-elements, created in steady-state, pre-supernova nucleosynthesis and ejected in the explosion. There is little iron produced, since the iron core is photo-disintegrated and largely forms the stellar remnant. Type Ia supernovae are produced by explosive nucleosynthesis in a white dwarf of mass greater than the Chandrasekhar limit, created by accretion from a binary companion. Such systems evolve on timescales that can be as long as a Hubble time and provide mostly iron, about ten times as much per supernova as a Type II supernova.

Stars formed in the earliest stages of a self-enriching system will have high levels of [α/Fe], reflecting the products of Type II (core-collapse) supernovae. There is negligible range in lifetime for Type II progenitors, compared to timescales of interest for star formation, and if there is good mixing of the supernova ejecta one expects to see an IMF-averaged yield of [α/Fe] in the next generation of stars. The most massive Type II progenitors produce the highest values of [α/Fe] (Gibson 1998; Kobayashi et al. 2007), so that a massive-star IMF biased towards the most massive stars will leave a signature, namely even higher values of IMF-averaged [α/Fe]. Some (model-dependent) time after the onset of star formation, white dwarfs will accrete sufficient material to explode as Type Ia supernovae. The first Type Ia are expected to result from progenitors with main sequence mass just below the threshold for core-collapse (~ 8 $M_\odot$), and these could in principle evolve to explosion in less than 10^8 yr in the double-degenerate model (e.g. Matteucci et al. 2009). It takes longer for sufficient Type Ia to explode and for their ejecta to be incorporated in the next generation of stars, and a typical lag seen in chemical evolution models is ~ 1 Gyr (Matteucci et al. 2009). Of course, the *iron abundance* corresponding to this time depends on the star formation rate and gas flows in or out. Inefficient enrichment, due, for example, to a slow rate of star formation or to loss of metals by winds, results in the signature of 'extra' iron from Type Ia being seen at lower values of [Fe/H]. High rates of star formation and efficient enrichment will produce a 'Type II plateau' that extends to higher [Fe/H] (e.g. Fig. 1 of Wyse & Gilmore 1993). One then expects that systems with low rates of star formation, over extended periods – such as the dwarf spheroidal satellites of the Milky Way – should show low values of [α/Fe] at low values of [Fe/H] (Unavane, Wyse & Gilmore 1996).

Observations have confirmed this expectation. A compilation of [Ca/Fe] for stars in the Milky Way and in representative gas-poor satellite galaxies is shown in Fig.4; the stars in satellite galaxies lie systematically below the field halo stars at the same value of [Fe/H], leading to the conclusion that accretion of stars from dwarf galaxies like those observed today could not play an important role in the formation of the stellar halo (Venn et al. 2004). Complementary, independent age information that the vast majority of halo stars are *old*, contrasting with the dominant intermediate age population of the luminous dSph, further constrains progenitors to consist only of stars formed early, in a short-lived burst of star formation that happened a long time ago (e.g. Unavane et al. 1996). A progenitor that would have been expected to have an extended star formation history, similar to a typical dSph, would have had to be accreted ~ 10 Gyr ago for its member stars to contribute to the bulk of the stellar halo (Unavane et al. 1996).

A new result is seen too, as observations have pushed to lower and lower metallicities within each system: we find, within the errors and limitations of small numbers, a consistent value for the enhanced 'Type II plateau' in [α/Fe] in all galaxies for their lowest metallicity member stars. These are the stars which one might expect to have formed in the earliest stages of star formation within each individual galaxy, and thus to show (pre)enrichment by core-collapse supernovae only. The fact that there is only moderate scatter in [α/Fe] in these stars, following the field halo stars,

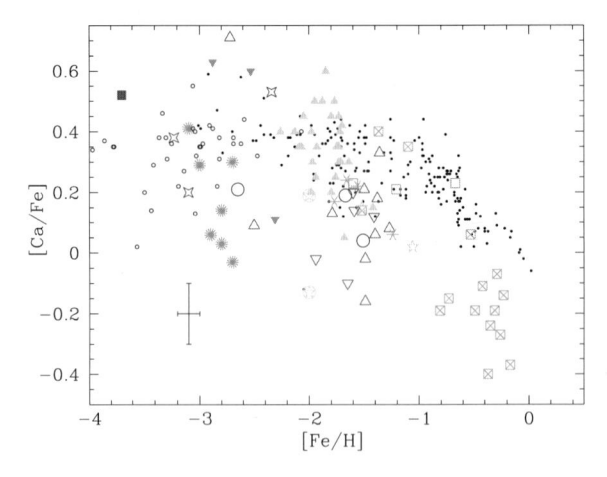

Figure 4: Abundance of calcium, an α-element, relative to iron, for field stars in the Milky Way Galaxy (small open and filled black circles, from Cayrel et al. 2004 and Fulbright 2000, respectively) and stars that are members of representative dwarf spheroidal galaxies, satellites of the Milky Way. Typical errors as indicated in the bottom right. The large black open circles represent stars in the Ursa Minor dSph (Sadakane et al. 2004), the magenta squares, enclosing crosses, represent stars in the Sgr dSph (Monaco et al. 2005), the red many-armed asterisks represent stars in the Sextans dSph (Aoki et al. 2009), the blue open triangles represent stars in the Carina dSph (usual orientation, Koch et al. 2008a; upside down, Shetrone et al. 2003), the green filled triangles represent stars in the Leo II dSph (Shetrone et al. 2009), the open red squares represent stars in the Fornax dSph (Shetrone et al. 2003), the six-armed magenta asterisks represent stars in the Sculptor dSph (Shetrone et al. 2003), the two green open star symbols represent stars in the Leo I dSph (Shetrone et al. 2003), the two cyan circles, enclosing star symbols, represent two stars in the Hercules dSph (Koch et al. 2008b), the blue 4-pointed star symbols represent stars in the Ursa Major I dSph (Frebel et al. 2009), the red upside-down filled triangles represent stars in the Coma Berenices dSph (Frebel et al. 2009) and the filled blue square represents a star in the Boötes I dSph (Norris et al. 2010). These last four dSph were all discovered within the SDSS imaging dataset and are very faint, fainter than absolute magnitude $M_V \sim -7$, well into the regime of globular star clusters (Belokurov et al. 2006, 2007b; Zucker et al. 2006).

implies that (i) the massive-star IMF is invariant and (ii) well-sampled and the ejecta well-mixed. These place rather stringent requirements on the early star formation. A further conclusion is that the stellar halo could form from any system(s) in which star-formation is short-lived, so that only Type II supernovae have time to enrich the star-forming gas, and enrichment is inefficient so that the mean metallicity is kept low. These could be star clusters, galaxies, or transient structures.

The old age and 'Type II' elemental abundances of the bulk of the field halo stars limits any late merging to be of systems that formed stars early, with short duration; we know of only one such luminous dwarf galaxy with no detectable age spread, that in Ursa Minor. Even for this system at least one star shows lower ratios of calcium to iron than does the bulk of the stellar halo, as shown in Fig. 4 (large black open circles; Sadakane et al. 2004). Recent models within the framework of ΛCDM argue that the field halo of the Milky Way formed from a few LMC-mass systems that were accreted early, and, unlike the LMC, had star-formation truncated abruptly at that time (e.g. Robertson et al. 2005). This has not yet been demonstrated within a self-consistent model of star formation and chemical enrichment.

The stars in the Milky Way in Fig. 4 are not identified by kinematics or spatial location into distinct stellar components. When this is done, further insight can be achieved. Fig. 5 shows the results from a large (by the standards of high-resolution spectroscopy) survey of metal-poor stars, with an emphasis on candidate disk members (Ruchti et al. 2010). The sample was selected from the Radial Velocity (RAVE) spectroscopic survey (Steinmetz et al. 2006) as having space motions more consistent with disk than halo, but being of low metallicity; the stellar parameters from the RAVE pipeline were used in this sample selection. The RAVE survey is obtaining moderate-resolution ($R \sim 8000$) spectra, around the IR Calcium triplet, for a magnitude-limited sample of bright stars, $I < 12$, using the multi-object fibre spectrograph (the 6dF instrument) on the UK Schmidt telescope. The aim is for a final sample of one million stars, allowing for a statistically significant sampling of the kinematic and metallicity distributions of each of the stellar components of the Milky Way. These stars are bright enough that follow-up echelle spectra of 'interesting' subsamples (such as metal-poor disk stars) can be fairly easily obtained on 4 m- or 8 m-class telescopes, while still probing distances of several kiloparsec from the solar neighbourhood (using giant stars).

The stars with elemental abundances shown in Fig. 5 have been re-classified into halo, thick disk and thin disk on the basis of the stellar parameters derived from the echelle observations, together with isochrone fitting for improved estimates of the gravity (particularly important for the distance estimates for low gravity, luminous red giant stars; we estimate that our distances for giants are accurate to 20–30%). The stars in the RAVE catalogue are sufficiently bright that proper-motion measurements are available, allowing full space motions to be derived once distance are estimated (the RAVE radial velocities are accurate to better than a few km/s). The component assignment is probabilistic by necessity, since the components have overlapping properties – and indeed characterising, then understanding, that overlap is a major science goal of the full RAVE survey. The different population assignments are indicated by different symbols in the Figure (see Ruchti et al. 2010 for details of the range of criteria used in the assignments). The elemental abundances were derived following the formalism of Fulbright (2000).

It is evident that the thick disk extends at least to [Fe/H] ~ -1.75 dex, and that these stars show the same enhanced values of [α/Fe] as do halo stars, for a range of α-elements. Most of the stars classified as thick disk or halo are giants, and probe distances of several kiloparsec from the Sun, the first time that elemental abundances for the metal-poor thick disk has been measured at these distances. Similar elemental abundance results for smaller samples of (high-velocity) thick disk and halo (dwarf) stars at [Fe/H] ~ -1 dex were reported previously (Nissen & Schuster 1997), and also for a local sample (distances less than 150 pc) of dwarf stars for the full range in [Fe/H] of (Reddy & Lambert 2008; Reddy, Lambert & Allende-Prieto 2006). The

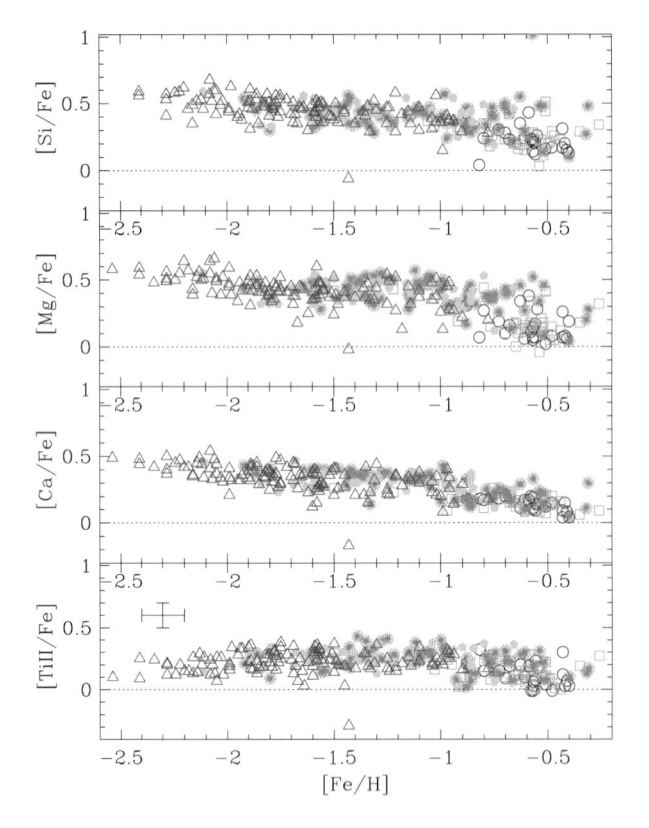

Figure 5: Elemental abundances, from analyses of high-resolution spectra, for stars selected from the RAVE survey database to be low metallicity but with disk-like kinematics. The distances and kinematics have been re-calculated based on the stellar parameters derived from the echelle spectra and isochrone fits. The blue triangles represent stars that have high probability to be halo stars, the red asterisks high probability thick-disk stars, with the green pentagons having similar probability of belonging to either the halo or thick disk. The black circles represent high probability thin-disk stars and the magenta open squares are either thin- or thick-disk stars.

enhanced values of [α/Fe] imply that core-collapse supernovae dominated the enrichment of the gas clouds from which these stars formed, which in turn implies that the stars formed within a short time after the onset of star formation. It could still be possible to incorporate an age range greater than $\sim$1 Gyr, should this be established by other means, if the stars were to form in distinct sub-units which each had a short duration of star formation, but different times of onset, but this requires some fine-tuning (see Gilmore & Wyse 1998). A simpler interpretation is that there is a narrow age range and that the metal-poor thick disk (and halo) stars formed during a short-lived epoch of star formation.

The similarity of the elemental abundances across the halo/thick disk transition implies that the core-collapse supernovae that pre-enriched the stars came from a very similar massive-star Initial Mass Function. Further, the low scatter implies good sampling of the massive-star IMF and good mixing within the interstellar medium. This would seem to imply large star-forming regions.

Several models for the thick disk invoke a minor merger and associated heating of a pre-existing thin disk; one then expects to find stars above the disk plane that came from the previous thin disk, together with stars from the now-shredded satellite that merged (e.g. Gilmore, Wyse & Norris 2002). Distinguishing these two contributions relies upon stellar population signatures. It is clear that the bulk of the thick-disk stars near the solar neighbourhood, for which the mean metallicity is relatively high, [Fe/H] ~ -0.6 dex, and the mean age old, ~ 10–12 Gyr, must have formed within a fairly deep potential well, since even the Large Magellanic Clouds did not self-enrich to this level until a few Gyr ago. The metal-poor end of the thick disk could more plausibly be associated with debris from accreted dwarf galaxies. The elemental abundance pattern seen in Fig. 5 requires that star formation in any such progenitor be very short-lived, which, compared to the typical extended star formation history, and early onset, in surviving dwarf galaxies implies that any such dwarf galaxy be accreted a long time ago. This is consistent with inferences from the old age of stars in the bulk of the thick disk, but not easy to reconcile with the late merging and accretion of satellites predicted in ΛCDM (e.g. Abadi et al. 2003).

Radial migration of stars and gas in disks, due to resonance with transient spiral arms (Sellwood & Binney 2002), perhaps in concert with the bar (Minchev & Famaey 2009) is potentially very important. Extremely efficient migration, with a probability that is independent of the amplitude of random motions of the stars, can create a metal-rich thick disk locally out of the inner regions of the disk (Schönrich & Binney 2009). However, we currently lack an understanding of the efficiency of migration, and this could be tested by large surveys of elemental abundances.

5 Concluding remarks

I have only two 'take-home' points:

- spectroscopy is critical in determining the astrophysics of stellar populations;
- large (planned) imaging surveys need matched spectroscopic capabilities.

Acknowledgements

I thank the organisers for their invitation to this stimulating conference. I acknowledge support from the US National Science Foundation (AST-0908326).

References

Aaronson, M.: 1983, ApJ 266, L11
Abadi, M., Navarro, J., Steinmetz, M., Eke, V.: 2003, ApJ 597, 21

Aoki, W., Arimoto, N., Sadakane, K., et al.: 2009, A&A 502, 569

Bell, E., Zucker, D.B., Belokurov, V., et al.: 2008, ApJ 680, 295

Belokurov, V., Zucker, D.B., Evans, N.W., et al.: 2006, ApJ 647, L111

Belokurov, V., Evans, N.W., Bell, E., et al.: 2007a, ApJ 657, L89

Belokurov, V., Zucker, D.B., Evans, N.W., et al.: 2007b, ApJ 654, 897

Belokurov, V., Walker, M., Evans, N.W., et al.: 2010, ApJ 712, L103

Bienaymé, O., Soubiran, C., Mishenina, T., Kovtyukh, V., Siebert, A.: 2006, A&A 446, 933

Binney, J., Mamon, G.: 1982, MNRAS 200, 361

Cayrel, R., Depagne, E., Spite, M., et al.: 2004, A&A 416, 1117

Ceverino, D., Dekel, A., Bournaud, F.: 2010, MNRAS, in press, astro-ph/0907.3271

Chapman, S., Ibata, R., Irwin, M., et al.: 2008, MNRAS 390, 1437

Chou, M.-Y., Cunha, K., Majewski, S.R., et al.: 2010, ApJ 708, 1290

Durrell, P., Harris, W., Pritchet, C..: 2004, AJ 128, 260

Elmegreen, B.G., Elmegreen, D.M.: 2006, ApJ 650, 644

Fakhouri, O., Ma, C.-P., Boylan-Kolchin, M.: 2010, MNRAS, submitted, astro-ph/1001.2304

Fardal, M.A., Babul, A., Guhathakurta, P., Gilbert, K.M., Dodge, C.: 2008, ApJ 682, L33

Fellhauer, M., Belokurov, V., Evans, N.W., et al.: 2006, ApJ 651, 167

Ferguson, A.M. N., Irwin, M.J., Ibata, R., et al.: 2002, AJ 124, 1452

Font, A., Johnston, K., Ferguson, A., et al.: 2008, ApJ 673, 215

Frebel, A., Simon, J., Geha, M., Willman, B.: 2010, ApJ 708, 560

Fulbright, J.: 2000, AJ 120, 1841

Genzel, R.: 2009, in: J. Andersen, J. Bland-Hawthorn, B. Nordström (eds.), *The Galaxy Disk in Cosmological Context*, IAU Symp. 254, p. 33

Gerhard, O.: 1993, MNRAS 265, 213

Gerhard, O., Jeske, G., Saglia, R., Bender, R.: 1998, MNRAS 295, 197

Gibson, B.K.: 1998, ApJ 501, 675

Gilbert, K., Guhathakurta, P., Kollipara, P., et al.: 2009, ApJ 705, 1275

Gilmore, G., Wyse, R.F.G.: 1998, AJ 116, 748

Gilmore, G., Wyse, R.F.G., Norris, J.E.: 2002, ApJ 574, L39

Gilmore, G., Wilkinson, M., Wyse, R.F.G., et al.: 2007, ApJ 663, 948

Governato, F., Brook, C.B., Brooks, A.M., et al.: 2009, MNRAS 398, 312

Grcevich, J., Putman, M.E.: 2009, ApJ 696, 385

Holmberg, J., Flynn, C.: 2004, MNRAS 352, 440

Huang, S., Carlberg, R.G.: 1997, ApJ 480, 503

Ibata, R.A., Gilmore, G.: 1995, MNRAS 275, 591

Ibata, R.A., Gilmore, G., Irwin, M.J.: 1994, Nat 370, 194

Ibata, R.A., Wyse, R.F.G., Gilmore, G., Irwin, M.J., Suntzeff, N.: 1997, AJ 113, 634

Ibata, R.A., Irwin, M., Lewis, G., et al.: 2001, Nat 412, 49

Ibata, R.A., Chapman, S., Ferguson, A.M.N., et al.: 2004, MNRAS 351, 117

Johnston, K.V., Bullock, J.S., Sharma, S., et al.: 2008, ApJ 689, 936

Kalirai, J., Beaton, R., Geha, M., et al.: 2010, ApJ 711, 671

Kazantzidis, S., Zentner, A.R., Kravtsov, A.V., et al.: 2009, ApJ 700, 1896

Klypin, A., Kravtsov, A.V., Valenzuela, O., Prada, F.: 1999, ApJ 522, 82

Kobayashi, C., Umeda, H., Nomoto, K., Tominaga, N., Ohkubo, T.: 2006, ApJ 653, 1145

Koch, A., Grebel, A., Gilmore, G., et al.: 2008a, AJ 135, 1580

Koch, A., McWilliam, A., Grebel, E., et al.: 2008b, ApJ 668, L13

Khochfar, S., Burkert, A.: 2006, A&A 445, 403

Komatsu, E., Dunkley, J., Nolta, M.R., et al. (WMAP team): 2009, ApJS 180, 330

Kuijken, K., Gilmore, G.: 1989, MNRAS 239, 605

Kuijken, K., Gilmore, G.: 1991, ApJ 367, L9

Kusenko, A.: 2009, Phys. Rep. 481, 1

Layden, A.C., Sarajedini, A.: 2000, AJ 119, 1760

Letarte, B., Chapman, S., Collins, M., et al.: 2009, MNRAS 400, 1472

McConnachie, A., Irwin, M.: 2006, MNRAS 365, 1263

McConnachie, A., Irwin, M., Ibata, R., et al.: 2009, Nat 461, 66

McWilliam, A., Smecker-Hane, T.A.: 2005, ApJ 622, L29

Martin, N., de Jong, J., Rix, H.-W.: 2008, ApJ 684, 1075

Martin, N., McConnachie, A., Irwin, M., et al.: 2009, ApJ 705, 758

Matteucci, F., Spitoni, E., Recchi, S., Valiante, R.: 2009, A&A 501, 531

Mihos, J.C., Hernquist, L.: 1996, ApJ 464, 641

Minchev, I., Famaey, B.: 2009, ApJ, submitted, astro-ph/0911.1794

Monaco, L., Bellazzini, M., Bonifacio, P., et al.: 2005, A&A 441, 141

Moore, B., Ghigna, S., Governato, F., et al.: 1999, ApJ 524, L19

Mould, J., Kristian, J.: 1986, ApJ 305, 591

Navarro, J., Frenk, C., White, S.D.M.: 1996, ApJ 462, 563

Nissen, P., Schuster, W.: 1997, A&A 326, 751

Norris, J.E., Yong, D., Gilmore, G., Wyse, R.F.G.: 2010, ApJ 711, L350

Ostriker, J.P., Steinhardt, P.: 2003, Sci 300, 1909

Peebles, P.J.E., Nusser, A.: 2010, astro-ph/1001.1484

Percival, W., Reid, B.A., Eisenstein, D.J., et al.: 2010, MNRAS 401, 2148

Prior, S., Da Costa, G., Keller, S.: 2009, ApJ 704, 1327

Putman, M., Peek, J., Muratov, A., et al.: 2009, ApJ 703, 1486

Quinn, P., Goodman, J.:1986, ApJ 309, 472

Read, J.I., Mayer, L., Brooks, A.M., Governato, F., Lake, G.: 2009, MNRAS 397, 44

Reddy, B., Lambert, D.: 2008, MNRAS 391, 95

Reddy, B., Lambert, D., Allende Prieto, C.: 2006, MNRAS 367, 1329

Robertson, B., Bullock, J.S., Font, A.S., Johnston, K.V., Hernquist, L.: 2005, ApJ 632, 872

Ruchti, G., Fulbright, J., Wyse, R.F.G., et al.: 2010, in preparation

Sadakane, K., Arimoto, N., Ikuta, C., et al.: 2004, PASJ 56, 1041

Sbordone, L., Bonifacio, P., Buonanno, R., Marconi, G., Monaco, L., Zaggia, S.: 2007 A&A, 465, 815

Scannapieco, C., White, S.D.M., Springel, V., Tissera, P.B.: 2009, MNRAS 396, 696

Schönrich, R., Binney, J.: 2009, MNRAS 399, 1145

Sellwood, J., Binney, J.: 2002, MNRAS 336, 785

Shetrone, M., Venn, K., Tolstoy, E., et al. : 2003, AJ 125, 684

Shetrone, M., Siegel, M.H., Cook, D., Bosler, T.: 2009, AJ 137, 62

Siegel, M.H., Dotter, A., Majewski, S.R., et al.: 2007, ApJ 667, L57

Stadel, J., Potter, D., Moore, B., et al.: 2009, MNRAS 398, L21

Steinmetz, M., Zwitter, T., Siebert, A., et al. (the RAVE collaboration): 2006, AJ 132, 1645

Stewart, K.R., Bullock, J. S., Wechsler, R.H., et al.: 2008, ApJ 683, 597

Tanaka, M., Chiba, M., Komiyama, Y., Guhathakurta, P., Kalirai, J., Iye, M.: 2010, ApJ 708, 1168

Tinsley, B.M., Larson, R.B.: 1979, MNRAS 186, 503

Toth, G., Ostriker, J.P.: 1992, ApJ 389, 5

Unavane, M., Wyse, R.F.G., Gilmore, G.: 1996, MNRAS 278, 727

Venn, K., Irwin, M., Shetrone, M., et al.: 2004, AJ 128, 1177

Walker, M., Mateo, M., Olszewski, E., et al.: 2009, ApJ 704, 1274

Watkins, L.L., Evans, N.W., Belokurov, V., et al.: 2009, MNRAS 398, 1757

White, S.D.M., Rees, M.J.: 1978, MNRAS 183, 341

Wyse, R.F.G.: 2001, in: J. Funes, E. Corsini (eds.), *Galaxy Disks and Disk Galaxies*, ASPC 230, p. 71

Wyse, R.F.G., Gilmore, G.: 1988, AJ 95, 1404

Wyse, R.F.G., Gilmore, G.: 1993, in: G. Smith, J. Brodie (eds.), *The Globular Cluster-Galaxy Connection*, ASPC 48, p. 727

Wyse, R.F.G., Gilmore, G., Franx, M.: 1997, ARA&A 35, 637

Zolotov, A., Willman, B., Brooks, A.M., et al.: 2009, ApJ 702, 1058

Zucker, D.B., Belokurov, V., Evans, N.W., et al.: 2006, ApJ 650, L41

Zurek, W.H., Quinn, P.J., Salmon, J.K.: 1988, ApJ 330, 519

Probing dark matter, galaxies and the expansion history of the Universe with Lyα in absorption and emission

Martin Haehnelt

Institute of Astronomy
Madingley Road, Cambridge CB3 0HA, UK
haehnelt@ast.cam.ac.uk

Abstract

Lyα radiation is an important diagnostic tool in a wide range of astrophysical environments. I will first describe here how measurements of the matter power spectrum on small scales from Lyα forest data constrain the mass of dark matter particles. I then will report on an ambitious program of searching for very faint spatially extended Lyα emission at $z \sim 3$ which has led to the discovery of a new population of faint Lyα emitters which I will argue should be identified with the long searched for host galaxies of damped Lyα absorbers. Finally, I will discuss the possibility of measuring the redshift drift of Lyα absorption features and therefore the change of the expansion rate of the Universe in real time with the ultra-stable high-resolution spectrograph CODEX proposed for the E-ELT.

1 Introduction

Astrophysical spectroscopy has a great tradition in Potsdam and very much in line with the conference theme this contribution will focus on spectroscopy of Lyα radiation absorbed and/or emitted by the intervening material between us and high redshift QSOs.

Figure 1 shows a nice example of Lyα forest absorption due to intervening neutral hydrogen superposed on the intrinsic emission spectrum of a QSO which itself is a superposition of a broad continuum and numerous strong and weak emission lines (see Rauch 1998 and Weinberg et al. 1999 for reviews). Note also the effect of a strong damped Lyα absorption system.

Figure 2 shows the same spectrum after it has been normalized by the continuum emission of the QSO. In the middle and bottom panel the wavelength scale has been stretched by factors of 10 and 100, respectively. The Lyα forest absorption is due to the neutral hydrogen in the warm photoionized intergalactic medium with a temperature of about 15 000 K and the width of typical absorption features ranges from about 10–100 km s^{-1}. Fully resolving these features requires high resolution spectroscopy which is performed with Echelle-type spectrographs. Some of the narrowest

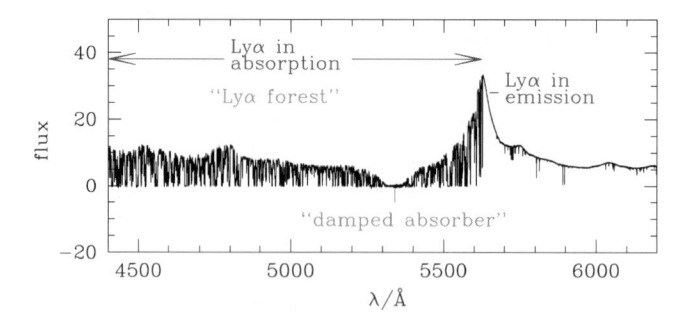

Figure 1: A high-resolution, high S/N spectrum of a QSO at $z \sim 3.6$. The spectrum shows the broad Lyα emission typical for QSOs. On the blue side of the emission line absorption due to intervening neutral hydrogen along the line-of sight leads to a characteristic Lyα "forest". Observed strong absorption systems exhibiting "damping wings" are presumably due to the neutral hydrogen in low-mass high-redshift galaxies akin to the typical building blocks of Milky Way like galaxies.

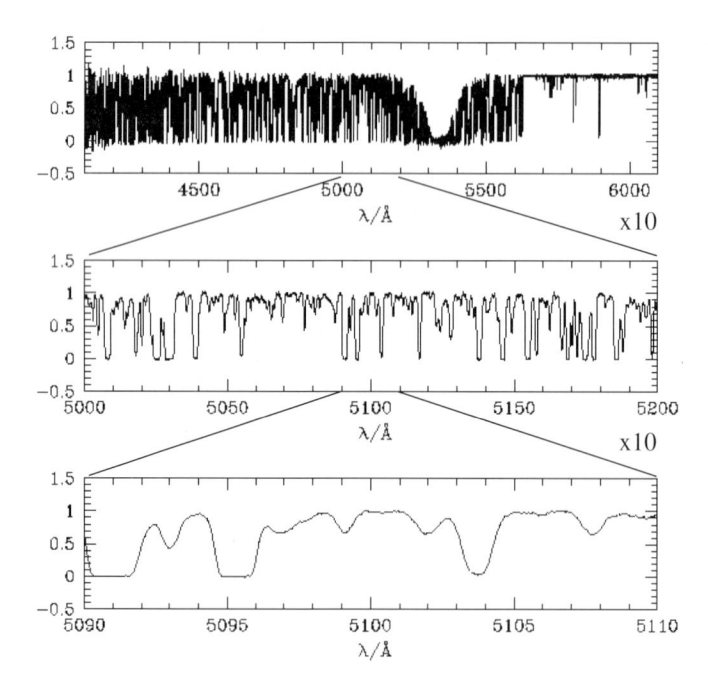

Figure 2: The "normalized" absorption spectrum after accounting for the wave length dependence of continuum emission. The middle and bottom panel show the central part of the spectrum in the panel above with the wavelength scale stretched by a factor of ten.

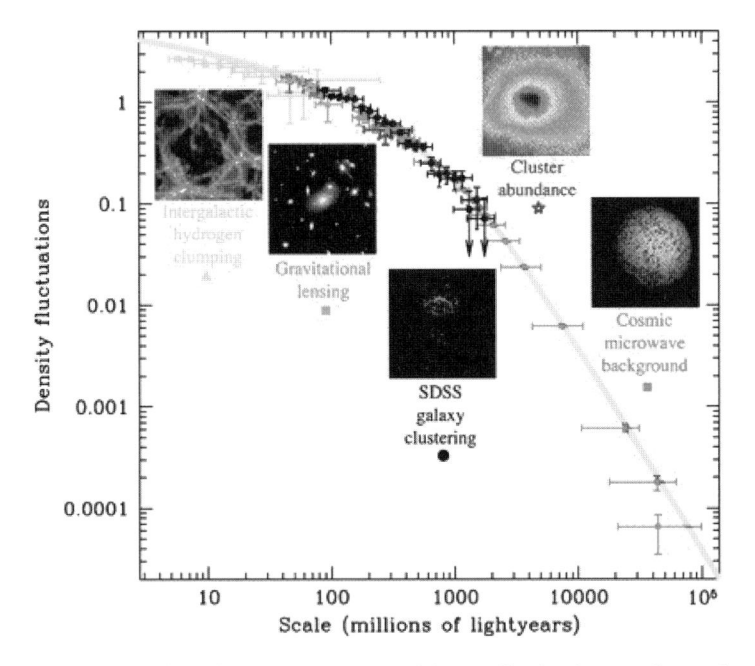

Figure 3: Illustration of the measurements of the amplitude of matter fluctuations with different observational methods across a wide range of scales. Figure courtesy of Max Tegmark.

features are not due to Lyα absorption but due to associated metal line absorption. Such associated metal line absorption is more easily visible on the red side of the broad Lyα emission line of the QSO where it is not confused with Lyα absorption. High-quality high-resolution absorption spectra like the one shown in Figs. 1 and 2 contain a tremendous amount of information most notably on high-redshift galaxies, the ionization and thermal state of the intergalactic medium, the enrichment of the intergalactic medium with metals, the epoch of reionization and the power spectrum of the matter distribution on small and intermediate scales.

I will not have space to review all of these here but will concentrate here on the three particular aspects mentioned in the abstract.

2 Constraining the mass of dark matter particles with the Lyα forest

2.1 Neutral hydrogen as a faithful tracer of the matter distribution

The last 10–15 years have seen the establishment of the ΛCDM paradigm for the emergence of structure in the Universe which is now based on overwhelming evidence from a wide range of observational methods. Figure 3 illustrates how the

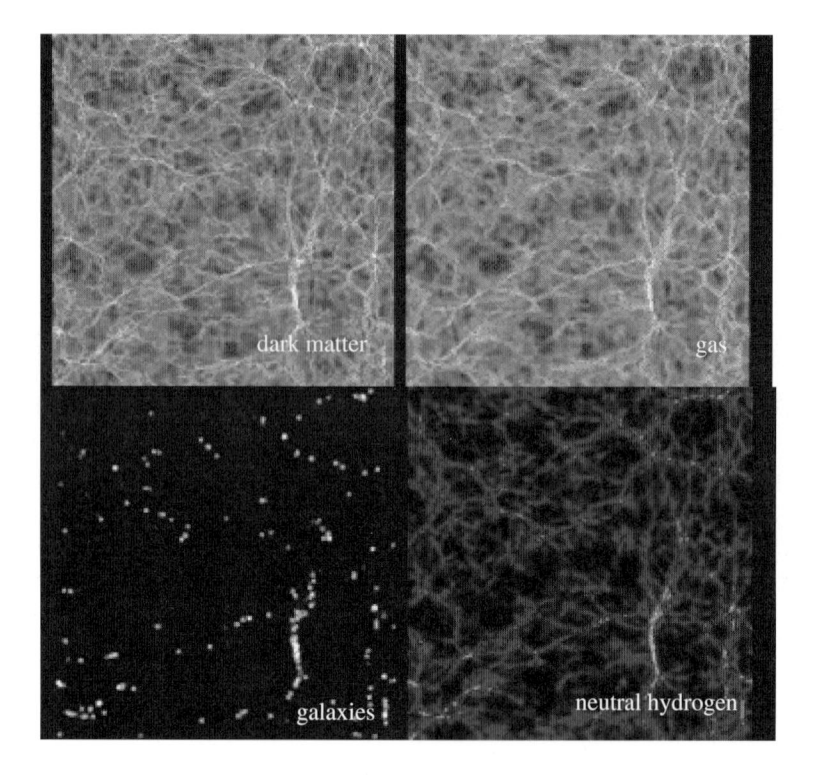

Figure 4: Hydrodynamical simulations of the matter distribution in a ΛCDM model. The upper left and right panel show the distribution of dark matter and gas, while the lower panels show the distribution of galaxies and neutral hydrogen. Figure courtesy of Matteo Viel.

combination of different observational methods allows the measurement of the amplitude of matter fluctuations across a wide range range of scales. On the largest scales the matter power spectrum is measured by CMB data while on intermediate scales galaxy cluster abundances, galaxy surveys, and weak gravitational lensing offer the best constraints.

I will focus here on the smallest scales for which the matter power spectrum is currently accessible for which the best measurements are based on Lyα forest data which probe the distribution of neutral hydrogen down to the the Jeans length of the warm photoionized IGM with a temperature of $\sim 15\,000$ K. Figure 4 shows a hydrodynamical simulation of the distribution of matter in a ΛCDM model of structure formation which follows the formation of galaxies and takes into account the physics governing the ionization and thermal state of the intergalactic medium. The upper right panel shows the distribution of dark matter which is faithfully traced by the distribution of gas and neutral hydrogen which are shown in the upper and lower right panel, respectively. The use of simulations like that shown in Fig. 4 allows us to obtain accurate measurements of the matter power spectrum on small scales from Lyα forest data.

2.2 How cold is dark matter?

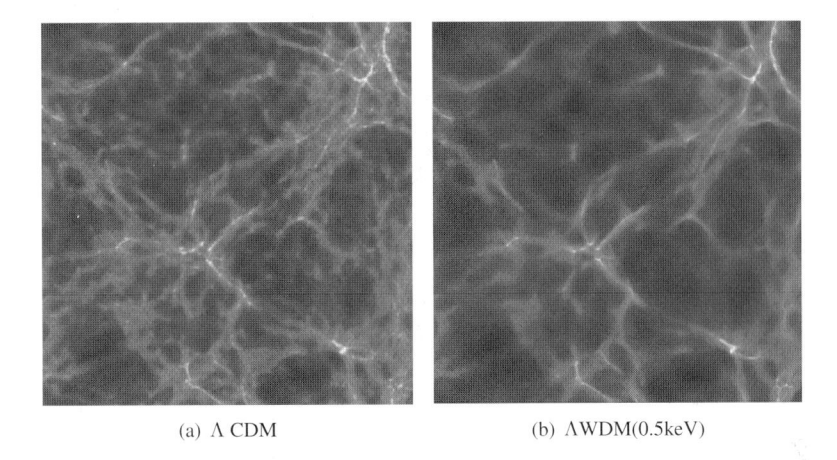

(a) ΛCDM

(b) ΛWDM(0.5keV)

Figure 5: Hydrodynamical simulations of the matter distribution of a cold/warm dark matter model. The reduced small scale structure due to the free-streaming of the warm dark matter ($m_{\rm WDM} = 0.5$ keV) is clearly seen and well probed by the Lyα forest. Simulations were run with the numerical hydrodynamical code Gadget-II on the COSMOS computer in Cambridge (box-size $30\,h^{-1}$ Mpc). Figure courtesy of Matteo Viel.

It is often tacitly assumed that the dark matter is cold (CDM) i.e. that the thermal motions of the dark matter particles are sufficiently small that they are not astrophysically relevant. It is thereby left somewhat vague what is meant by not astrophysically relevant. A large number of particle candidates has been suggested to be the dark matter and for many of these thermal velocities are, however, important. These are therefore generally called warm instead of cold dark matter. Prominent examples are early decoupled thermal relics and the sterile neutrinos which are an essential ingredient of the minimally extended standard model of particle physics proposed to account for the non-zero mass of active neutrinos required by the observed flavour oscillations of neutrinos.

As demonstrated in Fig. 5 the matter distribution is sensitive to the effect of free-streaming of dark matter particles due to thermal velocities. The signature of a cut-off in the matter power spectrum due to the free-streaming (or the lack thereof) gives interesting lower limits on the mass of dark matter particles. As already mentioned the power spectrum of the flux distribution in QSO absorption spectra offers the best constraint of the matter power spectrum on small scales and the strongest constraints on the free-streaming of dark matter particles are thus provided by Lyα forest data. The most accurate measurement of the Lyα flux power spectrum have been obtained from a large sample of SDDS QSO spectra. These spectra have, however, only moderate spectral resolution and the tightest lower limits on the mass of a putative warm dark matter particle have been obtained by combining the SDSS Lyα flux power spectrum with higher resolution Lyα data.

(a) $m_{\rm WDM} > 2.4$ keV

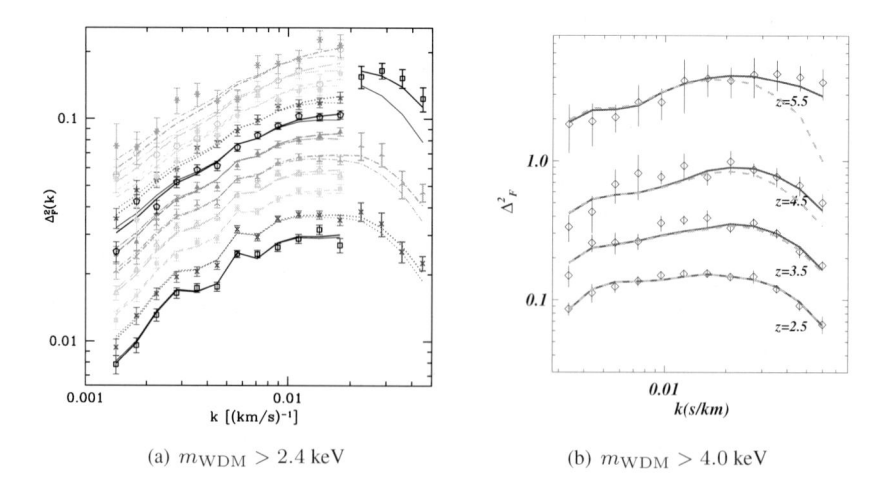

(b) $m_{\rm WDM} > 4.0$ keV

Figure 6: The Lyα flux power spectrum for the SDSS sample of low-resolution QSO absorption spectra and two different samples of high resolution spectra for a range of redshifts. The shape of the small scale cut-off of the flux-power spectrum is sensitive to the free-streaming of dark matter particles and therefore gives a lower limit for the mass of (warm) dark matter particles. The plots are reproduced from Seljak et al. (2006) and Viel et al. (2008) and the limits given below the plots are the 2σ lower limits for the mass of early decoupled thermal relics.

Figure 6 shows a comparison of the observed Lyα flux power spectrum with predictions for warm and cold dark matter models. In this way Seljak et al. (2006) obtained a 2σ lower limit of 2.5 keV using the SDSS data and a sample of Keck high resolution data. Using a larger sample of 55 high resolution spectra which extends to higher redshift where the strongest constraints can be obtained Viel et al. (2008) improved this lower limit to 4 keV. Note that what is actually constrained is the free streaming velocity of the dark matter particles and that the inferred masses are model dependent. The masses quoted above are applicable for early decoupled thermal relics. For sterile neutrinos the corresponding masses would be larger. These limits have virtually closed the parameter space for warm dark matter for which the free streaming of dark matter would have an effect on the structure of the dark matter haloes with masses $\geq 10^9$ M$_\odot$. Warm dark matter is thus very unlikely to be able to resolve what has been termed the small scale crisis of cold dark matter (see Kuzio de Naray et al. 2010 for a recent discussion of the CDM small scale crisis).

3 Going very faint in search for the building blocks of Milky Way type galaxies

3.1 Searching for faint spatially extended Lyα emission

I will now move to a very different topic, the host galaxies of damped Lyα absorbers. Detecting emission from the host galaxies of strong QSO absorption systems for which associated metal absorption suggests that they arise in the vicinity of galaxies is very difficult. There are two reasons for this. The first is that the emission has to be detected against the backdrop of what are generally very bright QSOs. The second is as I will argue here that the host galaxies of damped Lyα absorbers and Lyman limit systems are rather faint galaxies which are typically much fainter than typical (L_*) galaxies. I will begin here with reviewing the results of an ambitious attempt to push searches for spatially extended emission from damped Lyα absorption systems and Lyman limit systems to fainter surface brightness limits. The results are the outcome of a campaign of ultra-deep long-slit spectroscopy with FORS2 at the VLT where we have searched for extended Lyα emission in a mostly blank field (Rauch et al. 2008; Haehnelt et al. 2008).

The field contained a moderately bright QSO and was observed during 2004–2006 with the the volume-phased holographic grism 1400V on FORS2. The resulting spectrum which should be the deepest optical spectrum ever taken is shown in Fig. 7. The spatial direction is shown vertically and the spectral direction is shown horizontally. The spectrum ranges from 4457 to 5776 Å. Despite of the long exposure time Rauch et al. (2008) were able to reach the sky noise limit and the 1σ surface brightness detection limit for line photons of the spectrum is an unprecedented 8×10^{-20} erg cm^{-2} s^{-1} arcsec^{-2}. The white boxes in Fig. 7 mark faint emission line objects. Figure 8 shows a close up of a sample of 27 faint candidate Lyα emitters. About half of the profiles are extended, most likely owing to radiative transfer of Lyα photons from a central source, and there are candidates for both outflows and infall features. For a detailed description of the results see Rauch et al. (2008).

The original motivation was to search for fluorescent Lyα emission from the optical thick part of the gaseous cosmic web, induced by the general UV background (Hogan & Weymann 1987; Gould & Weinberg 1996). Line emission in response to recombinations caused by the impinging UV photons would cause any patches of optically thick gas to exhibit a universal "glow" of Lyα photons. Theoretical estimates of the UV background intensity had, however, gone down during the project, and so Rauch et al. (2008) were probably somewhat short of the signal-to-noise-ratio required to detect the fluorescent (re-)emission of the meta-galactic UV background from optically thick regions in a blank field. They nevertheless found 27 single line emitters on the slit, however with total fluxes typically a factor five to ten higher than their revised expectations for the signal from fluorescence alone.

Rauch et al. (2008) have argued that the high inferred space density and the clear signs of radiative transfer effects in many of the emitters leave Lyα at $2.67 < z < 3.75$ as the only plausible interpretation for the majority of the emitters.

If interpreted as Lyα the emitters present a steeply rising luminosity function with a total number density more than 20 times larger than the comoving density of

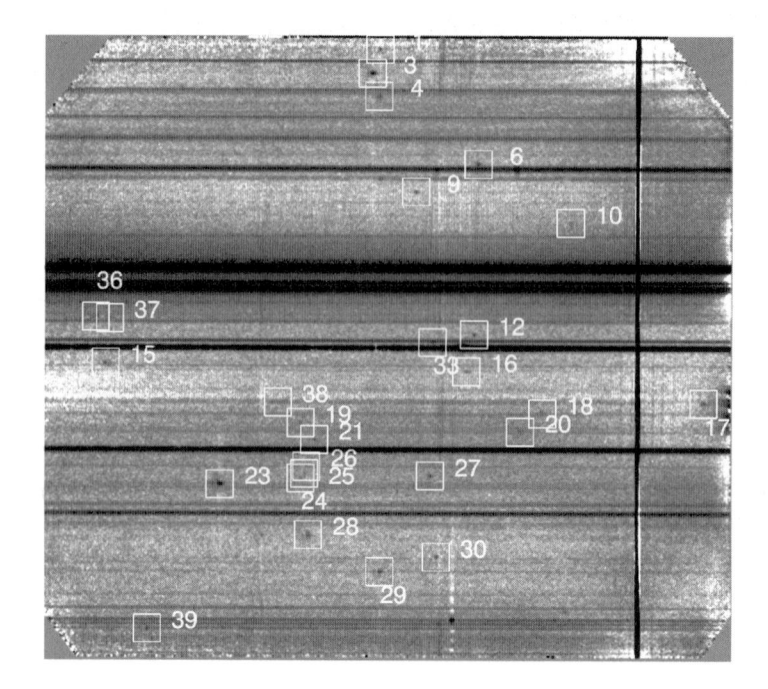

Figure 7: Two-dimensional spectrum obtained in 92 hours of exposure time with FORS2 at the VLT. The line emitter candidates for H I Lyα are enclosed by the numbered boxes. The dispersion direction is horizontal, with blue to the left and red to the right; the spatial direction along the slit is vertical (reproduced from Rauch et al. 2008).

Lyman break galaxies at comparable redshifts. As I will discuss in the next section theoretical modeling supports the suggestion that the emitters are the long-searched host galaxies of damped Lyα systems.

3.2 Modeling the emission and absorption properties of the faint Rauch et al. emitters

Building on the successful model for damped Lyα absorbers of Haehnelt, Steinmetz & Rauch (1998, 2000), Barnes & Haehnelt (2009) presented a model that simultaneously accounts for the kinematic properties and incidence rate of the observed damped Lyα absorbers *and* the luminosity function and the size distribution of the Rauch et al. (2008) emitters in the context of the ΛCDM model for structure formation. The model assumes a simple relation between the size of the damped absorption and Lyα emission regions, and proposes that the Lyα luminosity is proportional to the total halo mass. Barnes & Haehnelt (2009) further corroborated the suggestion of Rauch et al. that cooling radiation is not expected to contribute significantly to

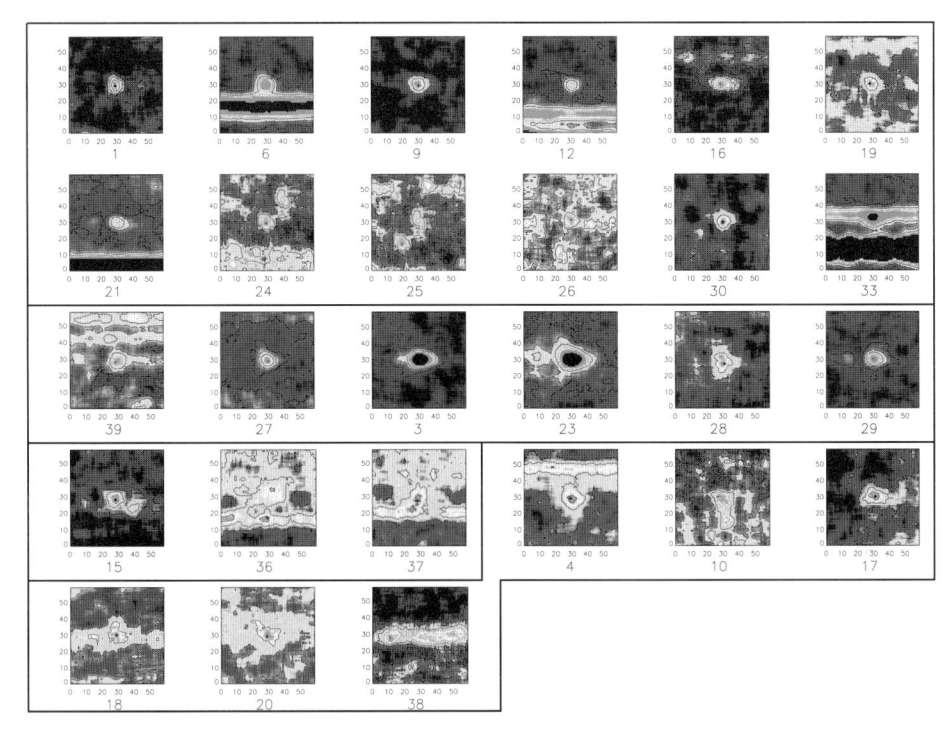

Figure 8: Spectra of individual line emitters (15.12″ or 116 proper kpc wide in the spatial direction and about 2266 km s^{-1} long in the spectral direction. The first 12 of the spectra show a single central peak; the next six show a clearly asymmetric red peak with a much weaker blue counter-peak; the following three either have a stronger blue than red peak (ID 15) or emission features blueward of an absorption line (36, 37); the remaining six spectra are not easily classifiable more amorphous objects (reproduced from Rauch et al. 2008).

the observed Lyα emission, and that the emitters are most likely powered by star formation at their centre.

Barnes & Haehnelt (2010) followed on from there and performed detailed Lyα radiative transfer calculations to show that the Lyα emission is indeed scattered to radii larger than those where the column density is sufficient for damped Lyα absorption as they had assumed in their earlier modeling without Lyα radiative transfer. The modeling with 1D Lyα radiative transfer simultaneously reproduces the observed properties of damped Lyα absorbers and the faint Lyα emitters, including the velocity width (Fig. 9) and column density distribution of damped Lyα absorbers and the large spatial extent of the emission of the faint emitters (Fig. 10). In the model damped Lyα absorbers are predominantly hosted by dark matter halos in the mass range $10^{9.5}$–10^{12} M$_\odot$, and are thus of significantly lower mass than those generally assumed to host L$_*$ Lyman break galaxies. The host galaxies of damped Lyα absorbers are therefore almost certainly the typical building blocks of Milky Way type galaxies. Note further, that in the model of Barnes & Haehnelt dark matter halos

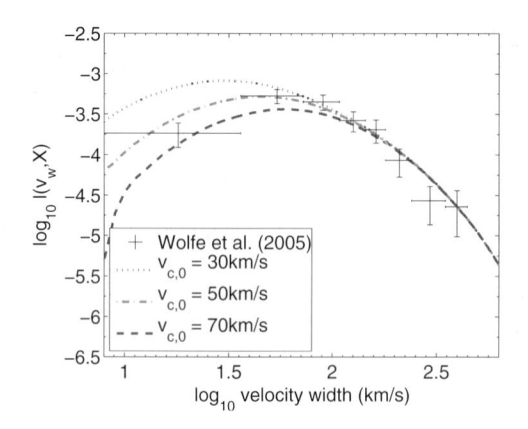

Figure 9: Velocity width of the associated low-ionization metal absorption for the joint modeling of the faint Rauch et al. (2008) emitters and damped Lyα absorption systems for a range of assumed minimum circular velocities for which a dark matter halo produces damped Lyα absorption together with data from Wolfe al. (2005). The figure is reproduced from Barnes & Haehnelt (2010).

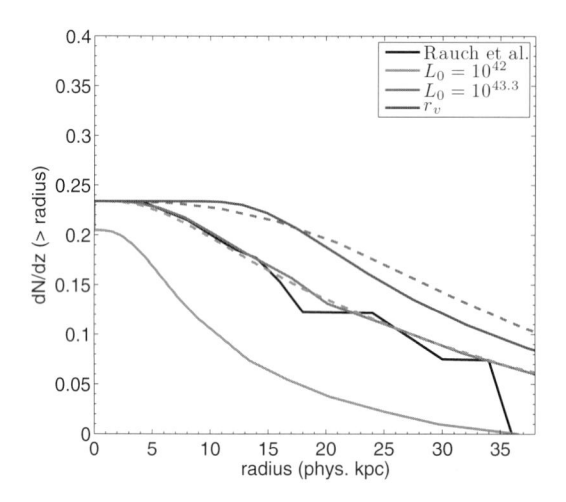

Figure 10: Cumulative size distribution for the joint modeling of the faint Rauch et al. (2008) emitters and damped Lyα absorption systems for different scalings of luminosity with halo mass together with the observed distribution of the faint Rauch et al. emitters. The figure is reproduced from Barnes & Haehnelt (2010).

hosting damped Lyα absorbers retain about 20% of the cosmic baryon fraction in the form of neutral hydrogen and that the Lyα emission due to star formation has a duty cycle of $\sim 25\%$.

In reality the spatial distribution and velocity field of the gas scattering the Lyα photons will not be spherically symmetric and more complex than assumed by Barnes & Haehnelt. Figure 11 shows 3D Lyα radiative transfer simulations for a dark

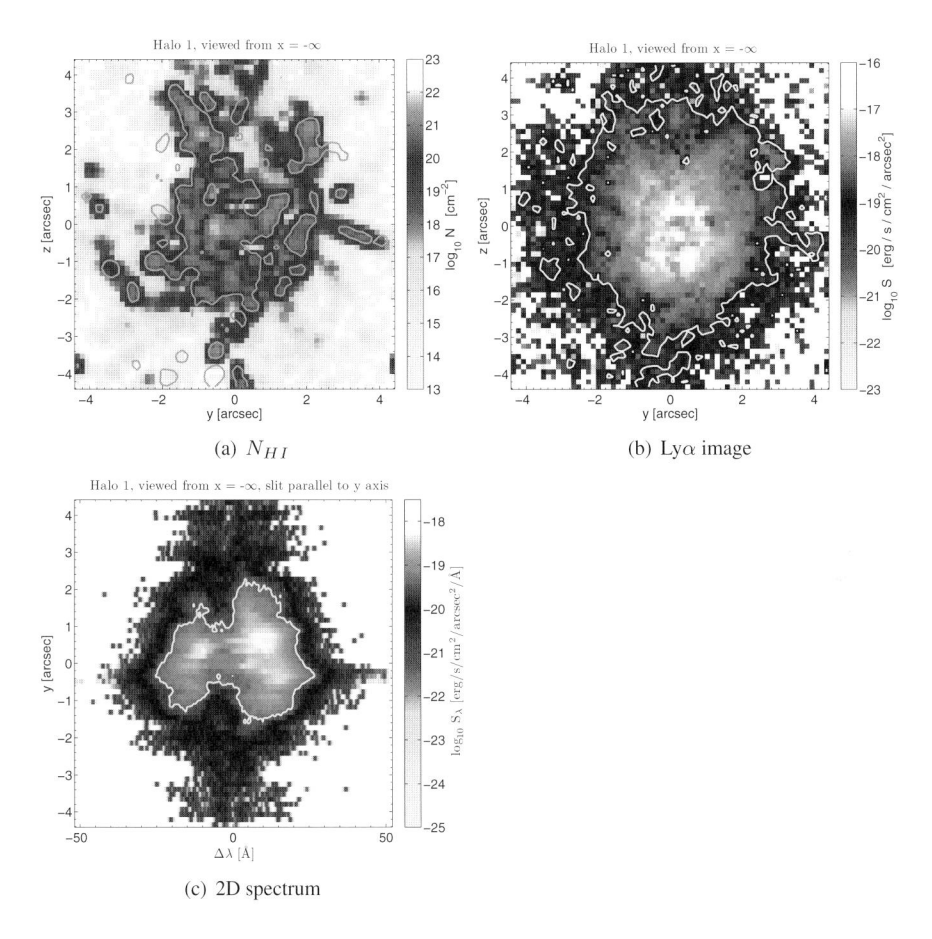

Figure 11: Simulation of the resonant scattering of Lyα radiation from a central source in a dark matter halo of a (ΛCDM) hydrodynamical simulation including star formation and a simple prescription for a supernova driven galactic wind. The *left panel* shows the projected column density of neutral hydrogen as integrated from the mid plane. The white contour corresponds to a total column density, $\log N_{HI} = 20.3$, the column density threshold for damped Lyα absorption. The *middle panel* shows the Lyα surface brightness and the white contour corresponds to the detection threshold of the Rauch et al. (2008) sample of faint emitters. The detectable emission extends to radii larger than 5 arcsec. The *right panel* shows a 2D spectrum for a slit along the y-axis in the left and middle panel. The white contour corresponds to the detection threshold of the Rauch et al. (2008) sample of faint emitters. Figure courtesy of Barnes (2009).

matter halo in a full hydrodynamical galaxy formation simulation aimed at modeling damped Lyα absorbers (Tescari et al. 2009). The simulation includes inflows due to gravitational infall and outflows due to a simple galactic wind description. The figure shows nicely that for a more realistic density distribution the cross section for damped absorption will be more irregular and that the resonant Lyα radiative transfer scatters the Lyα photons beyond those radii where the column density is sufficient for damped Lyα absorption. The 2D spectrum in the right panel shows the double-peak structure typical for Lyα emission. Note that the spectral shape thereby depends strongly on the direction from which the halo is viewed. For different directions stronger blue, stronger red peaks and peaks of equal strength all occur.

4 CODEX: an ultra-stable high-resolution spectrograph for the E-ELT

4.1 The rationale for CODEX

CODEX is a proposed ultra-stable optical high-resolution spectrograph for the E-ELT which is currently undergoing a Phase A study (PI: Luca Pasquini). CODEX will be the next step in pushing optical spectroscopy to new frontiers (Pasquini et al. 2005). CODEX will thereby use novel Laser Frequency Comb (LFC) calibration techniques. The core of an LFC is a mode-locked laser emitting a repetitive train of femtosecond pulses. In frequency space this results in thousands of equally spaced modes. The accuracy of the mode spacing is thereby ultimately related to the frequency accuracy of atomic clocks which provide the by far the most precise frequency measurements available (Steinmetz et al. 2008). The novelty of the LFC and its tremendous benefit for many areas of physics has been widely recognized with the award of the 2005 Nobel Prize in Physics to T. Hänsch and J. Hall. An LFC offers a number of advantages over traditional wavelength calibration sources. (i) the absolute position of each line is known a priori (i.e. without reference to any laboratory measurements) with relative precision better than 10^{-12}, which is limited only by the radio frequency clock; (ii) the density of lines may be as high as permitted by the spectrograph's resolution while at the same time guaranteeing the absence of blended lines; and (iii) the line density is perfectly uniform. The availability of such a precise and reproducible wavelength standard together with the enormous photon collecting power of the E-ELT will allow us to reach cm/s level accuracy stable over 30 years. This will enable a broad range of tremendously exciting science from planetary over stellar, galactic and extragalactic astronomy to cosmology and fundamental physics (Fig. 12) which I will have no space here to discuss in detail (see: http://www.eso.org/sci/facilities/eelt/instrumentation for more information).

In the next Sect. I will instead concentrate on a particular ambitious proposed experiment to measure the change of cosmological redshifts and thus the change of the expansion rate of the Universe in real time which CODEX at the E-ELT will bring for first time into reach.

4.2 The redshift drift experiment

The cosmological redshifts of spectroscopic features originating at large distances are the signature of an expanding Universe. As the expansion rate changes with time, the change of the expansion rate is expected to be mirrored by a corresponding change in redshift (Sandage 1962). The change in redshift at the present time t_0 is

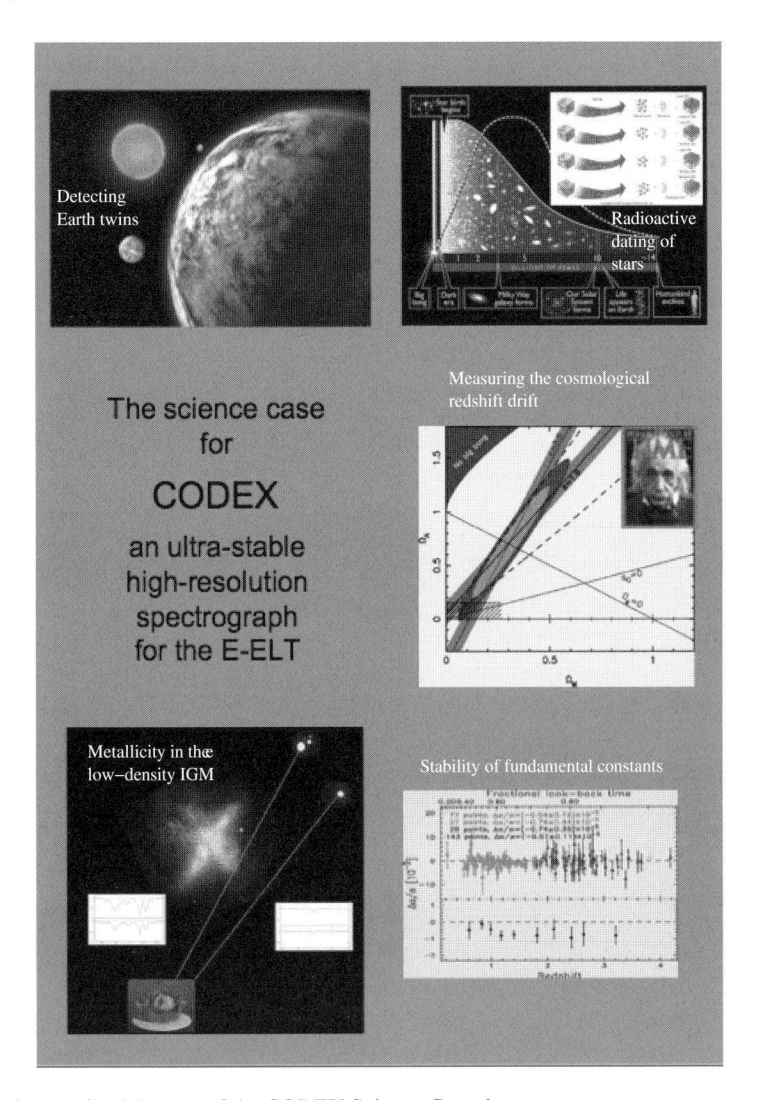

Figure 12: Title page of the CODEX Science Case document.

related to the change in expansion rate expressed in terms of the Hubble parameter as

$$\dot{z}(t_0) \equiv \frac{\mathrm{d}z}{\mathrm{d}t_{\mathrm{obs}}}(t_0) = (1+z)H_0 - H(z). \tag{1}$$

This is a very small effect, but should become measurable with CODEX if the collective signal in a large number of absorption features is monitored. Figure 13 shows the expected redshift drift which corresponds to less than one cm/s per year as a function of redshift for a range of cosmological parameters. The most favourable target for a measurement of the redshift drift is the multitude of absorption features making up the Lyα forest in QSO absorption spectra (Loeb 1998). The unprecedented wave-

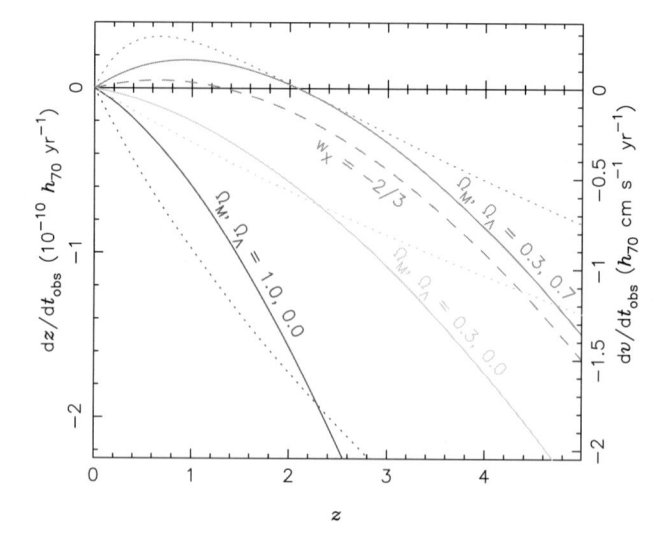

Figure 13: The expected redshift drift as a function of redshift for a range of cosmological parameters. The solid curve shows dz/dt_{obs} while the dotted curves shows dv/dt_{obs}. The figure is reproduced from Liske et al. (2008).

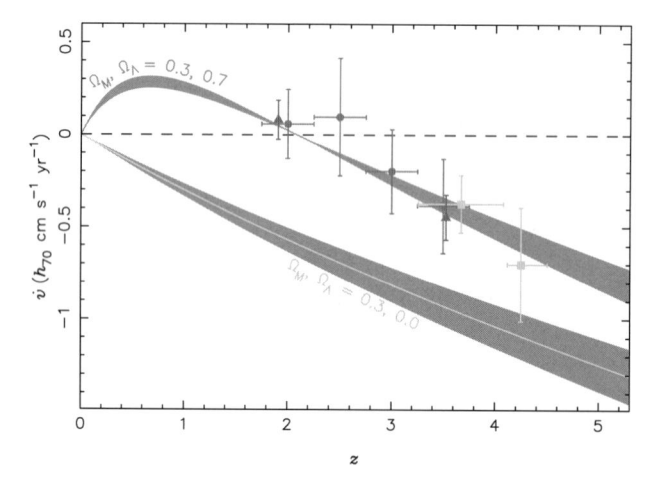

Figure 14: Monte Carlo simulation of three different implementations of a redshift drift experiment. The solid curves show the expected redshift drift for different cosmological parameters as indicated. The figure is reproduced from Liske et al. (2008).

length accuracy of CODEX means that photon noise should become the dominant source of error for a realistic redshift experiment with CODEX at the E-ELT. By monitoring a sample of bright QSOs for a decade or more it will be possible to measure the characteristic redshift with high statistical significance (Fig. 14). This will constitute an important model independent measurement of the expansion history of the Universe which does not require any cosmological assumptions or priors. Such a measurement may turn out to be crucial if a modification of 4D General Relativity is necessary as appears very likely.

I would like to end my contribution with a more general remark. The CODEX team had no difficulty to formulate an exciting Science case supporting the development of an ultra-stable optical high-resolution spectrograph. More importantly, however, such a spectrograph will open up substantial new discovery space which as we all know from the decisive role astrophysical spectroscopy has played in the development of quantum mechanics is often at the heart of revolutionary new developments in Fundamental Physics.

Acknowledgements

The work I have reviewed here has been a joined effort with a long list of collaborators who I hope to have duly acknowledged in the text.

References

Barnes, L.: 2009, PhD Thesis, Cambridge University

Barnes, L.A., Haehnelt, M.G.: 2009, MNRAS 397, 511

Barnes, L.A., Haehnelt, M.G.: 2010, MNRAS 403, 870

Gould, A., Weinberg, D.H.: 1996, ApJ 468, 462

Haehnelt, M.G., Steinmetz, M., Rauch, M.: 1998, ApJ 495, 647

Haehnelt, M.G., Steinmetz, M., Rauch, M.: 2000, ApJ 534, 594

Haehnelt, M.G., et al.: 2008, ESO Messenger 132, 41

Hogan, C.J., Weymann, R.J.: 1987, MNRAS 225, 1

Kuzio de Naray, R., Martinez, G.D., Bullock, J.S., Kaplinghat, M.: 2010, ApJ 710, L161

Liske, J., et al.: 2008, MNRAS 386, 1192

Loeb, A.: 1998, ApJ 499, L111

Pasquini, L., et al.: 2005, ESO Messenger 122, 10

Rauch, M.: 1998, ARA&A 36, 267

Rauch, M., et al.: 2008, ApJ 681, 856

Sandage, A.: 1962, ApJ 136, 319

Seljak, U., Makarov, A., McDonald, P., Trac, H.: 2006, Phys. Rev. Lett. 97, 1303

Steinmetz, T., et al.: 2008, Sci 321, 1335

Tescari, E., Viel, M., Tornatore, L., Borgani, S.: 2009, MNRAS 397, 411

Viel, M., Lesgourgues, J., Haehnelt, M.G., Matarrese, S., Riotto, A.: 2005, Phys. Rev. D 71, 063534

Viel, M., Becker, G.D., Bolton, J.S., Haehnelt, M.G., Rauch, M., Sargent, W.L.W.: 2008, Phys. Rev. Lett. 100, 041304

Weinberg, D.H., et al.: 1999, in: A.J. Banday, R.K. Sheth, L.N. da Costa (eds.) *Evolution of Large Sale Structure: From Recombination to Garching*, p. 346

Wolfe, A.M., Gawiser, E., Prochaska, J.X.: 2005, ARA&A 43, 861

Hypervelocity stars in the Galactic halo

Holger Baumgardt

Argelander-Institut für Astronomie
Auf dem Hügel 71, D-53121 Bonn, Germany
holger@astro.uni-bonn.de

Abstract

This article gives a brief overview of the formation mechanisms and properties of the recently discovered hypervelocity stars in the halo of the Milky Way. Hypervelocity stars (HVS) are stars that move with velocities up to 1000 km/sec and are unbound to the potential of the Milky Way. They are different from classical runaway stars which have maximum velocities of only a few hundred km/sec. The first HVS was discovered in 2005 and photometric and spectroscopic surveys have since then revealed a whole population of these stars in the Galactic halo. Different formation scenarios for hypervelocity stars are currently discussed, like the tidal disruption of stellar binaries by the Galactic center supermassive black hole, or the ejection of stars by a massive binary black hole. However, alternative scenarios like supernova explosions in close binary systems or the tidal disruption of a dwarf galaxy are also not ruled out. Precise proper-motion measurements will be necessary to explain the origin of hypervelocity stars and will allow to use hypervelocity stars to probe the mass and shape of the dark matter halo of the Milky Way.

1 Introduction

The Milky Way is a unique laboratory for studying stellar populations and stellar dynamical processes since it can be studied with the greatest resolution and down to the faintest magnitudes of all galaxies. One of the great successes of recent years has been the irrefutable discovery of a supermassive black hole at the Galactic centre through the orbital motion of the S-stars (Gillessen et al. 2009, Ghez et al. 2005). The proximity of stars in the Milky Way also allows to study large numbers of stars through automated surveys and thereby identify unusual stars in the disc or halo which would be very difficult to find in other galaxies. Examples include stars which belong to ultra-faint dwarf galaxies (e.g. Belokurov et al. 2007), very metal-poor stars (e.g. Christlieb et al. 2002) and hypervelocity stars (Brown et al. 2005).

Hypervelocity stars (HVS) are stars that move with velocities up to several 1000 km/sec and are unbound to the potential of the Milky Way. The first hypervelocity star was found by Brown et al. (2005) in the Sloan Digital Sky Survey. It

Table 1: Properties of so-far discovered hypervelocity stars (from Brown et al. 2009b).

Name	Spectral Type	V [mag]	R_{GC} [kpc]	M_V [mag]	$v_\odot$ [km s^{-1}]
SDSS J090744.99+024506.9	B	19.83	111	-0.3	840
US 708	sdO	19.05	26	$+2.6$	708
HE 0437-5439	B	16.20	62	-2.7	723
SDSS J091301.01+305119.8	B	18.50	82	-0.9	611
SDSS J091759.48+672238.3	B	17.70	45	-0.3	553
SDSS J110557.45+093439.5	B	19.11	78	-0.3	626
SDSS J113312.12+010824.9	B	17.80	60	-1.1	529
SDSS J094214.04+200322.1	B	18.09	53	-0.3	489
SDSS J102137.08-005234.8	B	18.76	68	-0.3	628
SDSS J120337.85+180250.4	B	19.36	87	-0.3	478
SDSS J095906.48+000853.4	A	19.70	70	$+0.6$	482
SDSS J105009.60+031550.7	A/BHB	19.76	70	$+0.6$	552
SDSS J105248.31-000133.9	B	20.16	125	-0.3	575
SDSS J104401.75+061139.0	B	19.89	112	-0.3	532
SDSS J113341.09-012114.2	B	19.33	85	-0.3	463
SDSS J122523.40+052233.8	B	19.49	90	-0.3	443

was a B-type star with a heliocentric radial velocity of 853 ± 12 km/sec and a heliocentric distance of 39 or 71 kpc, depending on whether the star is a main-sequence star or a blue horizontal-branch star. Corrected for the solar motion relative to the local standard of rest, the Galactic rest-frame velocity of this star is 709 km/sec, more than twice the escape speed from the Milky Way at the position of the star.

Since 2005, about 15 additional unbound hypervelocity stars and many slightly slower moving, bound stars have been found. Table 1 gives an overview of the properties of the unbound hypervelocity stars discovered so far.

Since hypervelocity star candidates are pre-selected by their colors and absolute magnitudes and later studied in more detail spectroscopically, all hypervelocity stars discovered so-far are B type stars in the halo of the Milky Way. Information on the mass function and true spatial distribution of hypervelocity stars is not available at the moment, although the present data already suggests that the distribution of hypervelocity stars is non-isotropic (Brown et al. 2009a, Lu et al. 2009). For the majority of hypervelocity stars it is also not yet clear if they are main sequence stars or blue horizontal branch stars. Detailed spectroscopic observations of HE 0437-5439 (Przybilla et al. 2008a), HVS 7 (Przybilla et al. 2008b) and US 708 (Hirsch et al. 2005) have shown that these stars are main-sequence stars. Most stars in Table 1 are therefore likely to be main-sequence stars, although López-Morales & Bonanos (2008) have reported evidence that SDSS J113341.09-012114.2 could be a blue horizontal branch star.

All stars in Table 1 have positive radial velocities, which cannot be attributed to a selection effect but must be a true feature of HVS. It implies that HVS are leaving the Milky Way and that there is a process operating inside the Milky Way which creates new HVS.

2 Mechanisms for creating hypervelocity stars

Several mechanisms have been suggested which can create hypervelocity stars and at the moment there is no consensus as to which process creates them, or, if there is more than one process involved, which of them is the dominant mode of formation. Most authors so far focussed on the Galactic centre as the birth place of HVS, and we will summarize their ideas below, however several alternatives to a Galactic centre origin exist.

Poveda et al. (1967) showed that stars can receive substantial velocity kicks as a result of close dynamical encounters during three or four-body interactions in the centers of star clusters and Leonard & Duncan (1990) found from the distribution of mass ratios and eccentricities of runaway binaries, that most OB runaway stars were created by dynamical encounters. Direct evidence for dynamical ejections was presented by Gualandris et al. (2004), who showed that several young stars in the vicinity of the Orion nebula cluster were ejected from it simultaneously with high velocities and in opposite directions.

The velocities that can be achieved by dynamical encounters between main-sequence stars in close binary systems are limited by the finite size of stars. Leonard (1991) showed that the maximum velocity which a runaway star can achieve in binary encounters is equal to the escape velocity from the surface of the most massive star involved in the interaction. This limits the velocities to ~ 1400 km/sec for interactions involving 100 $M_\odot$ stars. Typical ejection velocities are however much smaller. Gvaradmadze, Gualandris & Portegies Zwart (2009) found through 3-body scattering experiments that typical escape velocities of B-type stars in encounters with two O-type stars are around 200 km/sec and rarely exceed 400 km/sec. The escape velocities are thus large enough to explain runaway stars and bound hypervelocity stars but most likely not the unbound population listed in Table 1, which must have been created by a different mechanism.

Escape speeds of stars can be substantially higher if star clusters contain massive, so-called intermediate-mass black holes (IMBHs, Portegies Zwart et al. 2004). In such a case, escape speeds of up to 1000 km/sec are possible through the tidal disruption of stellar binaries by an IMBH. Indeed Gualandris & Portegies Zwart (2007) have argued, based on the proximity of HE0437-5439 to the LMC and the fact that its travel time is too long to have originated in the Milk Way, that HE0437-5439 was created by an IMBH in a young star cluster in the Large Magellanic Cloud. If true, the Milky Way likely also contains star clusters with IMBHs which could eject hypervelocity stars.

Another mechanism to give stars large velocity kicks are supernova explosions of one component of close binary systems, which was first suggested by Blaauw (1961) as a way to create runaway stars. Stone (1991) determined space frequencies of O and B type stars in the solar neighborhood and argued for supernova explosions as the main mechanism for creating runaway O stars. Wang & Han (2009) have recently re-investigated the supernovae scenario and studied SNe Ia supernovae explosions from close, mass-transferring WD-He star binary systems. They found that the companion star could survive the supernova explosion of the white dwarf. At the moment of supernova explosion, it would achieve a velocity kick in addition to its orbital velocity by the supernova ejecta which collide with it. Wang & Han (2009) found that velocities up to 650 km/sec could be achieved this way, enough to explain most hypervelocity stars.

Interestingly three high/hyper-velocity stars have been found for which travel times together with orbital information exclude a Galactic centre origin: HE 0437-5439 (Przybilla et al. 2008a), HD 271791 (Heber et al. 2008) and SDSS J013655.91+242546.0 (Tillich et al. 2009). Based on radial velocity and proper motion information an origin at the outer edge of the Galactic disc seems most likely for these stars, indicating that supernova explosions or dynamical ejections can contribute to the HVS population. We note however that Perets (2009) has recently argued that HE 0437-5439 is more likely a rejuvenated star following a merger of two stars in a binary system, in which case it could have reached its current position coming from the Galactic centre.

Abadi et al. (2009) finally investigated the tidal disruption of dwarf galaxies in the halo of the Milky Way and found that disrupting dwarf galaxies may contribute halo stars with velocities that sometimes exceed the nominal escape speed of the Milky Way. In such a case, the hypervelocity stars would be arranged in a thinly collimated tidal tail, which could explain the non-isotropic distribution of hypervelocity stars found by Brown et al. (2009a). However, except for the Magellanic Clouds, most dwarf galaxies in the Local Group are made up of mainly old stars, so the fact that many of the so-far discovered hypervelocity stars are young, B-type main-sequence stars is a potential problem for this scenario.

3 Hypervelocity stars from the Galactic centre

Hills (1988) was the first to suggest that hypervelocity stars can be created through encounters of stars with a supermassive black hole (SMBH) in the center of a galaxy. The main reasoning is like this: Consider a star moving on a nearly hyperbolic orbit around an SMBH. Near pericenter, the specific orbital energy of this star is given by

$$E = -\frac{GM_{SMBH}}{r_{Peri}} + \frac{1}{2}v_{Peri}^2 \qquad (1)$$

This energy is much smaller than either potential or kinetic energy at pericenter, hence $E \approx 0$. If the star receives a small velocity kick $v_k \ll v_{Peri}$ through a close encounter with another star in the vicinity of the black hole, its orbital energy is given by

$$\begin{aligned} E &= -\frac{GM_{SMBH}}{r_{Peri}} + \frac{1}{2}(v_{Peri} + v_k)^2 \\ &= -\frac{GM_{SMBH}}{r_{Peri}} + \frac{1}{2}v_{Peri}^2 + v_{Peri}v_k + \frac{1}{2}v_k^2 \\ &\approx v_{Peri}v_k \end{aligned} \qquad (2)$$

Close to the SMBH, the orbital velocity can reach $v_{Peri} = 10.000$ km/sec. For a small velocity kick of $v_k = 50$ km/sec one finds a velocity at infinity of $v_\infty = \sqrt{2 v_k v_{Peri}} = 1000$ km/sec, which is large enough to explain hypervelocity stars.

Under the assumption that HVS come from the Galactic centre, Brown et al. (2009b) showed that the ejection times of the so-far discovered hypervelocity stars have a broad distribution from 60 to 240 Myr. Hence, the ejection mechanism of hypervelocity stars must operate over long timescales. Three main mechanisms have been suggested which can give stars velocity kicks near a super-massive black hole: Encounters with other stars or black holes in the stellar cusp around the SMBH (Yu & Tremaine 2003, O'Leary & Loeb 2008), tidal disruption of stellar binaries by the SMBH (Hills 1988, Yu & Tremaine 2003, Bromley et al. 2006, Perets et al. 2007, Perets et al. 2009, Löckmann, Baumgardt & Kroupa 2009) and encounters of stars with an inspiraling IMBH (Yu & Tremaine 2003, Baumgardt, Gualandris & Portegies Zwart 2006, Sesana, Haardt & Madau 2006, Levin 2006, Lu, Yu & Lin 2007, Löckmann & Baumgardt 2008, Sesana, Madau & Haardt 2009).

3.1 Close encounters between single stars

Yu & Tremaine (2003) calculated the expected ejection rate of stars due to close encounters between single stars. They found that the finite size of stars strongly suppresses the ejection of stars with high velocities and estimated that the total ejection rate of stars with velocity $v > 500$ km/sec at the solar radius is only $3 \cdot 10^{-10}$/yr. The currently observed 15 HVS imply a formation rate of HVS of order 10^{-7}/yr already, so encounters between single stars should contribute a negligible amount of all HVS. In addition, no all sky survey looking for HVS exists so far and current surveys are limited to O/B type stars, so the total rate of HVS formation is likely to be far larger than 10^{-7}/yr.

O'Leary & Loeb (2008) pointed out that the stellar cusp in the innermost 0.1 pc of Sgr A* might be dominated by 10 $M_\odot$ to 15 $M_\odot$ black holes, which either spiraled into the center due to dynamical friction (Baumgardt et al. 2004, Freitag et al. 2006a), or formed in the center as a result of a top-heavy IMF. In such a case, the ejection rate of stars with high velocity can be considerably larger than estimated by Yu & Tremaine (2003). O'Leary & Loeb found that ejection rates of stars with velocities $v > 800$ km/sec can be as high as 10^{-6}/yr, which would be high enough to create all hypervelocity stars. Whether there is a sufficient number of stellar mass black holes close to Sgr A* is however uncertain since the mass function of stars in the Galactic centre is controversial (Maness et al. 2007, Buchholz, Schödel & Eckart 2009, Löckmann et al. 2009). In addition, recent near infrared observations of the Galactic centre indicate that the stellar cusp extends only down to radii of 0.4 pc and that the stellar distribution might have a core or even a hole at its centre (Buchholz, Schödel & Eckart 2009, Do et al. 2009). This could significantly reduce the number of ejected hypervelocity stars.

If larger HVS surveys become available in the future, it might be possible to discriminate between the single star scenario and other Galactic centre based ejection mechanisms by studying the binary properties of HVS. Ejection of stars by a massive black hole binary for example would leave close binaries intact, so a fraction of HVS should be in binaries in this case (Lu, Yu & Lin 2007). In addition, a close encounter of a star with a black hole would spin up the star (Löckmann & Baumgardt 2008), so HVS should be fast rotators if ejected by stellar mass black holes. In contrast, stars ejected by tidal disruption of stellar binaries by an SMBH are mainly slow rotators since stars in tight binaries show slower rotation compared to field stars (Hansen 2007).

3.2 Tidal disruption of binary stars by an SMBH

It is possible that binary stars are formed in the Galactic centre with similar frequency as in the solar neighborhood. Tidal disruption of a binary with the Galactic centre SMBH would then lead to an exchange reaction in which one of the components of the binary is replaced by the SMBH while the other component can escape from the Galactic centre. Yu & Tremaine (2003) showed that tidal disruption of binaries with separations $a \leq 0.3$ AU can create hypervelocity stars. An interesting feature of the binary disruption scenario is that the component which gets bound to the SMBH would orbit it at distances of around 0.01 pc, thereby offering the possibility to explain both hypervelocity stars in the Galactic halo and the S-stars close to the SMBH by one mechanism. Since the S-stars are significantly younger than the hypervelocity stars, multiple generations of both types of stars have to exist in this case.

The exact rate of hypervelocity stars escaping from the Galactic centre due to tidal disruption of binaries is hard to quantify since it depends on several unknown parameters like the number and semi-major axis distribution of binary stars in the Galactic centre and the rate with which these binaries approach the SMBH.

One problem is to produce a large enough number of binaries that have close encounters with the SMBH, since star formation in the direct vicinity of the SMBH is thought to be suppressed by the strong gravitational field (Vollmer & Duschl 2001), while star formation at larger distances likely produces mostly stars on more or less circular orbits. Stars and stellar binaries can in principle change their orbital eccentricities through standard two-body and resonant relaxation (Rauch & Tremaine 1996) as a result of encounters with other stars. However, the timescale over which these changes happen in the Galactic centre is of order 100 Myr or larger for resonant relaxation and of order 1 Gyr for two-body relaxation (Hopman & Alexander 2006). This is too slow for binaries containing B-type stars, since these would turn into white dwarfs before acquiring sufficiently high eccentricities to produce hypervelocity stars.

Perets et al. (2007) suggested that massive perturbers like giant molecular clouds or star clusters could scatter binary stars from distances of a few pc towards Sgr A*. The tidal disruption of such binaries would lead the bound component in a close and initially highly eccentric orbit around the SMBH. The eccentricity would thermalise due to encounters with other stars within a few 10s of Myr and Perets et al. (2009) showed that the eccentricity distribution of the S-stars is compatible with such a scenario and too high for a scenario in which the S-stars migrated inwards on low eccentricity orbits.

An alternative origin of both S-stars and hypervelocity stars was suggested by Löckmann et al. (2008) and Löckmann et al. (2009), who found that they might have originated in the disc(s) of young stars which have been found at distances of a few 0.1 pc from Sgr A* in near infrared surveys of the Galactic centre (Paumard et al. 2006, Lu et al. 2009, Bartko et al. 2009). As a result of encounters between disc stars themselves, and between disc stars and stars in the cusp, and due to orbital precession and Kozai resonance with the discs, some stars can acquire high orbital eccentricities within a few Myr. Binaries can stay bound long enough until they are tidally disrupted by the SMBH, leading to the formation of S-stars and hypervelocity stars. Löckmann & Baumgardt (2009) found good agreement between the observed warping of the Galactic centre discs and the eccentricity distribution of stars in the discs and their simulations. Madigan et al. (2009) showed that S-stars and hypervelocity stars can also be created in an eccentric stellar disc through a new type of secular instability. Interestingly, Lu et al. (2009) found that almost all the discovered HVSs are located in two thin planes and that the orientation of one of these planes is consistent with the orientation of the more massive disk in the Galactic centre, supporting a connection between the disks in the Galactic centre and hypervelocity stars.

One problem with the disc scenario could be that the observed mass function of stars in the young stellar discs is strongly top-heavy: Bartko et al. (2009) find a power law mass function $dN/dm \sim m^{-0.45\pm0.30}$ for stars in the discs, which is in quite strong contrast to the Salpeter like mass function $dN/dm \sim m^{-2.15\pm0.30}$ which they find for the S-stars. How much this is a problem is not clear though, since, if originating from a stellar disc, the hypervelocity stars would have come from previous generations of discs as the currently observed ones are only a few Myr old. Since star formation in the Galactic centre is not well understood yet, there is no a priori reason against the idea that previous discs had a different stellar mass function.

3.3 Ejection of stars by an inspiraling IMBH

Intermediate-mass black holes might form in young, dense star clusters through runaway merging of stars (Portegies Zwart et al. 2004, Baumgardt et al. 2006, Freitag et al. 2006, Suzuki et al. 2007, Chatterjee et al. 2009). Portegies Zwart et al. (2006) estimated that up to 10% of all star clusters in the inner 100 pc of the Milky Way could form IMBHs and that they will merge with the central SMBH at a rate of about 1 every 10 Myr.

Baumgardt, Gualandris & Portegies Zwart (2006), Matsubayahsi et al. (2007) and Löckmann & Baumgardt (2008) followed the orbital evolution of inspiraling IMBHs in the Galactic centre and found that the evolution is first dominated by a quick, dynamical friction driven phase. The inspiral stalls at radii of a few mpc from the SMBH since the stellar cusp runs out of stars and dynamical friction becomes inefficient. The orbital evolution at later times is driven by close encounters between the IMBH and cusp stars, which can excite high eccentricities in the orbit of the IMBH. As a result, gravitational wave emission becomes important and the IMBH merges with the SMBH within a few Myr after arriving in the Galactic centre.

Once an IMBH has arrived in the Galactic centre, hypervelocity stars are efficiently created due to the high relative velocities. Levin (2006) and Baumgardt, Gualandris & Portegies Zwart (2006) found that hypervelocity stars are generated in short bursts which last only a few Myr until the IMBH is swallowed by the SMBH. Since the observed hypervelocity stars show a spread in ejection times, several IMBHs must have been present in the Galactic centre within the last 200 Myrs. Baumgardt et al. (2006) also found that HVS are created nearly isotropically due to rapid changes in the orbital plane of the IMBH, unless the IMBH is very massive ($M_{IMBH} \geq 10^4 \, M_\odot$). The anisotropic spatial distribution of HVS is therefore more consistent with relatively massive IMBHs. An interesting feature of the IMBH scenario is that the central stellar cusp becomes strongly depleted in stars, Baumgardt et al. (2006) found that a core develops in the stellar distribution due to the scattering of stars by the IMBH, This offers an explanation for the core seen in the density profile of Galactic centre stars (Buchholz, Schödel & Eckart 2009). The scattering of stars could also be an explanation why the currently observed number of B type stars is too low to generate a significant population of hypervelocity stars (Perets 2009). One problem might be that in the IMBH scenario a significant fraction of stars acquires very high velocities $v > 1000$ km/sec (Sesana, Haardt & Madau 2007, Löckmann & Baumgardt 2008), while no such hypervelocity stars have been observed so far. If larger surveys of hypervelocity stars become available, this might help to verify or disprove the IMBH ejection scenario.

4 Conclusions

Hypervelocity stars were first predicted by Hills (1988) as a consequence of the fact that galaxies contain massive black holes in their centers and we now know about ~ 16 stars in the Galactic halo which are unbound to the Milky Way. Their ejection mechanism has not yet been identified. They might originate in the Galactic centre but could also form in supernova explosions in tight binary stars and come from the Galactic disc. In order to identify their place of origin, high-precision proper motion information would be most useful and should be possible to obtain with HST or perhaps *GAIA*. Such data would also be useful to constrain the shape of the Galactic potential field (Gnedin et al. 2005). If coming from the Galactic centre, information on the spatial distribution and binary properties of hypervelocity stars is also necessary to constrain the ejection mechanism.

References

Abadi, M.G., Navarro, J.F., Steinmetz, M., 2009, ApJ 691, L63
Bartko, H. et al.: 2009, ApJ 697, 1741
Baumgardt, H., Gualandris, A., Portegies Zwart, S.: 2006, MNRAS 372, 174
Baumgardt, H., et al.: 2006, MNRAS 372, 467
Baumgardt, H., Makino, J., Ebisuzaki, T.: 2004, ApJ 613, 1143
Belokurov, V., et al.: 2007, ApJ 654, 897

Blaauw, A.: 1961, Bull. Astron. Inst. Netherlands, 15, 265

Bromley, B.C. et al.: 2006, ApJ 653, 1194

Brown, W.R., Geller, M.J., Kenyon, S.J., Kurtz, M.J.: 2005, ApJ 622, 33

Brown, W.R., Geller, M.J., Kenyon, S.J., Bromley, B.C.: 2009a, ApJ 690, L69

Brown, W.R., Geller, M.J., Kenyon, S.J.: 2009b, ApJ 690, 1639

Buchholz, R.M., Schödel, R., Eckart, A.: 2009, A&A 499, 483

Chatterjee, S., et al.: 2009, ApJ submitted, arXiv:0911.1483

Christlieb, N., et al.: 2002, Nature 419, 904

Do, T., et al.: 2009, ApJ in press, arXiv:0908.0311

Freitag, M., Amaro-Seoane, P., Kalogera, V.: 2006a, ApJ 649, 91

Freitag, M., Gürkan, M.A., Rasio, F.A., 2006b, MNRAS 368, 141

Ghez, A.M., et al.: 2005, ApJ 620, 744

Gillessen, S., et al.: 2009, ApJ 692, 1075

Gnedin, O.Y., et al.: 2005, ApJ 634, 344

Gualandris, A., Portegies Zwart, S.: 2007, MNRAS 376, L29

Gualandris, A., Portegies Zwart, S., Eggleton, P.P.: 2004, MNRAS 350, 615

Gvaramadze, V.V., Gualandris, A., Portegies Zwart, S.: 2009, MNRAS 396, 570

Hansen, B.M.S.: 2007, ApJ 671, L133

Heber, U., et al.: 2008, A&A 483, L21

Hills, J.G.: 1988, Nature 331, 687

Hirsch, H.A., Heber, U., O'Toole, S.J., Bresolin, F.: 2005, A&A 444, L61

Hopman, C., Alexander, T.: 2006, ApJ 645, 1152

Leonard, P.J.T.: 1991, AJ 101, 562

Leonard, P.J.T., Duncan, M.J.: 1990, AJ 99, 608

Löckmann, U., Baumgardt, H.: 2008, MNRAS 384, 323

Löckmann, U., Baumgardt, H., Kroupa, P.: 2008, ApJ 683, L151

Löckmann, U., Baumgardt, H.: 2009, MNRAS 394, 1841

Löckmann, U., Baumgardt, H., Kroupa, P.: 2009a, MNRAS 398, 429

Löckmann, U., Baumgardt, H., Kroupa, P.: 2009b, MNRAS in press, arXiv:0910.4960

Levin, Y.: 2006, ApJ 653, 1203

Lopéz-Morales, M., Bonanos, A.Z.: 2008, ApJ 685, L47

Lu, J., et al.: 2009, ApJ 690, 1463

Lu, Y., Yu, Q., Lin, D.N.C.: 2007, ApJ 666, L89

Lu, Y., Zhang, F., Yu, Q.: 2009, ApJ submitted, arXiv:0910.3260

Madigan, A.M., Levin, Y., Hopman, C.: 2009, ApJ 697, L44

Maness, H., et al.: 2007, ApJ 669, 1024

Matsubayashi, T., Makino, J., Ebisuzaki, T.: 2007, ApJ 656, 879

O'Leary, R.M., Loeb, A.: 2008, MNRAS 383, 86O
Paumard, T. et al.: 2006, ApJ 643, 1011
Poveda, A., Ruiz, J., Allen, C.: 1967, Bol. Obs. Tonantzintla Tacubaya, 4, 86
Perets, H.B.: ApJ 690, 795
Perets, H.B., Hopman, C., Alexander, T.: 2007, ApJ 656, 709
Perets, H.B. et al.: 2009, ApJ 702, 884
Portegies Zwart, S., et al.: 2004, Nature 428, 724
Portegies Zwart, S., et al.: 2006, ApJ 641, 319
Przybilla, N. et al.: 2008a, A&A 480, L37
Przybilla, N. et al.: 2008b, A&A 488, L51
Rauch, K.P., Tremaine, S., 1996, New Astronomy, 1, 149
Sesana, A., Haardt, F., Madau, P.: 2006, ApJ 651, 392
Sesana, A., Haardt, F., Madau, P.: 2007, MNRAS 379, L45
Sesana, A., Madau, P., Haardt, F.: 2009, MNRAS 392, 31
Stone, R.C.: 1991, AJ 102, 333
Suzuki, T.K., et al.: 2007, ApJ 668, 435
Tillich, A., et al.: 2009, A&A 507, L37
Vollmer, B., Duschl, W.J.: 2001, A&A 377, 1016
Wang, B., Han, Z.: 2009, A&A in press, arXiv:0911.3316
Yu, Q., Tremaine, S.: 2003, ApJ 599, 1129

Schwarzschild modelling of elliptical galaxies and their black holes

Jens Thomas

Universitäts-Sternwarte München
Scheinerstr. 1, D-81679 München, Germany
jthomas@mpe.mpg.de

Abstract

This article describes the Schwarzschild orbit superposition method. It is the state-of-the-art dynamical modelling tool for early-type galaxies. Tests with analytic models show that masses and orbital anisotropies of not too face-on galaxies can be recovered with about 15 percent accuracy from typical observational data. Applying Schwarzschild models to a sample of Coma galaxies their dark matter halos were found to be 13 times denser than those of spirals with the same stellar mass. Since denser halos assembled earlier, this result indicates that the formation redshift $1 + z_{form}$ of ellipticals is about two times higher than of spirals. Roughly half of the sample galaxies have halo assembly redshifts in agreement with their stellar-population ages. Galaxies where stars appear younger than the halos show strong phase-space density gradients in their orbital structure, indicative for dissipational evolution and possibly connected with secondary star-formation after the main halo assembly epoch. The importance of considering dark-matter in dynamical models aimed to measure black-hole masses is briefly discussed.

1 Introduction

Early-type galaxies are characterised by an overall smooth and featureless spheroidal morphology and a dynamically hot system of stellar orbits. This is thought to be the result of a dynamically violent assembly process. However, when and how exactly these galaxies have formed is still not well known. Their mostly old and α-enhanced stellar populations imply a relatively short star-formation period in the distant past. The epoch when those stars assembled to form the early-types seen today cannot be directly deduced from stellar population ages. For example, the stars might be born at high redshift, in progenitors that only recently merged into spheroidal galaxies. If these mergers are mostly collisionless (without significant amounts of gas and star-formation), then stellar-population ages do not change, and the main assembly epoch is delayed with respect to the star-formation epoch. In contrast, a spheroidal galaxy formed early in the universe could have grown a disky stellar subcomponent recently, e.g. triggered by a gas-rich minor merger. The main halo assembly would then precede star formation (along the disk).

Pure dark-matter, collisionless N-body simulations predict a close relationship between the main assembly redshift of dark matter halos (when, say, half of the mass had been assembled) and their average density. The infall of baryons might change the dark matter density as it enforces an extra gravitational pull on the halo. Still, measuring the dark matter density in elliptical galaxy halos is a valuable tool to gain information about their assembly epoch. In comparison with stellar-population ages it also gives indirect information about the evolution of these systems. According to the above considerations one can expect that galaxies in which stars appear younger than the halos have likely experienced some secondary star-formation episodes. These systems have evolved dissipatively. In contrast, when stars appear older than the halo, this might indicate a more gasless evolution (e.g. by gas-free or collisionless or dry, respectively, mergers).

Since the system of stars in galaxies is collisionless it preserves some information about how the stars have assembled. A dynamically violent formation – characterised by violent relaxation in phase-space – is supposed to result in a highly mixed orbit distribution with strong phase-space density gradients likely being washed out. In contrast, dissipational evolution (e.g. through gas-rich mergers) likely results in disky subsystems with high phase-space density peaks on near-circular orbits. Consequently, the analysis of the orbital structure provides additional information about the assembly mechanism of early-type stellar systems.

In the last years we have collected photometric and kinematic observations for a sample of 19 early-type galaxies in the Coma cluster: 2 central cD galaxies, ten giant ellipticals and seven S0 or E/S0 galaxies, respectively, with $-18.8 \geq M_B \geq -22.26$ (Mehlert et al. 2000; Wegner et al. 2002; Corsini et al. 2008; Thomas et al. 2009a). For all galaxies ground-based and HST photometry are available. Long-slit stellar absorption-line spectra have been taken along at least the major and minor axes. In many cases additional data along other position angles has been collected as well. The spectra extend to $1 - 4\,r_{\mathrm{eff}}$, entering the regions where dark matter becomes noticeable.

By means of dynamical models we have measured the dark matter content and orbital structure of these galaxies. The results are summarised below and implications for the formation process of early-type galaxies are discussed.

2 The Schwarzschild method

A complete description of a stellar system is provided by its phase-space distribution function f, i.e. the density of stars in 6-dim phase-space. Unlike for a collisional gas the distribution function of collisionless dynamical systems like galaxies is not known in advance. However, for steady-state objects Jeans theorem ensures that the phase-space density is constant along individual orbits. Orbits, in turn, are identified by the integrals of motion they respect. Thus, for systems in a steady state the distribution function (DF) reads

$$f = f(I_1, \ldots, I_n), \tag{1}$$

where $I_1, \ldots, I_n$ are integrals of motion (e.g. Binney & Tremaine 1987). The simplest symmetry assumption consistent with the flattening and rotation of elliptical galaxies is that they are axisymmetric. Orbits in typical axisymmetric galaxy potentials respect three integrals of motion: energy E, angular momentum along the symmetry axis L_z (the rotation axis being parallel to the z-axis) and the so-called third integral I_3 (e.g. Contopoulos 1963). The last integral is usually not known explicitly and the distribution function can thus not be written in terms of elementary functions.

Schwarzschild (1979) introduced an orbit-superposition technique (now called Schwarzschild method) to construct collisionless DFs. In brief, one assumes a trial potential for a given galaxy and composes a library of several thousand orbits. The light distribution and projected kinematics of each orbit are stored. Then, a galaxy model is constructed as the superposition of all orbits. The unknown weight or total amount of light, respectively, of each orbit is chosen to match the model as good as possible with the given constraints (see below). This method corresponds to the approximation $f \approx \sum_i f_i$, where the f_i are single-orbit DFs (Vandervoort 1984; Thomas et al. 2004). The accuracy of the method only depends on the density of the orbit grid.

The Schwarzschild method provides very general dynamical models. In contrast to Jeans models there is no necessity for any a priori restriction upon the orbital structure. Moreover, Schwarzschild models are easily constructed to be everywhere positive in phase space (i.e. physically meaningful) – a property that is not guaranteed in Jeans models. The main challenge is to ensure that the orbit library is representative for all orbital shapes.

In axisymmetric potentials it is straight forward to sample the energy E and the angular momentum L_z. The issue related to our ignorance about I_3 is overcome by a systematic sampling of orbital initial conditions, which implicitly guarantees the inclusion of all orbital shapes. More specifically, orbits with given E and L_z but different I_3 follow distinct invariant curves in appropriate surfaces of section (SOS). Launching orbits at constant E and L_z from various initial starting points on a grid in such SOSs ensures a representative sampling of the unknown I_3 (Thomas et al. 2004).

In practice, one goes through the following steps to construct a Schwarzschild model of a galaxy:

- The photometry of the galaxy is deprojected to obtain the 3d light distribution ν.

- A trial mass distribution ρ is set up, combining the various mass components, e.g.

$$\rho = \Upsilon \times \nu + \rho_{\rm DM} + M_{\rm BH} \times \delta(r). \tag{2}$$

    Here, Υ is the stellar mass-to-light ratio (typically assumed to be radially constant in ellipticals), $\rho_{\rm DM}$ is the dark matter density (see below) and $M_{\rm BH}$ represents a supermassive central black hole. Given ρ, the gravitational potential follows by solving Poisson's equation and an orbit library can be assembled.

- The orbits are superposed and the orbital weights that best match with the observations are determined. In our implementation this is done by maximising

$$S - \alpha \chi^2, \qquad (3)$$

where χ^2 quantifies deviations between observed and modelled kinematics (the observed light profile is used as a boundary condition in the maximisation). The function

$$S = -\int f \ln f \, \mathrm{d}^3 r \, \mathrm{d}^3 v \qquad (4)$$

is the Boltzmann entropy of the orbit distribution and α is a regularisation parameter (Richstone & Tremaine 1988; see also below).

- The parameters determining the potential (e.g. Υ, halo parameters, $M_{\rm BH}$) are systematically varied and the final best-fit is obtained from a χ^2 analysis.

Various implementations of this method exist for spherical (Romanowsky et al. 2003), axisymmetric (Cretton et al. 1999; Cappellari et al. 2006; Chanamé, Kleyna & Van der Marel 2008; Valluri, Merritt & Emsellem 2004; Gebhardt et al. 2003; Thomas et al. 2004) as-well as triaxial potentials (van den Bosch et al. 2008). Gebhardt et al. (2003) measured supermassive black-holes in the centres of galaxies (ignoring any contribution of dark matter). Cappellari et al. (2006) modelled a subsample of the SAURON galaxies, again ignoring dark matter because their data cover only the inner regions $< r_{\rm eff}$.

Beyond $r_{\rm eff}$ dark matter becomes important in early-type galaxies. The Coma sample is by now the only larger sample of generic (e.g. flattened and rotating) ellipticals that has been modelled with Schwarzschild's method including dark matter. Previous attempts to measure the dark matter content of early-types via stellar dynamics focussed on round and non-rotating systems and assumed spherical symmetry (Kronawitter et al. 2000; Gerhard et al. 2001; Magorrian & Ballantyne 2001).

In the Coma galaxies we probed for two different parametric halo profiles: (1) logarithmic halos with a constant-density core and asymptotically constant circular velocity and (2) so-called NFW-profiles which are good fits to dark matter halos in cosmological N-body simulations (Navarro, Frenk & White 1996). The majority of Coma galaxies are better fit with logarithmic halos, but the significance over NFW halo profiles is marginal (Thomas et al. 2007b).

3 Tests of the method

A unique feature of our Schwarzschild models is the incorporation of orbital phase-space volumes V (Thomas et al. 2004). Together with the total amount of light w on the orbit it allows to calculate the phase-space density $f = w/V$ on each orbit. This offers several new applications. Firstly, one can project analytic phase-space DFs $f_{\rm analytic}$ through orbit libraries. In this case, the orbital weights are not determined

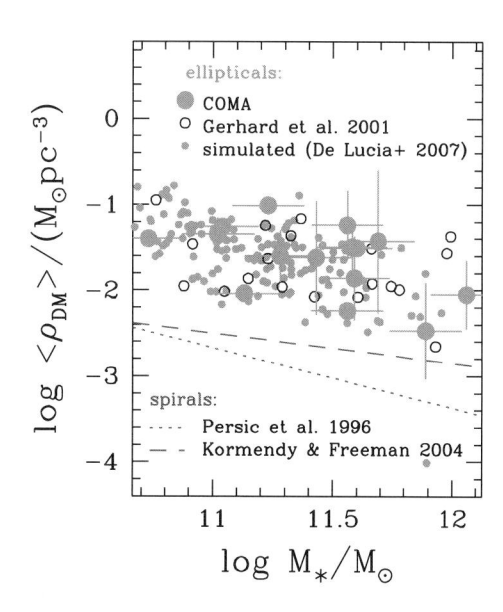

Figure 1: Dark matter density $\langle \rho_{DM} \rangle$ (averaged within $2\,r_{\rm eff}$) versus stellar mass M_*. Red dots: Coma early-type galaxies; open circles: round and non-rotating galaxies from Gerhard et al (2001); small dots: semi-analytic galaxy formation models (De Lucia & Blaizot 2007). Blue lines: spiral galaxies.

from a fit to photometric and kinematic constraints, but directly from the analytic DF, i.e. the weight w_i of orbit i reads $w_i = f_{\rm analytic} \times V_i$. The projected kinematics, spatial density profile, and intrinsic velocity dispersion profiles of the so-constructed orbit superposition can be compared with direct phase-space integrations of $f_{\rm analytic}$. Both methods agree very well (Thomas et al. 2004), confirming that the orbit sampling is representative.

In addition, mock data from analytic model galaxies can be used to measure how accurate the orbital DF can be reconstructed. In this Monte-Carlo approach photometric and kinematic data with spatial resolution and coverage similar to real data are simulated and modelled exactly as real galaxies. For the Coma data we have shown that with the appropriate choice of the regularisation parameter α the orbital structure and mass distribution of not too face-on galaxies can be recovered with an accuracy of about 15 percent (Thomas et al. 2005). Similar tests have been presented in Cretton et al. (1999), Krajnović et al. (2005) and Siopis et al. (2009).

The input models for the above tests had the same symmetry as the dynamical models (axial symmetry). Isophotal twists and kinematic misalignment suggest that at least the most massive early-type galaxies are slightly triaxial. In order to explore the systematic errors arising from too restrictive symmetry assumptions, we applied our models also to mock data sets created from collisionless N-body binary disk merger remnants. These remnants are strongly triaxial in the central, box-orbit dominated regions (Jesseit, Naab & Burkert 2005). While the central mass measured with axisymmetric models then underestimates the true mass on average (depending on projection angle and box-orbit content) the enclosed mass within r_eff is still recovered mostly with better than 20 percent accuracy unless for highly flattened, face-on systems (Thomas et al. 2007a).

4 The dark matter density and assembly epoch of early-type galaxies

Figure 1 shows dark matter densities $\langle \rho_\mathrm{DM} \rangle$ as a function of stellar mass. Stellar masses $M_* = \Upsilon \times L$ of early-types are derived from the best-fit (dynamical) stellar mass-to-light ratio Υ and the observed total luminosity L. Dark matter densities $\langle \rho_\mathrm{DM} \rangle$ are averaged within $2\,r_\mathrm{eff}$.

The dark matter densities of Coma ellipticals (from axisymmetric modelling) and round and non-rotating galaxies (from spherical models) match well. With increasing stellar mass dark matter densities tend to decrease. Spiral galaxy dark matter densities are also shown in Fig. 1 (Persic, Salucci & Stel 1996a,b; Kormendy & Freeman 2004, respectively). Stellar masses for spirals have been derived using the Tully-Fisher and stellar-mass Tully-Fisher relations of Bell & De Jong (2001) (cf. Thomas et al. 2009a for details). As in early-types, dark matter densities in spirals decrease with increasing stellar mass. Most important, the dark matter in ellipticals is about 13 times denser than in spirals of the same stellar mass (compared at the same B-band luminosity, early-type galaxies have 7 times denser halos than late-types).

Finally, Fig. 1 also includes semi-analytic galaxy formation models of De Lucia & Blaizot (2007). They are in good agreement with the observations in Coma galaxies. This is surprising because the N-body cosmological simulation underlying the semi-analytic models was performed without baryon dynamics. Thus, either the net effect of baryons on dark matter around early-types in the given mass interval vanishes or there is some discrepancy between the semi-analytic models and the observations. Note that baryons can also cause a halo expansion (e.g. by dynamical friction).

The light distribution in late-type galaxies is less centrally concentrated than in early-types. Then, even if the net effect of baryons on elliptical galaxy halos would be negligible, the stronger gravitational pull in early-type galaxy halos could still contribute to the observed overdensity of dark matter in ellipticals relative to spirals. Within the adiabatic contraction approximation this can at most explain a factor of two between average halo densities around ellipticals and spirals, respectively (Thomas et al. 2009a). The remaining over-density indicates that ellipticals have assembled earlier – at a time where the universe was denser.

The simplest analytic models, as well as pure dark matter cosmological N-body simulations predict a scaling of the average dark matter density $\langle \rho_\mathrm{DM} \rangle$ with halo-assembly redshift z_form of $\langle \rho_\mathrm{DM} \rangle \propto (1 + z_\mathrm{form})^3$ (e.g. Gerhard et al. 2001). Accordingly, the density contrast between elliptical and spiral galaxy halos translates into $1 + z_\mathrm{form}$ being about two times higher for ellipticals than for spirals. Absolute assembly redshifts can only be estimated with an additional assumption upon the formation redshift of spirals. Supposed that spirals assemble typically at $z_\mathrm{form} \approx 1$

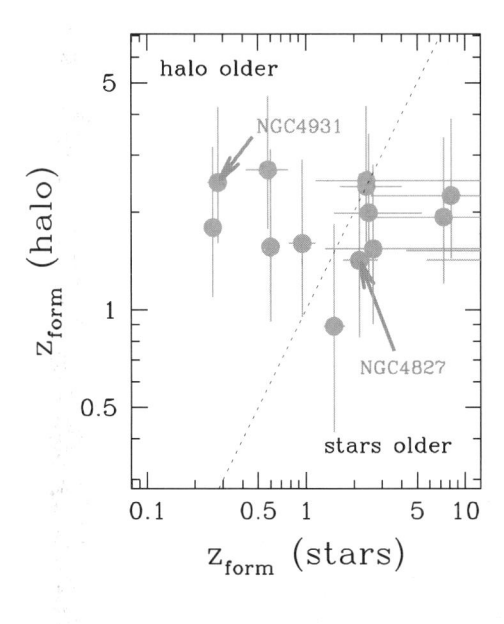

Figure 2: Halo assembly redshifts (y-axis) versus star-formation redshift (x-axis) for Coma early-type galaxies. The one-to-one relation is indicated by the dotted line. The phase-space DFs of the two galaxies NGC4827 and NGC4931 are shown in Fig. 4.

(at higher redshifts spirals become rare, e.g. Conselice et al. 2005) then Coma early-types have assembled around $z_{\rm form} \approx 2 - 3$. These dark-matter based formation redshifts are shown in Fig. 2 against stellar-population ages. In about one third of Coma galaxies the stars appear to be younger than the halos but many galaxies are consistent with equal assembly and star-formation redshifts.

5 The orbital structure of early-type galaxies

As outlined above a galaxy in which the stars appear to be younger than the halo is a candidate dissipative system, while objects in which stars are equally old or older than the halo have likely formed monolithically or through mostly collisionless mergers. In dissipative systems one would expect a high phase-space density on near-circular disk orbits while the phase-space DF of systems which have undergone violent relaxation should lack strong phase-space density gradients.

Classically, the orbital structure is measured in terms of anisotropy parameters, i.e. ratios of intrinsic velocity dispersions. Let f denote the phase-space DF of a galaxy, then the intrinsic dispersions σ_{ij} read

$$\sigma_{ij}^2 = \frac{1}{\rho} \int f\,(v_i - \overline{v_i})(v_j - \overline{v_j})\,{\rm d}^3 v, \qquad (5)$$

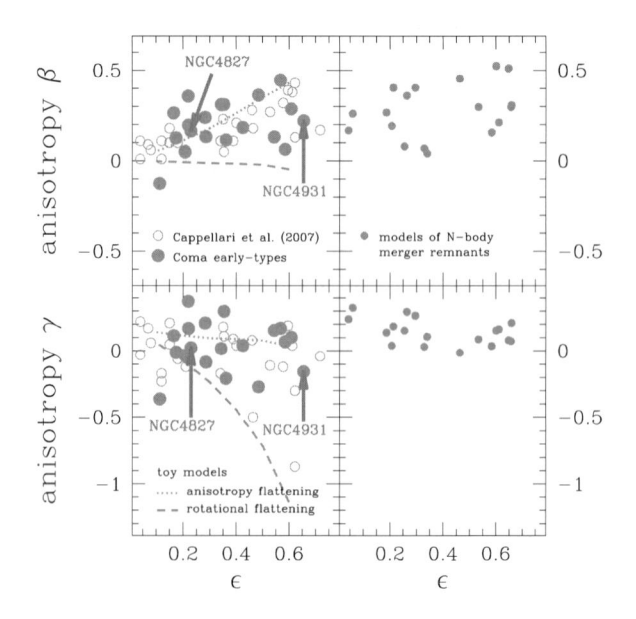

Figure 3: Anisotropy β and γ versus intrinsic ellipticity ϵ for observed galaxies (left-hand panels) and N-body binary disk merger remnants (right-hand panels). Dotted lines: maximum-entropy toy-models, dashed lines: 2I models (details in the text). The phase-space DFs of the two galaxies NGC4827 and NGC4931 are shown in Fig. 4.

$$\overline{v_i} = \frac{1}{\rho} \int f \, v_i \, \mathrm{d}^3 v. \tag{6}$$

In the following we assume $i, j \in \{R, \phi, z\}$, where z is the symmetry axis (cylindrical coordinates). The unordered kinetic energy along coordinate direction i is

$$\Pi_{ii} = \int \rho \sigma_{ii}^2 \, \mathrm{d}^3 r \tag{7}$$

and the anisotropy in the velocity dispersions can be quantified by $\beta \equiv 1 - \Pi_{zz}/\Pi_{RR}$ and $\gamma \equiv 1 - \Pi_{\phi\phi}/\Pi_{RR}$ (Cappellari et al. 2007). An isotropic system has $\beta = \gamma = 0$.

Fig. 3 shows β and γ versus intrinsic ellipticity ϵ. In observed galaxies $\beta > 0$ and, on average, $\gamma \approx 0$, but with significant scatter. The range of anisotropies found in Coma early-types is similar as in SAURON galaxies but the Coma galaxies do not exhibit a trend of increasing β with increasing ϵ as observed by Cappellari et al (2007). Neither the inclusion of dark matter in the Coma models nor regularisation can explain this difference (Thomas et al. 2009b). Instead it most likely reflects differences in the selections of the two samples (Thomas et al. 2009b).

In order to clarify the relationship between the classical anisotropy and the intrinsic orbital structure two proto-typical orbital compositions are shown by the lines in Fig. 3: (1) The case of flattening by rotation, i.e. by extra-light on near-circular

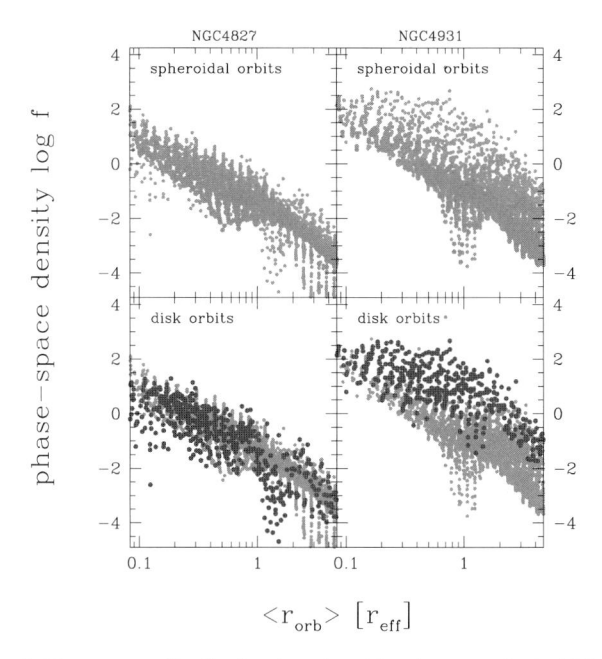

Figure 4: Phase-space distribution function f for the two galaxies NGC4827 (left-hand panels) and NGC4931 (right-hand panels). Each dot represents the phase-space density on a single orbit (in solar masses per cubed kpc and cubed km/s) plotted against the mean orbital radius. In the top panels spheroidal orbits are highlighted; in the bottom panel disk orbits are highlighted (details in the text). Only prograde orbits are shown.

orbits. Circular orbits mostly contribute to the kinetic energy in ϕ direction and γ becomes negative. The more, the flatter the galaxy is (dashed lines). (2) The dotted lines show toy models with maximum entropy[1]. Without any other conditions the maximisation of Boltzmann's entropy yields a flat DF $f = $ const. This is altered as soon as the luminosity distribution is used as a boundary condition. Still, maximising the entropy yields, in a sense, the smoothest phase-space DF compatible with a given density profile. It turns out that the classical notion of flattening by anisotropy (i.e. a suppression of vertical versus horizontal energy, with isotropy between R and ϕ) is closely related to the maximisation of entropy. Accordingly, systems along the maximum-entropy line, while having varying anisotropy β, are similar in that their phase-space distribution functions are smooth.

The phase-space DFs of two Coma galaxies shown in Fig. 4 (both galaxies are flagged by the arrows in Figs. 2 and 3). The two galaxies have similar β and γ but different intrinsic flattening: NGC4827 is an E2 early-type while NGC4931 is highly flattened. In addition, in NGC4827 stars are about as old as the halo while the stars in NGC4931 appear younger than its halo.

---

[1] For both toy models 75 percent of the light was distributed on prograde orbits; cf. Thomas et al. (2009b) for details.

In the upper panels of Fig. 4 spheroidal orbits are highlighted. Disk orbits are highlighted in the lower panels. We identify spheroidal orbits by $|\vartheta|_{\max} > 70°$, where ϑ is the angle between a point on the orbit and the equatorial plane of the galaxy. So-defined spheroidal orbits come close to the galaxy's minor-axis and contrast disk orbits, which are instead selected by $|z|_{\max} < r_{\mathrm{eff}}/4$, where z is the vertical height of a point on the orbit with respect to the equatorial plane.

Fig. 4 shows that phase-space densities on spheroidal orbits are similar in both galaxies. Moreover, in NGC4827 the phase-space densities of disk orbits are comparable to those of spheroidal orbits. Yet, the stellar density on disk orbits in NGC4931 is up to two orders of magnitude higher than on spheroidal orbits. These findings support the interpretation that galaxies left to the one-to-one relation in Fig. 2 (stars younger than halo) are dissipative systems, while galaxies close to the one-to-one relation or right of it are systems which formed in a dynamically violent process and have smoother phase-space DFs. A more detailed investigation of phase-space densities will be presented in Thomas et al. (in preparation).

The right-hand panels of Fig. 3 display dynamical models of collisionless N-body binary disk merger remnants. Mock data sets with similar spatial resolution and coverage as for Coma galaxies were constructed from projections along the three principal axes of six remnants selected representatively from the N-body sample of Naab & Burkert (2003). While the distribution of β versus ϵ is like in observed galaxies, none of the analysed merger remnants has $\gamma < 0$. In view of the above discussion this is plausible since galaxies with $\gamma < 0$ are likely dissipative systems (large energy in ϕ caused by extra-light on circular orbits), while the analysed N-body merger simulations are gas-free.

6 Summary

The Schwarzschild technique is the state-of-the-art tool to model early-type galaxies. For typical observational data masses and orbital anisotropies can be recovered with about 15 percent accuracy (for not too face-on systems). We have applied Schwarzschild models to 19 Coma early-type galaxies to measure the dark matter content and distribution of stellar orbits. By today, it is the largest sample of generic early-type galaxies with dynamical models including dark matter. Dark matter densities in early-types are larger than in spirals of the same luminosity or mass. Extra gravitational pulling by the more concentrated baryons in ellipticals is not sufficient to explain this over-density. Instead, the formation redshift $1 + z_{\mathrm{form}}$ of early-types is about two times higher than of spirals. Under the assumption $z_{\mathrm{form}} \approx 1$ for spirals the Coma early-types have formed around $z_{\mathrm{form}} \approx 2 - 3$. Observed dark matter densities are in good agreement with recent semi-analytic galaxy formation models. In about half of the sample galaxies halo assembly redshifts match with stellar population ages. In galaxies where stars appear younger than the halo, the orbit distribution indicates dissipational evolution (i.e. strong phase-space density peaks on near-circular disk orbits). This suggest that these galaxies had some secondary star-formation after the main halo assembly epoch.

When modelling the very central regions of early-types to study their supermassive central black-holes it is important to include dark matter in the models, even if the central parts itself are not dominated by dark matter. Neglect of the halo can result in an overestimation of the stellar mass-to-light ratio which subsequently leads to an overestimation of the central *stellar* mass. The latter is degenerate with the black-hole mass and neglect of dark matter might then yield a too small black-hole. This has been illustrated for M87, where the black-hole mass more than doubled after including dark matter in the models (Gebhardt & Thomas 2009). The effect is supposed to be strongest for the most massive galaxies since their light profiles are shallow. Further galaxies have to be modelled in order to establish if the inclusion of dark matter in black-hole models can reduce the discrepancy between the $\approx 10^9 \, M_\odot$ solar mass black-holes in the most massive nearby galaxies and the $\approx 10^{10} \, M_\odot$ solar mass black-holes in high-redshift quasars.

References

Bell, E.F., & De Jong, R.S.: 2001, ApJ, 550, 212

Binney, J., & Tremaine, S.: 1987, Galactic Dynamics (Princeton: Princeton University Press)

Cappellari, M., et al.: 2006, MNRAS, 366, 1126

Cappellari, M. et al.: 2007, MNRAS, 379, 418

Chanamé, J., Kleyna, J., & van der Marel, R.: 2008, ApJ, 682, 841

Conselice, C.J., Blackburne, J.A., & Papovich, C.: 2005, ApJ, 620, 564

Contopoulos, G.: 1963, AJ, 68, 1

Corsini, E.M., Wegner, G., Saglia, R.P., Thomas, J., Bender, R., Thomas, D.: 2008, ApJS, 175, 462

Cretton, N., de Zeeuw, P.T., van der Marel, R.P., & Rix, H.W.: 1999, ApJS, 124, 383

De Lucia, G., & Blaizot, J.: 2007, MNRAS, 375, 2

Gebhardt, K., et al.: 2003, ApJ, 583, 92

Gebhardt, K., & Thomas, J.: 2009, ApJ, 700, 1690

Gerhard, O.E., Kronawitter, A., Saglia, R.P., Bender, R.: 2001, AJ, 121, 1936

Jesseit, R., Naab, T. & Burkert, A.: 2005, MNRAS, 360, 1185

Kormendy, J., & Freeman, K.C.: 2004, in IAU Symp. 220, Dark Matter in Galaxies, ed. S.D. Ryder et al. (San Francisco: ASP), 377

Krajnović, D., Cappellari, M., Emsellem, E., McDermid, R.M., de Zeeuw, P.T.: 2005, MNRAS, 357, 1113

Kronawitter, A., Saglia, R.P., Gerhard, O.E., Bender, R.: 2000, A&AS, 144, 53

Magorrian, J., Ballantyne, D.: 2001, MNRAS, 322, 702

Mehlert, D., Saglia, R.P., Bender, R., Wegner, G.: 2000, A&AS, 141, 449

Mehlert, D., Thomas, D., Saglia, R.P., Bender, R., Wegner, G.: 2003, A&A, 407, 423

Naab, T. & Burkert, A.: 2003, ApJ, 597, 893

Navarro, J.F., Frenk, C.S., & White, S.D.M.: 1996, ApJ, 462, 563

Persic, M., Salucci, P., & Stel, F.: 1996a, MNRAS, 281, 27

Persic, M., Salucci, P., & Stel, F.: 1996b, MNRAS, 283, 1102

Richstone, D. O., & Tremaine, S.: 1988, ApJ, 327, 82

Romanowsky, A.J., Douglas, N.G., Arnaboldi, M., Kuijken, K., Merrifield, M.R., Napolitano, N.R., Capaccioli, M., Freeman K.C.: 2003, Sci, 301, 1696

Schwarzschild, M.: 1979, ApJ, 232, 236

Siopis, C. et al.: 2009, ApJ, 693, 946

Thomas, J., Saglia, R.P., Bender, R., Thomas, D., Gebhardt, K., Magorrian, J., Richstone, D.: 2004, MNRAS, 353, 391

Thomas, J., Saglia, R.P., Bender, R., Thomas, D., Gebhardt, K., Magorrian, J., Corsini, E.M., Wegner, G.: 2005, MNRAS, 360, 1355

Thomas, J., Jesseit, R., Naab, T., Saglia, R. P., Burkert, A., & Bender, R.: 2007a, MNRAS, 381, 1672

Thomas, J., Saglia, R.P., Bender, R., Thomas, D., Gebhardt, K., Magorrian, J., Corsini, E.M., Wegner, G.: 2007b, MNRAS, 382, 657

Thomas, J., Saglia, R.P., Bender, R., Thomas, D., Gebhardt, K., Magorrian, J., Corsini, E.M., Wegner, G.: 2009a, ApJ, 691, 770

Thomas, J., et al.: 2009b, MNRAS, 393, 641

Valluri, M., Merritt, D., & Emsellem, E.: 2004, ApJ, 602, 66

Vandervoort, P.O.: 1984, ApJ, 287, 475

Wegner, G., Corsini, E. M., Saglia, R. P., Bender, R., Merkl, D., Thomas, D., Thomas, J., Mehlert, D.: 2002, A&A, 395, 753

van den Bosch, R.C.E., van de Ven, G., Verolme, E.K., Cappellari, M., & de Zeeuw, P.T.: 2008, MNRAS, 385, 647

Star and protoplanetary disk properties in Orion's suburbs

Roy van Boekel[1], Min Fang[1,2], Wei Wang[1], Andrès Carmona[3,1], Aurora Sicilia-Aguilar[1], and Thomas Henning[1]

[1] Max Planck-Institut für Astronomie, Königstuhl 17,
D-69117 Heidelberg, Germany
boekel@mpia.de

[2] Purple Mountain Observatory, West Beijing Road,
210008 Nanjing, PR China

[3] ISDC & Geneva Observatory, University of Geneva, chemin d'Ecogia 16,
1290 Versoix, Switzerland

Abstract

This contribution aims to advance our knowledge of the physical mechanisms driving accretion onto pre-main sequence stars and the dissipation of their circumstellar disks, which are currently not well understood. We present results of a survey of the L1630N and L1641 star-forming clouds in Orion. For a sample of $\approx$400 YSOs we determine their fundamental stellar parameters, disk geometry and accretion rates from a combined data set of optical and infrared photometry and optical spectroscopy. We show that stars residing in clusters or aggregates disperse their disks faster than those formed in isolation, even in very small clusters with low number densities, and in the absence of OB stars. We exclude primordial binarity as a mechanism causing a "transition disk" appearance. We argue that our observations provide observational support for a scenario put forward by Hartmann et al. (2006), in which disks around low-mass stars and brown dwarfs are fully viscous, yielding a steep dependency of the accretion rate on stellar mass of $\dot{M}_{\mathrm{acc}} \propto M_*^{5/2}$, and stars in the $\sim$solar to intermediate mass regime have substantial dead zones in the disk interior resulting in a shallower relation of approximately $\dot{M}_{\mathrm{acc}} \propto M_*$.

1 Introduction

The disks around young stars are pivotal to our understanding of the star- and planet formation process. The physical mechanisms governing the accretion of material onto the central star and the ultimate dissipation of the disks are not well understood.

---

[1] Based on data from VLT, CAHA, SDSS, 2MASS, and Spitzer.

Here, we investigate these processes by comparing the stellar properties, accretion rates, and disk geometries of a large sample of young stars. We start, however, by briefly outlining the formation process of low mass stars (e.g. Shu et al. 1977) from an observer's point of view.

Based on the emerging spectral energy distribution, we can distinguish four phases in the low-mass star formation process (Lada & Wilking 1984; Andrè et al. 1993). Class 0 sources, only visible at far-infrared and millimeter wavelengths, represent the brief first phase of mass gathering in the center of a collapsing molecular cloud core, lasting on order of 10^4 yr. Due to the large angular momentum of the contracting cloud core a disk forms. Accretion through the disk is accompanied with strong bi-polar outflow activity typical for the Class I phase, during which the sources are visibile at infrared through millimeter wavelengths for a period on order of 10^5 yr. During the Class 0 and I phases, mass is accreted onto the central object through the accretion disk, which is fed by infalling material from the surrounding envelope. At some point the supply of fresh cloud material onto the disk comes to a halt. This marks the beginning of the Class II phase during which the central stars become optically visible but are still surrounded by a gas-rich circumstellar disk. The central stars in Class II objects have already gathered essentially their entire final mass but are still contracting and will become substantially smaller and hotter before reaching the main sequence. It is during this phase that planet formation is thought to take place, and at least the giant gas planets must have completely assembled before the disk is dissipated, after typically several million years (Haisch et al., 2001). The accretion process, which continues at rates several orders of magnitude lower than during the earlier phases, is no longer important for the mass build-up of the star during the Class II phase but may be highly relevant for the disk dissipation process. After the disk has been drained of most of its material, an essentially gasless "debris disk" consisting of dust particles that likely originate from the collisional grind-down of larger bodies may remain. Such "Class III" objects show only small amounts of infrared excess emission and represent the final stages of the pre-main sequence evolution of a low-mass star.

In this contribution we focus on the properties of objects in the Class II phase, and on the transition to Class III.

The Lynds 1630N and 1641 clouds are located in the Orion molecular cloud complex at a distance of approximately 450 pc (e.g. Anthony-Twarog, 1982). In L1630, the young stars are mainly formed in two clusters (NGC 2068 and NGC 2071), whereas in L1641 there is a population of young stars located in small clusters (or "aggregates") as well as a more distributed population of young stars that are forming in isolation.

The work presented here has recently appeared in print Fang et al. (2009). In this contribution we briefly outline the observational basis and main methodology of that work, and highlight a subset of its results. In section 5 we supplement the aforementioned paper with additional discussion on the dependency of the average accretion rate on the stellar mass.

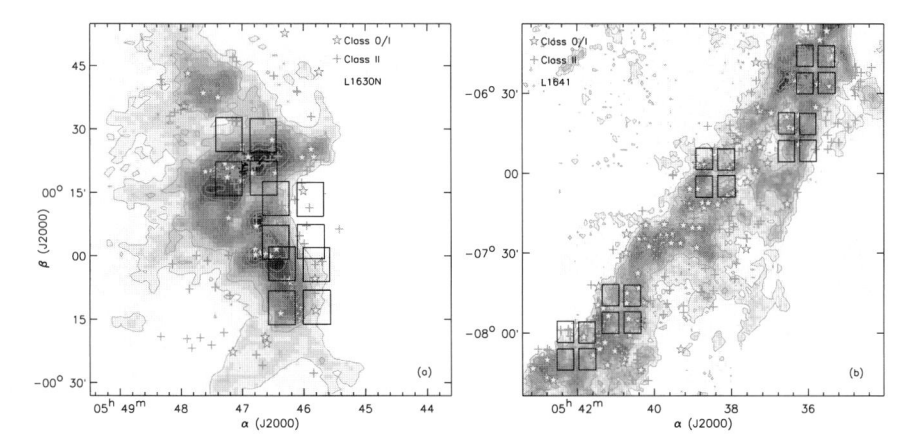

Figure 1: The Lynds 1630N (left) and 1641 (right) star forming clouds in Orion. The positions of Class I and Class II sources detected in the Spitzer data are indicated. Also drawn are the fields of the optical spectroscopic observations with VIMOS.

2 Observations

The data used in this work can be roughly divided into two categories: (1) infrared photometry from which the presence of circumstellar disks and measures for their geometry and evolutionary status can be derived, and (2) optical photometry and optical spectroscopy yielding the fundamental properties of the central stars as well as the accretion behavior of individual objects.

The L1630N and L1641 regions were observed with the Spitzer Space Telescope's IRAC and MIPS cameras, these data were previously presented by Megeath et al. (2005). IRAC images in the 3.6, 4.5, 5.8 and 8.0 μm bands with exposure times of 0.4 and 10.4 seconds were combined to achieve high sensitivity and dynamic range. Several mosaics were made of each cloud, resulting in average exposure times of ≈25 s per band. The MIPS mosaics in the 24 μm band had an effective expourue time of 80 s. We ran point source finding and PSF photometry routines and detected ≈25 000 and ≈40 000 sources in the L1630N and L1641 fields, respectively. Near infrared photometry in the JHK bands, centered on 1.25, 1.65 microns, and 2.17 μm, was taken from the 2MASS catalogue.

Both clouds were fully imaged in the $g'r'i'z'$ bands centered on 0.48, 0.62, 0.76, and 0.91 μm. For L1630N and the north-west half of L1641 we used data from the Sloan Digital Sky survey (SDSS). We imaged the south-west half of L1641 with the Large Area Imager for Calar Alto 3.5 m telescope (LAICA). The magnitudes of the detected stars were taken from the survey catalogues (SDSS) or determined using PSF photometry (LAICA).

For a subset of the optically detected stars, we have obtained medium resolution spectra with the Visible Multi-Object Spectrograph (VIMOS) at the VLT (spectral range covered ≈ 4100–8750 Å, spectral resolution $R \approx 2000-2500$). In total we observed 715 targets distributed over 8 VIMOS pointings (see Fig. 1), with three 25-minute exposures each.

3 Analysis

In this section we describe the methods that we applied to extract the physically interesting information from our data. The main goals are the identification of young stars, the determination of their fundamental stellar parameters, the characterization of their disk geometry and the estimation of the rate at which disk material is accreted onto the central star.

3.1 YSO selection criteria

A star in our sample observed with VIMOS is classified as a young star if it obeys any of the following criteria:

1. IR excess
2. Li I absorption
3. Hα emission

3.2 Fundamental stellar parameters

Robust determination of the stellar effective temperature T_{eff} and luminosity L_{bol} is a prerequisite for a meaningful correlation of star and disk properties. If the reddening by intervening dust towards a star is negligible, its broad-band optical colors are a good measure for T_{eff}. However, young stars usually suffer substantial extinction due to dust residing in the circumstellar environment or in the parent molecular cloud. In this case, a degeneracy arises between T_{eff} and the foreground extinction A_{V} if only broad-band photometry is available. This degeneracy is broken if the stellar temperatures are determined using spectroscopic methods, since these rely on the equivalent widths of photospheric absorption features or on ratios of neighboring lines. These quantities are unaffected by reddening.

We have performed spectral typing using the classification scheme developed by Hernández et al. (2004). We could reliably classify $\approx$75% of the stars observed with VIMOS, the remaining spectra having insufficient SNR. The resulting spectral types are accurate to $\sim$1 sub-type. Effective temperatures were derived from the spectral types using the relation given by Kenyon & Hartmann (1995). We then fitted reddened model atmospheres to the optical and, for those stars showing no near-infrared excess emission, near-infrared photometry, see Fig. 2. In this fit T_{eff} was kept fixed at the spectroscopically determined value, and the only free parameters were the angular diameter θ and the optical extinction A_{V}. The extinction law by Cardelli et al. (1989) was used with a total to selective extinction value typical of ISM dust ($R_{\text{V}} = 3.1$). The bolometric luminosity of each star L_{bol} can now easily be calculated if the distance is known: $L_{\text{bol}} = \pi \theta^2 d^2 \sigma T_{\text{eff}}^4$, where d is the distance and

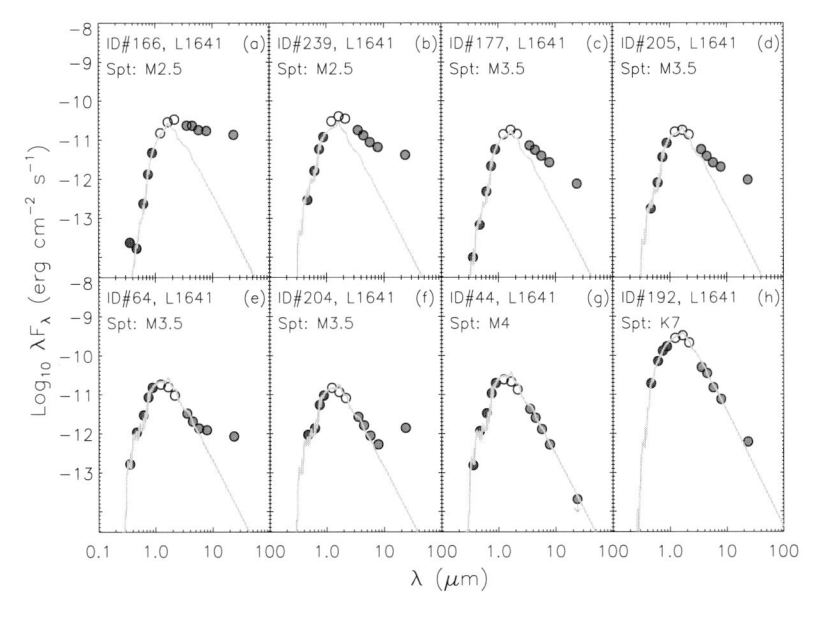

Figure 2: Example spectral energy distributions illustrating YSO disk geometry evolution and the fitting of stellar model atmospheres to photometric data (See sections 3.2 and 3.3 for a description).

σ denotes the Stefan-Boltzmann constant. We have adopted a distance of 450 pc for both L1630N and L1641. Note that we have implicitly corrected for extinction by calculating the bolometric luminosities in this way.

With $T_{\rm eff}$ and $L_{\rm bol}$ known, the mass and age of each star can be estimated by placement in the Hertzsprung-Russel diagram and comparison with theoretical pre-main sequence stellar evolutionary tracks, for which we adopt those of Dotter et al. (2008). We note, though, that several sets of such theoretical tracks exist and that there are substantial systematic differences in the derived masses and ages between these (see e.g. Hillenbrand et al. 2008 for a discussion). Within a specific set of evolutionary tracks, however, the *relative* masses and ages are more robust.

3.3 Disk geometry

The infrared SED can be used to constrain the geometry of a circumstellar disk. The basic idea is that in a young star+disk system the main energy source of the disk material is absorption of radiation from the central object in the disk surface[2]. The heated disk material re-radiates the absorbed energy at longer wavelengths. The disk material attains a temperature such that the absorbed energy equals the re-emitted energy. The hottest material close to the central star will radiate mostly at near-infrared wavelengths, while cooler material residing at larger radii is brightest at longer in-

---

[2] In the very early phases when the accretion rate is high, the release of gravitational potential energy of accreting material in the disk interior may contribute substatially close to the central star.

frared or millimeter wavelengths. Thus, the amount of infrared excess seen at different wavelengths provides a rough measure of how much material we see at certain temperatures, and thus certain distances to the star, i.e. it traces the disk structure, even if only in an approximate sense[3].

In Fig. 2 we illustrate how the disk geometry reflects on the emerging SED, by showing representative example SEDs from our survey. The stars shown in Fig. 2(a-c) have an infrared excess starting around 2 μm and continuing through longer wavelengths, indicating the presence of an optically thick, dusty circumstellar disk that reaches inward all the way to the dust sublimation radius, where the temperature is ≈1500 K. Going from Fig. 2(a) to (c) the strength of the excess is reduced by roughly the same factor at all wavelengths probed. This indicates that a decreasing fraction of the stellar radiation is absorbed and re-emitted, i.e. the disk evolves from a "flared" to a "flat" geometry (see e.g. Dullemond & Dominik, 2004), possibly due to dust settling. From Fig. 2(d) to (f) we see that the near-infrared excess disappears whereas the excess at longer wavelengths remains strong. This indicates that the hot material close to the star is no longer present whereas the cooler material at larger radii is still in place, i.e. an "inside-out" disk clearing. Such objects with partially dissipated disks are thought to represent a transition phase between the gas-rich disks of Class II sources and the much less massive, gas-deprived "debris disks" of Class III sources. Hence, they are referred to as *transition disks*. In Fig. 2(g) we show a young cloud member that shows no evidence of infrared excess emission at wavelengths of ≤24 μm and may have lost its disk entirely. An example of a Class III source, showing no excess in the near-infrared and only a weak excess at 24 μm is shown in Fig. 2(h).

3.4 Accretion rates

The rate at which material is accreted onto the young stars in our sample is deduced from emission lines detected in the VIMOS spectra, specifically Hα, He I 5876Å, and Hβ. We have compiled a large set of measurements of the line luminosities and the accretion luminosity as derived from the UV continuum excess from the literature. By establishing relations between the line luminosity and the accretion luminosity, we can use the line strengths as extracted from our VIMOS spectra to estimate the accretion rate of our stars. We note that the correlations between line luminosity and accretion luminosity show substantial scatter around the average relation, in particular for the Hα line, and that the accretion rate estimates for individual stars are uncertain by at least a factor of ≈3 but often more (See Fang et al. 2009 for a detailed report). However, for larger samples of stars the observed line luminosities form a good measure of the accretion properties in a statistical sense.

---

[3]The equilibrium temperature of the disk material depends on the distance to the star, the dust properties, and the disk geometry itself. The disk geometry cannot be accurately reconstructed from SED measurements alone since usually multiple geometries, optionally with different dust properties, yield equally good fits to the observed SED. Observations that spatially resolve the disks can lift these degeneracies.

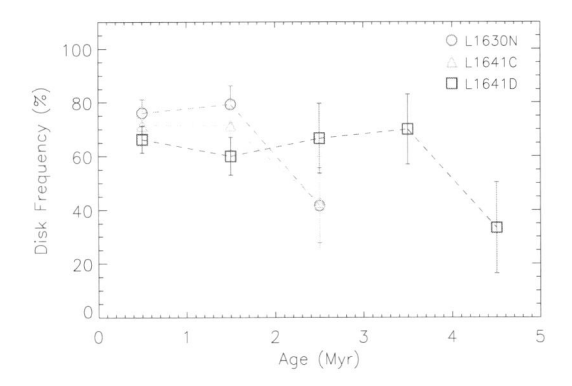

Figure 3: The fraction of stars that exhibit circumstellar disks as a function of stellar age. The circles and triangles represent "clustered" populations, the squares indicate a "distributed" or "isolated" population.

4 Highlighted results.

In this section we highlight a subset of the results from our survey. For a complete overview and discussion we refer to Fang et al. (2009).

4.1 Disk survival in different environments

Here we investigate how the average lifetime of a circumstellar disk depends on its nearby environment, in particular on whether a star resides in a cluster and thus has neighbours or has formed in relative isolation. From a theoretical vantage point, it is expected that disk dispersal is faster in the center of dense clusters due to tidal disruption of the disks during close encounters with other cluster members (e.g. Adams 2008), and due to enhanced UV fields from nearby OB stars photo-evaporating the disks. However, the volume density of stars (# stars pc^{-3}) in the clustered environments that we investigated is too low to expect substantial gravitational disruption of disks, and no OB stars are present. Therefore, we would not expect to see a difference in average disk lifetime between our clustered and distributed populations.

In Fig. 3 we show the disk frequency as a function of stellar age for the clustered populations "L1630N" and "L1641C" as well as the distributed population "L1641D". In the latter population we see a relatively constant disk frequency initially, but a $\sim 2\sigma$ decrease in the last age bin (4-5 Myr). In both clustered populations we see a similar decrease, however already after 2-3 Myrs. Thus, contrary to expectations, disk dispersal does seem to be faster if a star has close neighbors, despite the low stellar density of our clusters and the absence of OB stars.

4.2 Accretion rate as a function of stellar mass

The rate at which disk material is accreted onto young stars increases with stellar mass, decreases with stellar age, and shows large source-to-source variations within each group of objects with similar mass and age. Previous studies have revealed a

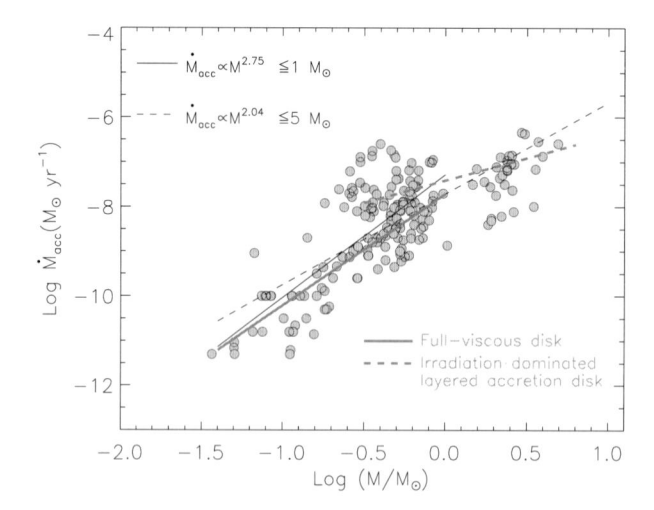

Figure 4: The mass accretion rate $\dot{M}_{\rm acc}$ as a function of stellar mass for a large sample of literature data (see sections 4.2 and 5).

dependence of the *average* accretion rate on the stellar mass of the form $\dot{M}_{\rm acc} \propto M_*^\alpha$, with $\alpha \approx 2$ (e.g. Natta et al. 2006). From our analysis, however, we find a steeper dependence of $\alpha = 3.1 \pm 0.3$ in the mass regime that we probe ($M_* \lesssim 1\,{\rm M}_\odot$).

In order to investigate this apparent discrepancy, we turned back to the literature, gathering stellar mass and accretion rate estimates. We included only those stars whose fundamental parameters have been determined robustly, i.e. *through spectroscopy*. The result of this exercise is shown in Fig. 4. We immediately see the usual correlation of increasing accretion rate with increasing stellar mass, as well as the large scatter at any given mass. If we make a power law fit over the whole mass range, we find $\alpha \approx 2$, in agreement with previous studies. However, if we consider only the mass range $M_* \leq 1\,{\rm M}_\odot$, the literature data show a substantially steeper dependence of $\alpha \approx 2.8$, in agreement with the value we find in our present study (Fang et al. 2009). Thus, the dependency of the accretion rate on stellar mass appears steeper in the sub-solar mass regime than for stars of $\approx 1\,{\rm M}_\odot$ or more. We will discuss this further in section 5.

4.3 Transition disks

The nature of transition disks, i.e. YSO systems with an SED similar to that shown in Fig. 2(f), is still poorly understood. The weak or even undetectable near-infrared excess emission indicates the absence of optically thick material in the inner disk, whereas at larger radii the optically thick disk still remains intact. The physical mechanism driving the clearing of the inner disk is yet unclear and the importance of various proposed mechanisms, i.e. viscous evolution, photoevaporation, grain growth, planet-formation, or (close) binarity, remains to be established.

In our survey we have identified 28 transition disk systems (≈11% of the disk population), of which 20 were previously unknown. We found 47 additional candiate systems that show the typical transition disk SED but were not in our spectroscopic sample.

When comparing the ages of the transition disk systems to those objects which still have optically thick disks, and thus presumably represent an earlier evolutionary phase, we found the following. The age dispersion within each group is much larger than can be accounted for by observational uncertainties, and must be intrinsic. The median ages of the populations are 0.8 Myr for the stars with optically thick disks and 1.9 Myr for the transition objects, and a KS test yields a very low probability ($P \approx 0.003$) for both groups to be drawn randomly from the same population. Thus, in a statistical sense *the transition disk systems are clearly older*. This is direct observational support for the hypothesis that transition disks represent an evolutionary phase that comes *after* the optically thick disk phase. It also excludes close binarity, i.e. transition disks are actually circumbinary disks, as the cause for a transition disk appearance: such a mechanism would take effect from the very earliest phases of the pre-main sequence evolution, and would not result in an "old" transition disk population.

Some transition disks still show signs of active accretion. This implies that, although their inner disks are optically thin, they are not entirely empty, and must at least contain some gaseous material. When comparing the accretion behavior of transition disk systems to that of stars optically thick disks, we find the following: the fraction of stars that show substantial accretion is much lower in transition disks (26±11% vs. 57±6%), but those transition disk systems that *are* actively accreting do so at roughly the same median rate as their optically thick disk bearing counterparts, namely $\approx 3-4 \times 10^{-9}$ M$_\odot$ yr^{-1}.

5 Discussion

For an extensive discussion of our results, including those highlighted here, we refer to Fang et al. (2009). Here we will discuss the dependence of the accretion rate $\dot{M}_{\rm acc}$ on the stellar mass M_* in more depth than was done in that work.

As we have argued in section 4.2, our data indicate that the scaling law of the average accretion rate with stellar mass appears to be steeper in the sub-solar mass regime than at higher masses. The available literature data, considering only those objects whose fundamental stellar parameters were robustly determined using spectroscopy, yield the same result (Fig. 4): instead of a single power law with an exponent of ≈2 from the brown dwarf to the intermediate mass (HAeBe) regime, an exponent of 2.5 to 3 in the sub-solar regime and of ≈1 at higher masses seems more appropriate. What are the implications of this behavior for the physical mechanism, or mechanisms, driving the accretion process?

The first concept that predicted a clear relation between the mass of a YSO system and the rate at which it acquires material from its surroundings was put forward by Bondi & Hoyle (1944), see Edgar (2004) for a review. This formalism describes at what rate an object of mass M_* gathers material from a homegenous medium with density ρ and sound speed c_s through which it is moving at constant velocity v:

$$\dot{M}_{\text{BH}} = \frac{4\pi G^2 \rho}{(c_s^2 + v^2)^{3/2}} M_*^2 \qquad (1)$$

Thus, it reproduces the M_*^2 that is found when the whole range of masses upto ≈ 5 M$_\odot$ is fitted simultaneously. Padoan et al. (2005) noted that for plausible input parameters accretion rates on order of 10^{-8} M$_\odot$ yr^{-1} are obtained, typical for actively accreting T Tauri stars. However, as pointed out by Hartmann et al. (2006), the Bondi-Hoyle formalism ignores the non-zero angular momentum of accreting material and thus describes only the rate at which material is fed onto the outer disk regions. Inward transport through the disk is not considered and it is not clear why $\dot{M}_{\text{acc}}$ should match $\dot{M}_{\text{BH}}$. Moreover, T Tauri stars are known that lie within H II regions, and yet show accretion rates very similar to those of YSOs residing in molecular clouds, whereas $\dot{M}_{\text{BH}}$ is ≈ 5 orders of magnitude lower in an H II region due to the low ambient density and high sound speed Hartmann et al. (2006); Sicilia-Aguilar et al. (2005)

Hartmann et al. (2006) provide an insightful discussion of the angular momentum transport in disks. The key quantity that needs to be understood is the viscosity of the disk material. The prime candidate mechanism, the magneto-rotational instability (MRI, e.g. Balbus & Hawley, 1998), requires a minimum degree of ionization of the disk material ere it couples to the magnetic field and the MRI can take effect (see e.g. Chiang & Murray-Clay 2007). Various means of ionization have been proposed, including absorption of cosmic rays (Gammie, 1996) or stellar X-rays (Glassgold et al., 1997) in the disk surface, and thermal ionization driven by either viscous dissipation within the disk or irradiation of the disk surface by the central star (Hartmann et al. 2006). The surface densities of typical T Tauri star disks are high and the irradiation by the central star or cosmic rays can typically only ionize to a limited depth, causing an "active layer" on the disk surface of column density Σ_a in which the MRI can be sustained. Deeper in the disk, a neutral "dead-zone" may exist, except close to the star where also the disk interior is hot enough to be sufficiently ionized or at very large radii where the surface density is low and cosmic rays can reach the midplane.

Two limiting cases are considered by Hartmann et al. (2006): 1) that of a low mass disk in which the entire column is sufficiently ionized for an active MRI, and which thus are subject to fully viscous evolution, and 2) that of a higher mass disk in which thermal ionization by irradiation from the central star is dominant, causing an MRI-active surface layer of column density Σ_a.

The former case may be appropriate for the disks around very low mass stars and brown dwarfs. Assuming a fully viscous disk whose initial mass scales with M_*, a scaling of the luminosity with stellar mass appropriate for low-mass pre-main sequence stars ($L_* \propto M_*^2$), and that we observe the disks at a time when substantial viscous evolution has already taken place, which is appropriate for Class II sources, Hartmann et al. (2006) predict the following relation for the mass accretion rate (eq. 17 in that work):

$$\dot{M}_{\text{acc}} \propto t^{-3/2} \alpha_\nu^{-1/2} M_*^{5/2} \qquad (2)$$

where α_ν denotes the viscosity parameter (Shakura & Sunyaev 1973). Thus, a steep dependency on M_* is derived, which matches the observed slope in the sub-solar mass regime fairly well (Fig. 4, and Fang et al. 2009).

In the latter case, i.e. a more massive disk with an MRI-active surface layer that is ionized by irradiation from the central star and a dead zone in the disk interior, the dependency of $\dot{M}_{\rm acc}$ on M_* depends on whether the temperature required for sufficient ionization is reached interior to the dust evaporation radius $R_{\rm d}$ or at larger radii. If the critical radius $R_{\rm c}$ is smaller than $R_{\rm d}$, Hartmann et al. (2006) derive:

$$\dot{M}_{\rm acc} \propto \alpha_\nu \Sigma_a M_*^{1/2} \qquad (3)$$

In case $R_{\rm c}$ is larger than $R_{\rm d}$,

$$\dot{M}_{\rm acc} \propto \alpha_\nu \Sigma_a M_* \qquad (4)$$

is obtained. This shallower dependency is seen in the observations as well (Fig. 4). We once more stress, however that the source-to-source scatter in the observations is very large. This makes it difficult to accurately determine the appropriate slope empirically. Furthermore, the relations in equations 2, 3, and 3 do not comprise the full complexity of the accretion process, and the input parameters likely vary substantially from source to source.

We conclude that our observations (Fang et al. 2009) as well as the available literature provide observational support for the scheme put forward by Hartmann et al. (2006), in which the dependency of $\dot{M}_{\rm acc}$ on M_* is steeper for low mass disks around low-mass stars and brown dwarfs than for higher mass disks around ∼solar to intermediate mass stars, due to the presence of dead zones in the higher mass disk, which are not present or less important in the lower mass disks.

References

Adams, F. C.: 2008, Physica Scripta Volume T 130, 014029

Andre, P., Ward-Thompson, D., & Barsony, M.: 1993, ApJ 406, 122

Anthony-Twarog, B. J.: 1982, AJ 87, 1213

Balbus, S. A. & Hawley, J. F.: 1998, Reviews of Modern Physics 70, 1

Bondi, H. & Hoyle, F.: 1944, MNRAS 104, 273

Cardelli, J. A., Clayton, G. C., & Mathis, J. S.: 1989, ApJ 345, 245

Chiang, E., Murray-Clay, R.: 2007, NatPh 3, 604

Dotter, A., Chaboyer, B., Jevremović, D., et al.: 2008, ApJS 178, 89

Dullemond, C. P. & Dominik, C.: 2004, A&A 417, 159

Edgar, R.: 2004, New Astronomy Reviews 48, 843

Fang, M., van Boekel, R., Wang, W., Carmona, A., Sicilia-Aguilar, A., Henning, Th.: 2009, A&A 504, 461

Gammie, C. F.: 1996, ApJ 457, 355

Glassgold, A. E., Najita, J., & Igea, J.: 1997, ApJ 485, 920

Haisch, Jr., K. E., Lada, E. A., & Lada, C. J.: 2001, ApJ 553, L153

Hartmann, L., D'Alessio, P., Calvet, N., & Muzerolle, J.: 2006, ApJ 648, 484

Hernández, J., Calvet, N., Briceño, C., Hartmann, L., & Berlind, P.: 2004, AJ 127, 1682

Hillenbrand, L. A., Bauermeister, A., & White, R. J.: 2008, ASP 384, 200

Hoyle, F. & Lyttleton, R. A.: 1939, Proc. Cam. Phil. Soc. 35, 405

Kenyon, S. J. & Hartmann, L.: 1995, ApJS 101, 117

Lada, C. J. & Wilking, B. A.: 1984, ApJ 287, 610

Natta, A., Testi, L., & Randich, S.: 2006, A&A 452, 245

Megeath, S. T., Flaherty, K. M., Hora, J., et al.: 2005, IAU Symposium 227, 383

Padoan, P., Kritsuk, A., Norman, M. L., & Nordlund, Å.: 2005, ApJ 622, L61

Shu, F. H.: 1977, ApJ 214, 488

Sicilia-Aguilar, A., Hartmann, L. W., Hernández, J., Briceño, C., & Calvet, N. 2005, AJ, 130, 188

Molecular gas at high redshift

Fabian Walter[1], Chris Carilli[2], and Emanuele Daddi[3]

[1]Max Planck Institut für Astronomie
Auf dem Hügel 71, 69117 Heidelberg, Germany
walter@mpia.de

[2]National Radio Astronomy Observartory
Socorro, NM 87801, USA

[3]Laboratoire AIM, CEA Saclay
91191 Gif-sur-Yvette Cedex, France

Abstract

In order to understand galaxy evolution through cosmic times it is critical to derive the properties of the molecular gas content of galaxies, i.e. the material out of which stars ultimately form. The last decade has seen rapid progress in this area, with the detection of massive molecular gas reservoirs at high redshifts in submillimeter–selected galaxies and quasars. In the latter case, molecular gas reservoirs have been quantified out to redshifts z>6, i.e. towards the end of cosmic reionization when the universe was less than one Gyr old. The recent discovery of molecular gas in more normal galaxies have extended these studies from the most extreme objects in the universe (SFR$\sim$1000 $M_\odot$ yr^{-1}; quasars and submillimeter galaxies) to more 'normal' starforming systems at redshifts 1.5–2.5 (with SFR$\sim$100 $M_\odot$ yr^{-1}). However, detecting the molecular gas reservoirs of high–redshift galaxies that only have moderate star formation rates ($\sim$10 $M_\odot$ yr^{-1}, similar to the faint galaxies seen in the Hubble Ultra Deep Field) will likely have to await the completion of ALMA.

1 Recent progress in understanding galaxy formation

The last decade has seen dramatic advances in our understanding of galaxy formation. Cosmic geometry, the mass-energy content of the Universe, and the initial density fluctuation spectrum, have been constrained to high accuracy (e.g., Spergel et al. 2007). Given the underlying distribution of dark matter, the main challenge of galaxy formation studies is to understand how the baryons assemble into the observed Universe. In this context, structure formation through gravitational instabilities has been calculated in detail through numerical studies, and observationally verified through studies of galaxy distributions (e.g., Springel et al. 2006). The history of star formation, and the build up of stellar mass, as a function of galaxy type and mass, are now constrained out to the tail-end of cosmic reionization, within 1 Gyr of the Big Bang, and pushing toward first light in the Universe (e.g., Bouwens et al. 2010).

The principle results of these studies can be summarized as follows. The comoving cosmic star formation rate density increases by more than an order of magnitude from $z \sim 0$ to 1, peaks around $z \sim 2$ to 3, and likely drops gradually out to $z \sim 8$ (e.g., Hopkins & Beacom 2006, Bouwens et al. 2010). The build-up of stellar mass follows this evolution, as does the temporal integral (for z<2, e.g. Ilbert et al. 2010). The redshift range $z \sim 1.5$ to 3 has been called the 'epoch of galaxy assembly', when about half the stars in the Universe form in spheroidal galaxies. It has been shown that star formation shifts systematically from lower luminosity galaxies ($L_{FIR} \sim 10^{10}$ $L_\odot$) at low redshift, to high star formation rate galaxies ($L_{FIR} \geq 10^{11}$ $L_\odot$) at $z \sim 2$ ('downsizing', e.g., Le Floc'h et al. 2005, Smolčić et al. 2009). Another key finding is the tight relation between the star formation rate and the stellar mass for star forming galaxies up to redshifts z~3 (e.g. Noeske et al. 2007; Elbaz et al. 2007; Daddi et al. 2007; Magdis et al. 2010). High–redshift starforming galaxies are typically selected through optical/IR studies (e.g., Steidel et al. 1999, Daddi et al. 2004, van Dokkum et al. 2006)

Besides the star forming galaxies, a population of passively evolving, relatively massive (stellar masses $\sim 10^{10}$ to 10^{11} $M_\odot$) galaxies ('red and dead'), have been detected at $z > 1$, comprising roughly 50% of the galaxies selected in near-IR surveys (see, e.g., review by Renzini 2006). These galaxies must form the majority of their stars quickly at even earlier epochs (e.g., Wiklind et al. 2008).

Lastly, substantial populations of supermassive black holes and galaxies are now being detected routinely back to first light and cosmic reionization ($z \sim 6.5$). Populations include normal star forming galaxies, such as the Ly-α selected galaxies, with star formation rates of order $10\,M_\odot$ year^{-1} (Taniguchi et al. 2005, Ouchi et al. 2009) as well as quasar host galaxies, often with star formation rates that are 100 times higher (e.g., Wang et al 2008). The latter likely represent a major star formation epoch for very massive galaxies within 1 Gyr of the Big Bang. Most recently, GRBs are showing great potential to probe galaxy formation out to $z > 8$ (Tanvir et al. 2009, Salvaterra et al. 2009).

These observations have led to a new model for galaxy formation. As opposed to either cooling of virialized, hot halo gas, or major, gas-rich mergers, the dominant mode of star formation during the epoch of galaxy assembly may be driven by cold mode accretion, or stream-fed galaxy formation (e.g., Kereš et al. 2005, Dekel et al. 2009). Simulations suggest that gas accretion in early massive galaxies occurs along cold streams from the filamentary intergalactic medium (IGM) that never shock-heat, but streams onto the galaxy at close to the free-fall time. This cool gas forms a thick, turbulent rotating disk which very efficiently forms stars in a few giant regions in the disk (e.g., Bournaud et al. 2007). The star forming regions then migrate to the galaxy center via dynamical friction and viscosity, forming compact stellar bulges. The process leads to relatively steady, active (~ 100 $M_\odot$ year^{-1}) star formation in galaxies over timescales of order 1 Gyr. Subsequent dry mergers at $z < 2$ lead to continued total mass build up, and morphological evolution, but little subsequent star formation (e.g., Skelton et al. 2009, Van der Wel et al. 2009, Robaina et al. 2010).

2 Molecular gas: the key to testing galaxy formation models

While progress has been impressive, the model above is based almost exclusively on optical and near-IR studies of the stars, star formation, and ionized gas. There remains a major gap in our understanding of galaxy formation, namely, observations of the cool gas, the fuel for star formation in galaxies. In essence, current studies probe the products of the process of galaxy formation, but miss the source. Numerous observational and theoretical papers have pointed out the crucial need for observations of the cool molecular gas feeding star formation in galaxies (e.g., Dressler et al. 2009; Obreschkow & Rawlings 2009).

The molecular gas density history of the Universe: The key to future studies of galaxy formation is to determine the evolution of the molecular gas density of the Universe. Over the last two decades, the star formation history of the Universe (SFHU) plot has been perhaps the most fundamental tool for studying galaxy formation. At the same time, very detailed studies of star formation in nearby galaxies have reached two critical conclusions. First, star formation relates closely with the molecular gas content, but has little relation to the atomic neutral gas content (e.g., Bigiel et al. 2009). This fact is verified at high z by the lack of evolution of the cosmic HI mass density, as determined from damped Lyα systems, over the same redshift range where the star formation rate density is increasing by more than an order of magnitude (Prochaska & Wolfe 2009). Second, once molecular gas forms, it forms stars, to first order, according to a star formation law (e.g., Kennicutt 1998, Bigiel et al. 2009, Leroy et al. 2009, Krumholz et al. 2009, Daddi et al. 2010b, Genzel et al. 2010). In low–density environments, the general idea is that, once a giant molecular cloud becomes self-gravitating, star formation proceeds via local processes inherent to the GMC. If the universal star formation law were to hold to high redshift, then the classic SFHU plot would just be a reflection of the molecular gas density history of the Universe.

The stellar to gas mass density ratio: A related, and similarly fundamental issue is to compare the stellar and gas mass of galaxies versus redshift. Two extremes scenarios can be envisioned. At one extreme, the gas builds up to large values (approaching 10^{11} M$_\odot$), and then a dramatic starburst is triggered via, e.g., a major gas rich merger, when the gas is rapidly converted into stars on dynamical timescales $\sim 10^8$ years. At the other extreme, a continuous supply of gas comes from the IGM, cooling into molecular clouds and forming stars at a rate comparable to the free-fall rate of gas onto the galaxy over $\sim$ Gyr timescales. This latter case corresponds to the cold mode accretion model. This point is currently addressed by a number of observational programs that aim at studying star forming galaxies at z$\sim$ 2 (Daddi et al. 2010a, 2010b, Tacconi et al. 2010, Genzel et al. 2010, as discussed in Sec. 4.3 and 4.4 below).

More generally, the study of the cool molecular gas phase in 'typical' high high–redshift galaxies (with SFR$\sim$100 M$_\odot$ yr^{-1}) provide key insight into the process of galaxy formation in a number of ways:

- A universal star formation law: The low order transitions of CO provide the most direct means of measuring the total molecular gas mass. These measurements can then be compared to star formation rates to determine the evolution of the fundamental star formation laws relating cool gas and star formation. Is there a universal relationship that governs the efficiency of conversion of molecular gas to stars in galaxies of various types through cosmic time? What are the fundamental physical parameters behind the relationship? For a first attempt to address this topic see Sections 4.3 and 4.4 below.

- How does the ratio of molecular gas to stellar mass evolve? This ratio is typically < 0.1 in the nearby Universe, even for gas rich galaxies such as the Milky Way. Observations of a few the sBzK, and BX/BM galaxies provide evidence that this ratio may increase to > 1 at $z \sim 2$ (Daddi et al. 2010a, Tacconi et al. 2010).

- Galaxy dynamics and star formation: Molecular line observations are the most direct means of studying galaxy dynamics. Are these systems dominated by rotating disks, as predicted by cold mode accretion, or major mergers? What is the relative magnitude of ordered motion vs. turbulence? What are the stability criteria, and how do they compare to the distribution of star formation?

- Black hole – bulge mass relation: Establishing a galaxy's mass through molecular gas dynamics can be used to test the evolution of the black hole – bulge mass relation back to the first galaxies. Indeed, in many cases, the molecular line observations are the only means to get galaxy dynamics for galaxies with optically bright AGN (e.g., Walter et al. 2003, Shields et al. 2006, Riechers et al. 2008, 2009). The local relation suggests that the formation of black holes and galaxies are closely tied, but the origin of this relationship remains a mystery (see Sec. 4.2).

- ISM physics and chemistry: Detailed studies of the physical conditions in the ISM of primeval galaxies can be performed using numerous molecular line transitions, including excitation studies of CO, observations of dense gas tracers like CN, HCN and HCO^+, and other key astrochemical tracers. Such observations determine the temperature and density of the ISM, molecular abundances and the fraction of dense gas. To date, only a few extreme high–redshift systems have been detected in these dense gas tracers (e.g. Vanden Bout et al. 2003, Riechers et al. 2006, 2007a, 2007b, Garcia–Burillo et al. 2006)

3 Recent progress on molecular lines observations in early galaxies

3.1 Submillimeter Galaxies (SMGs)

Surveys at submm wavelengths have revealed highly obscured, extreme starburst galaxies (SFR $>$ 1000 $M_\odot$ year^{-1}) at high redshift, the so–called submillimeter galaxies (SMGs, e.g., Smail et al. 2007, Blain et al. 2002). Subsequent studies have detected large reservoirs of molecular gas in these systems ($> 10^{10}$ $M_\odot$; e.g., Frayer et al. 1998, 1999, Genzel et al. 2003, Greve et al. 2005, Tacconi et al. 2006, 2008, Knudsen et al. 2009, Ivison et al. 2010). General properties of the SMGs are summarized in the review by Solomon & Vanden Bout (2005) and the detections to date are summarized in Figure 1 (as discussed in Sec. 4.3).

High–resolution imaging of SMGs by Tacconi et al. (2006, 2008) have shown that their molecular gas reservoirs are compact, with a median diameter of $\sim$4 kpc. This morphology can be explained if these galaxies are the results of mergers, in which the molecular gas settles in the centre of two interacting galaxies (leading to a starburst). Recent CO imaging of one of the most distant SMGs suggests that cold mode accretion may also play a role in powering the ongoing star formation (Carilli et al. 2010).

Chapman et al. (2005) have shown that the median redshift of SMGs is z$\sim$2.3 — the existence of a substantial high–redshift (z$>$4) tail of the submillimeter galaxy population that host molecular gas has recently been established by a number of studies (Schinnerer et al. 2009, Daddi et al. 2009a, 2009b, Carilli et al. 2010, Coppin et al. 2010).

While the study of these SMGs has opened a new window on the optically obscured Universe at high redshift, the fact remains that these sources are rare, likely representing the formation of the most massive galaxies in merger-driven hyper-starbursts at high redshift, with gas depletion timescales < 0.1Gyr,

3.2 Quasars

Likewise, surveys of molecular gas in quasar host galaxies have revealed the presence of massive reservoirs of molecular gas in these objects ($> 10^{10}$ $M_\odot$, i.e. comparable to the submillimeter galaxies, e.g., Solomon & Vanden Bout 2005, Coppin et al. 2008). Quite remarkably, such reservoirs have now been detected in a sizable sample of quasars at redshift 6 and greater, i.e. at the end of cosmic reionization when the Universe was less than one Gyr old (Walter et al. 2003, 2004, Bertoldi et al. 2003, Maiolino et al. 2007, Carilli et al. 2007, Wang et al. 2010). Resolved imaging of a few quasar hosts (Walter et al. 2004, Riechers et al. 2008a, b, Riechers et al. 2009) imply that the molecular gas is extended on kpc scales around the central black hole. It remains puzzling how such amounts of *cold* molecular gas, rich in C and O, can be distributed over many kpc scales on short (few 100 Myr) timescales.

The high FIR luminosity of the quasars, in conjunction with their massive molecular gas reservoirs imply high star formation rates in the host galaxies of SFR > 1000 $M_\odot$ year^{-1}. This finding is corroborated by the detection of [CII] emission in a quasar host at z=6.42 (Maiolino et al. 2005, Walter et al. 2009). In fact, spatially resolved measurements of the [CII] emission were able to constrain the size of the starburst to be emerging from the central $\sim$1.5 kpc of the quasar host. This implies extremely high star formation rate surface densities ($\sim$1000 $M_\odot$ yr^{-1} kpc-2), at the maximum what is allowed theoretically based on Edington–limited star formation of a radiation–pressure–supported starburst (Walter et al. 2009a).

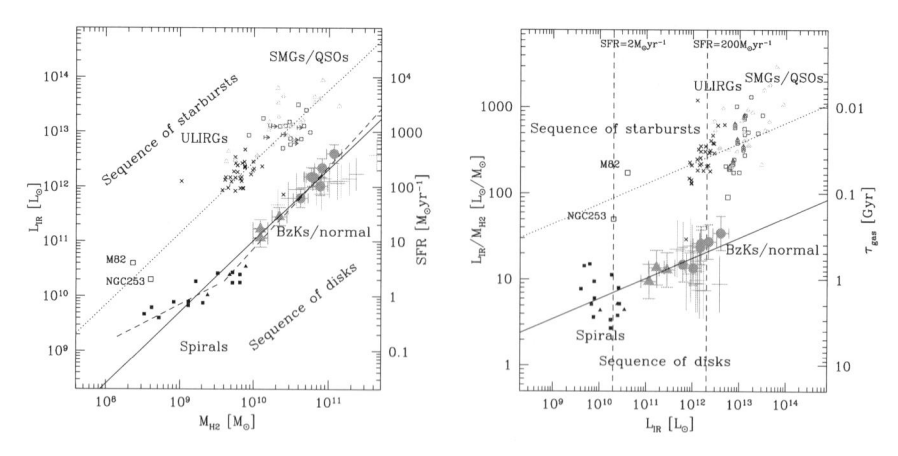

Figure 1: Comparison of molecular gas masses and total IR bolometric luminosities (taken from Daddi et al. 2010b) for all high–redshifted systems detected in CO to date: BzK galaxies (red filled circles; Daddi et al. 2010a), $z \sim 0.5$ disk galaxies (red filled triangles; Salmi et al. in prep.), $z = 1$–2.3 normal galaxies (Tacconi et al. 2010; brown crosses), SMGs (blue empty squares; Greve et al. 2005, Frayer et al. 2005, Daddi et al. 2009a, 2009b), QSOs (green triangles; Riechers et al. 2006, Solomon & Vanden Bout 2005), local ULIRGs (black crosses; Solomon et al. 1997), local spirals (black filled square: Leroy et al. 2009; black filled triangles: Wilson et al. 2009). The two nearby starbursts M82 and the nucleus of NGC 253 are also shown (see Daddi et al. 2010b for references). The solid line (slope of 1.31 in the left panel) is a fit to local spirals and BzK galaxies and the dotted line is the same relation shifted in normalization by 1.1 dex. For guidance, two vertical lines indicate $SFR = 2$ and 200 M$_\odot$ yr^{-1} in the right panel. For more details see Daddi et al. (2010b).

The central black hole masses of the quasars that are currently detectable reach masses of many 10^9 M$_\odot$. If the local relation seen between the central black hole and surrounding bulge mass (e.g., Magorrian et al. 1998, Ferrarese et al. 2000, Gebhardt et al. 2000) were to hold at these high redshifts, one would expect bulge masses of order 10^{12} M$_\odot$. Such high masses exceed dynamical mass estimates of a few sources by more than an order of magnitude (Walter et al. 2004, Riechers et al. 2008a,b, 2009), which in turn indicates that black holes built up their masses more quickly than the stellar bulges (at least in some quasars).

3.3 'Normal' starforming galaxies

More recently, major progress has been made on detecting CO emission from more normal star forming galaxies at high z. These come in the form of near-IR and color selected galaxy samples at $z \sim 1.5$ to 3. The BzK color selection technique in particular has revealed a sample of star forming galaxies at $z \sim 1.5$ to 2.5, selected at near-IR wavelengths, with stellar masses in the range 10^{10} to 10^{11} M$_\odot$, and star formation rates ~ 100 M$_\odot$ yr^{-1} (Daddi et al. 2004). These galaxies have a space density > 10 times those the submm galaxies.

Daddi et al. (2008) showed that star forming BzK galaxies (sBzK) uniformly contain molecular gas reservoirs of comparable mass to the submm galaxies, and yet they are forming stars at ~ 10 times lower rates. This leads to high gas depletion times of up to ~ 1 Gyr (see Fig. 1). Follow–up observations showed that massive gas reservoirs have been detected in each BzK galaxy that was targeted (Daddi et al. 2010a). The implied gas fractions are very high and the gas mass in these galaxies is comparable to or larger than the stellar mass, and the gas accounts for 50–65% of the baryons with the galaxies' half–light radius. New observations of an even larger sample of 'normal' starforming galaxies have confirmed and extended these findings (Tacconi et al. 2010).

In a few cases, the molecular gas emission in these objects could even be spatially resolved (see Tacconi et al. 2010, Fig. 2, for a particularly striking example) allowing for a determination of the dynamical mass. A good constraint on the dynamical mass also allows one to derive an independent measurement of the CO–to–H$_2$ conversion factor (as done in the BzK sample by Daddi et al. 2010a). The interesting result of such an analysis is that the BzK's have a conversion factor similar to the Galaxy, i.e. a factor of ~ 5 larger than what is found in local ULIRGs.

In conclusion these systems likely represent the major star formation epoch in early massive galaxies (presumably driven by cold mode accretion). In the current picture these galaxies may turn into passively evolving galaxies at $z \sim 1$, which may eventually turn into 'red and dead' cluster elliptical galaxies seen today. They are substantially different from the quasar and submillimeter population discussed above in terms of their larger spatial sizes, lower star formation efficiencies, and lower molecular gas excitation (e.g. Dannerbauer et al. 2009, Aravena et al. 2010).

3.4 Different Star Formation Laws at High Redshift?

The detection of molecular gas in systems spanning a wide range of parameters now allows one to investigate the star formation law (i.e. the dependence of the star formation rate on the available gas) at high redshift. In Figure 1 (left), we show the summary plot of Daddi et al. 2010b. Here the infrared luminosities (a proxy for the SFR) is plotted as function of molecular gas mass (M$_{H2}$) of all high–redshift objects detected in CO to date, including submillimeter galaxies, quasars and BzK,BM/BX galaxies (see figure caption for references). As discussed in Daddi et al. (2010b), two different conversion factors (to derive M$_{H2}$) have been used for the high redhift galaxy populations in this plot: the Galactic one for the local spiral galaxies and the BzK/normal galaxies at z$\sim$1.5–2.5 and a lower 'ULIRG' conversion factor for the ULIRS, submillimeter galaxies and quasars. Plotted this way, there are two sequences emerging (each of which can be fitted with a power law of slope $\sim$1.3), one for the 'starbursts' (ULRIGS, SMGs and QSOs) and one for the 'disk' galaxies. It should be stressed that the offset in sequences seen here is not only due to the choice of conversion factors (e.g., Fig. 13 in Daddi et al. 2010a). The right hand panel shows the same data, but the y–axis shows the ratio of infrared luminosity to molecular gas mass (i.e. the the gas depletion time) plotted as a function of total infrared luminosity.

It is obvious from this figure that the gas depletion times in the BzKs/normal galaxies are about an order of magnitude longer than in the systems that are presumably undergoing mergers (such as the ULIRGs, SMGs and QSOs; see also discussion in Genzel et al. 2010). These different star formation efficiencies appear to be regulated by the dynamical timescales of the systems (Daddi et al. 2010b, Genzel et al. 2010).

These initial molecular gas observations of the different population of star forming galaxies during the epoch of galaxy assembly present a remarkable opportunity to study the formation of normal galaxies at critical early epochs, when most of their stars form, and when the dominant baryon mass constituent was gas, rather than stars. The galaxies are typically extended on scales $\sim 1"$ in the optical, providing ideal targets for follow-up high resolution imaging of the gas using the EVLA and ALMA.

4 Outlook: EVLA and ALMA

Major progress in the studies of the molecular gas content of galaxies throughout cosmic times is currently limited by the following technological restrictions:

- *Sensitivity:* Although remarkable progress has been made in recent years, the sensitivities of current interferometers do only allow to study the molecular gas content in objects that have high star formation rates and/or high molecular gas masses (of order $10^{10}\,M_\odot$, SFR$\sim 100\,M_\odot\,\mathrm{yr}^{-1}$). To push this to more typical object at high redshift (with SFR that are an order of magnitude lower) much higher sensitivities are required. ALMA, with its unprecedented collecting area, will without doubt revolutionize the field of quantifying the molecular gas content in high–z systems, e.g. through dedicated studies of molecular line deep fields.

- *Resolution:* Spatially resolved imaging of galaxies remains an important task to derive the extent of the molecular gas emission and to constrain the dynamical masses of the systems (which in turn give independent estimates of the conversion factor to derive molecular gas masses from CO observations), and to search for observational evidence for interactions and/or cold mode accretion. Both EVLA and ALMA will enable such studies at high spatial resolution – but even with these new facilities such studies will be restricted to small samples due to the intrinsic faintness of the sources.

- *Spectral Coverage:* As galaxies are observed at higher and higher redshifts, their molecular lines are redshifted. Since the molecular gas excitation has proven to be different in quasars, submillimeter galaxies and more normal galaxies (e.g., Weiss et al. 2005, 2007a, 2007b, Dannerbauer et al. 2009), it is important that the same transitions (e.g. rotational J transitions of carbon monoxide) are observed to compare the various galaxy populations. The EVLA will enable detailed studies of the ground transitions of CO up to high redshift. On the other hand ALMA will enable studies of molecular gas tracers that are currently very difficult to observe (e.g. [CII], [NII], [OIII] and other fine structure lines, e.g., Walter et al. 2009b).

Acknowledgements

It is our pleasure to acknowledge our collaborators, in particular Frank Bertoldi, Pierre Cox, Helmut Dannerbauer, Roberto Maiolino, Roberto Neri, Dominik Riechers, Ran Wang, and Axel Weiss.

References

Aravena, M., et al. 2010, ApJ, in press

Bigiel, F., Leroy, A., Walter, F., Brinks, E., de Blok, W. J. G., Madore, B., & Thornley, M. D. 2008, AJ, 136, 2846

Blain, A. W., Smail, I., Ivison, R. J., Kneib, J.-P., & Frayer, D. T. 2002, PhysRep, 369, 111

Bournaud, F., Elmegreen, B. G., & Elmegreen, D. M. 2007, ApJ, 670, 237

Bouwens, R. J., et al. 2010, ApJL, 709, L133

Carilli, C. L., et al. 2010, arXiv:1002.3838

Carilli, C. L., et al. 2007, ApJL, 666, L9

Chapman, S. C., Blain, A. W., Smail, I., & Ivison, R. J. 2005, ApJ, 622, 772

Coppin, K. E. K., et al. 2008, MNRAS, 389, 45
  Coppin et al. 2010, MNRAS, to be submitted.

Daddi, E., et al. 2010a, ApJ, 713, 686

Daddi, E., et al. 2010b, arXiv:1003.3889

Daddi, E., Dannerbauer, H., Krips, M., Walter, F., Dickinson, M., Elbaz, D., & Morrison, G. E. 2009b, ApJL, 695, L176

Daddi, E., et al. 2009a, ApJ, 694, 1517

Daddi, E., et al. 2007, ApJ, 670, 156

Daddi, E., Dannerbauer, H., Elbaz, D., Dickinson, M., Morrison, G., Stern, D., & Ravindranath, S. 2008, ApJL, 673, L21

Daddi, E., Cimatti, A., Renzini, A., Fontana, A., Mignoli, M., Pozzetti, L., Tozzi, P., & Zamorani, G. 2004, ApJ, 617, 746

Dannerbauer, H., Daddi, E., Riechers, D. A., Walter, F., Carilli, C. L., Dickinson, M., Elbaz, D., & Morrison, G. E. 2009, ApJL, 698, L178

Dekel, A., et al. 2009, Nature, 457, 451

Dressler, A., Oemler, A., Gladders, M. G., Bai, L., Rigby, J. R., & Poggianti, B. M. 2009, ApJL, 699, L130

Elbaz, D., et al. 2007, A&A, 468, 33

Ferrarese, L., & Merritt, D. 2000, ApJL, 539, L9

Frayer, D. T., et al. 2008, ApJ, 680, L21

Frayer, D. T., et al. 1999, ApJ, 514, L13

Frayer, D. T., Ivison, R. J., Scoville, N. Z., Yun, M., Evans, A. S., Smail, I., Blain, A. W., & Kneib, J.-P. 1998, ApJ, 506, L7

García-Burillo, S., et al. 2006, ApJ, 645, L17

Gebhardt, K., et al. 2000, ApJL, 539, L13

Genzel, R., Baker, A. J., Tacconi, L. J., Lutz, D., Cox, P., Guilloteau, S., & Omont, A. 2003, ApJ, 584, 633

Genzel, R., et al. 2010, arXiv:1003.5180

Greve, T. R., et al. 2005, MNRAS, 359, 1165

Hopkins, A. M., & Beacom, J. F. 2006, ApJ, 651, 142

Ilbert, O., et al. 2010, ApJ, 709, 644

Ivison, R. J., Smail, I., Papadopoulos, P. P., Wold, I., Richard, J., Swinbank, A. M., Kneib, J.-P., & Owen, F. N. 2010, MNRAS, 261

Kennicutt, R. C., Jr. 1998, ApJ, 498, 541

Kereš, D., Katz, N., Weinberg, D. H., & Davé, R. 2005, MNRAS, 363, 2

Knudsen, K. K., Neri, R., Kneib, J.-P., & van der Werf, P. P. 2009, A&A, 496, 45

Krumholz, M. R., McKee, C. F., & Tumlinson, J. 2009, ApJ, 699, 850

Le Floc'h, E., et al. 2005, ApJ, 632, 169

Leroy, A. K., Walter, F., Brinks, E., Bigiel, F., de Blok, W. J. G., Madore, B., & Thornley, M. D. 2008, AJ, 136, 2782

Magdis, G. E., Rigopoulou, D., Huang, J.-S., & Fazio, G. G. 2010, MNRAS, 401, 1521

Magorrian, J., et al. 1998, AJ, 115, 2285

Maiolino, R., et al. 2007, A&A, 472, L33

Maiolino, R., et al. 2005, A&A, 440, L51

Noeske, K. G., et al. 2007, ApJL, 660, L43

Obreschkow, D., & Rawlings, S. 2009, ApJL, 696, L129

Ouchi, M., et al. 2009, ApJ, 696, 1164

Prochaska, J. X., & Wolfe, A. M. 2009, ApJ, 696, 1543

Renzini, A. 2006, ARA&A, 44, 141

Riechers, D. A., et al. 2006, ApJ, 650, 604

Riechers, D. A., Walter, F., Carilli, C. L., Weiss, A., Bertoldi, F., Menten, K. M., Knudsen, K. K., & Cox, P. 2006, ApJL, 645, L13

Riechers, D. A., Walter, F., Cox, P., Carilli, C. L., Weiss, A., Bertoldi, F., & Neri, R. 2007a, ApJ, 666, 778

Riechers, D. A., Walter, F., Carilli, C. L., & Bertoldi, F. 2007b, ApJL, 671, L13

Riechers, D. A., Walter, F., Carilli, C. L., Bertoldi, F., & Momjian, E. 2008a, ApJL, 686, L9

Riechers, D. A., Walter, F., Brewer, B. J., Carilli, C. L., Lewis, G. F., Bertoldi, F., & Cox, P. 2008b, ApJ, 686, 851

Riechers, D. A., et al. 2009, ApJ, 703, 1338

Robaina, A. R., Bell, E. F., van der Wel, A., Somerville, R. S., Skelton, R. E., McIntosh, D. H., Meisenheimer, K., & Wolf, C. 2010, arXiv:1002.4193

Salvaterra, R., et al. 2009, Nature, 461, 1258

Schinnerer, E., et al. 2008, ApJL, 689, L5

Shields, G. A., Menezes, K. L., Massart, C. A., & Vanden Bout, P. 2006, ApJ, 641, 683

Skelton, R. E., Bell, E. F., & Somerville, R. S. 2009, ApJ, 699, L9

Solomon, P. M., Downes, D., Radford, S. J. E., & Barrett, J. W. 1997, ApJ, 478, 144

Solomon, P. M., & Vanden Bout, P. A. 2005, ARA&A, 43, 677

Spergel, D. N., et al. 2007, ApJS, 170, 377

Springel, V., Frenk, C. S., & White, S. D. M. 2006, Nature, 440, 1137

Steidel, C. C., Adelberger, K. L., Giavalisco, M., Dickinson, M., & Pettini, M. 1999, ApJ, 519, 1

Smail, I., Ivison, R. J., & Blain, A. W. 1997, ApJL, 490, L5

Smolčić, V., et al. 2009, ApJ, 690, 610

Tacconi, L. J., et al. 2010, Nature, 463, 781

Tacconi, L. J., et al. 2008, ApJ, 680, 246

Tacconi, L. J., et al. 2006, ApJ, 640, 228

Taniguchi, Y., et al. 2005, PASJ, 57, 165

Tanvir, N. R., et al. 2009, Nature, 461, 1254

Vanden Bout, P. A., Solomon, P. M., & Maddalena, R. J. 2004, ApJ, 614, L97

van der Wel, A., Rix, H.-W., Holden, B. P., Bell, E. F., & Robaina, A. R. 2009, ApJ, 706, L120

van Dokkum, P. G., et al. 2006, ApJ, 638, L59

Walter, F., Riechers, D., Cox, P., Neri, R., Carilli, C., Bertoldi, F., Weiss, A., & Maiolino, R. 2009, Nature, 457, 699

Walter, F., Weiß, A., Riechers, D. A., Carilli, C. L., Bertoldi, F., Cox, P., & Menten, K. M. 2009, ApJL, 691, L1

Walter, F., Carilli, C., Bertoldi, F., Menten, K., Cox, P., Lo, K. Y., Fan, X., & Strauss, M. A. 2004, ApJL, 615, L17

Walter, F., et al. 2003, Nature, 424, 406

Wang, R., et al. 2008, ApJ, 687, 848

Wang, R., et al. 2010, arXiv:1002.1561

Weiss, A., Downes, D., Walter, F., & Henkel, C. 2007a, From Z-Machines to ALMA: (Sub)Millimeter Spectroscopy of Galaxies, 375, 25

Weiß, A., Downes, D., Neri, R., Walter, F., Henkel, C., Wilner, D. J., Wagg, J., & Wiklind, T. 2007b, A&A, 467, 955

Weiß, A., Downes, D., Walter, F., & Henkel, C. 2005, A&A, 440, L45

Wilson, C. D., et al. 2009, ApJ, 693, 1736

Wiklind, T., Dickinson, M., Ferguson, H. C., Giavalisco, M., Mobasher, B., Grogin, N. A., & Panagia, N. 2008, ApJ, 676, 781

X-ray spectroscopy and mass analysis of galaxy clusters

Robert W. Schmidt

Astronomisches Rechen-Institut
Zentrum für Astronomie der Universität Heidelberg
Mönchhofstrasse 12–14, D-69120 Heidelberg, Germany
rschmidt@ari.uni-heidelberg.de

Abstract

An overview of the subject of X-ray imaging-spectroscopy of galaxy clusters and its application to mass analysis is given. It is demonstrated how cluster masses can be measured using the hydrostatic method for dynamically relaxed galaxy clusters. X-ray observations have resulted in a detection of the mass-concentration relation that had been predicted for dark matter haloes by simulations of cold dark matter dominated structure formation. The gas mass fraction of galaxy clusters obtained from X-ray observations provides strong constraints on the distance scale of the universe and yields independent evidence for the presence of a cosmological constant or dark energy in the universe. It is also shown that galaxy clusters feature strongly in the science goals of the already approved and planned future X-ray observatories.

1 X-ray observations of galaxy clusters

Galaxy clusters are the largest collapsed structures in the universe. They consist of galaxies, large amounts of intracluster gas and dark matter.

When viewed with an X-ray telescope galaxy clusters appear as large gaseous objects with with a typical angular extent of up to several arcminutes and temperatures of several keV. The dominant emission process is thermal bremsstrahlung with gas densities of the order of 10^{-3} atoms cm^{-3}. The emission from denser central regions is particularly enhanced since bremsstrahlung is proportional to the square of the gas density (a recent review is Böhringer & Werner 2010).

The X-ray morphology of galaxy clusters varies widely, in many cases substructure is clearly observed. This can be well understood with simulations of cold dark matter dominated structure formation: after the smallest objects formed in the universe at high redshift they continue to merge constantly (e.g., Davis et al., 1985; Springel et al., 2006). Especially since the advent of NASA's Chandra X-ray observatory and its superb spatial resolution of 0.5 arcsec, much detail has been detected in the X-ray morphology of galaxy clusters. The range of phenomena extends from bi-

Figure 1: Archival Chandra image of the galaxy cluster Abell 1835 ($z = 0.252$). The exposure time was 120 ks. The image is 7.5×6.3 arcmin. Note the regular shape of the cluster emission.

or multi-modal cluster structures over shock-heated regions (e.g., Markevitch et al., 2002; Russell et al., 2010), to the so-called cold fronts which separate cold regions from hotter surrounding gas (Markevitch et al., 2000). Objects with clear substructure or multi-modality are often called "unrelaxed" because in these systems the gas is unlikely to have settled into an equilibrium.

In other cases, however, the galaxy clusters appear undisturbed and regular in the X-rays, circular or perhaps slightly elliptical (see Fig. 1). Such clusters are often termed dynamically relaxed or short "relaxed" galaxy clusters because much of the former substructure has virialized and the gas is thought to be in hydrostatic equilibrium.

There is a close connection between relaxed clusters and the so-called cool-core clusters. In such systems, the gas is rather cool in the centres (see Fig. 2, likely due to the high central densities and the short cooling times) and may even accrete into the central galaxies to fuel star formation there (so called cooling-flows, e.g. Crawford et al. 2009; Fabian 1994). In the outer parts of galaxy clusters the temperature is also observed to drop (Leccardi & Molendi, 2008; Pratt et al., 2007). This temper-

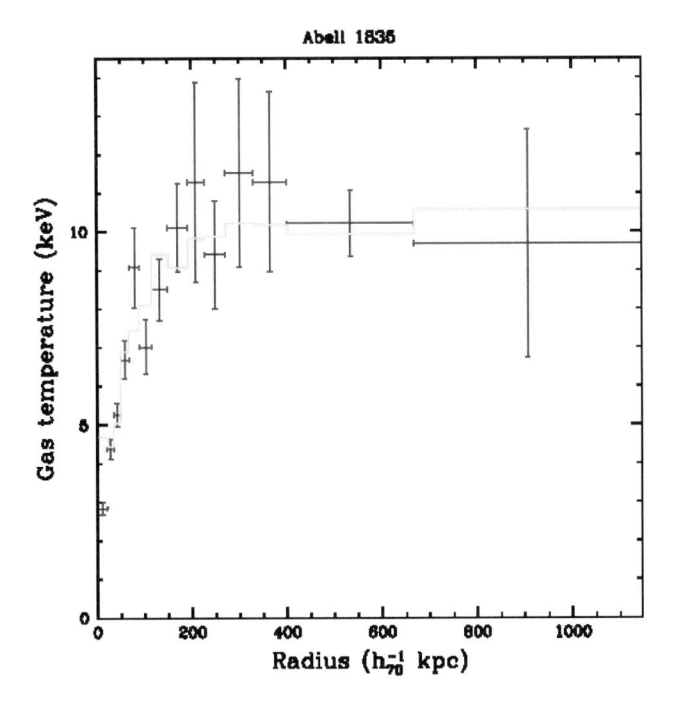

Figure 2: Deprojected temperature profile of the galaxy cluster Abell 1835. The profile was obtained by analysing the data set shown in Fig. 1. The cyan line corresponds to the best-fitting temperature profile and mass model. The mass model used is an NFW mass model.

ature drop towards the outside has recently been particularly well observed with the Japanese/US satellite Suzaku that can trace the cluster emission out to and beyond the virial radius (Bautz et al., 2009; George et al., 2009; Reiprich et al., 2009).

An important result from the grating spectrometer onboard ESA's XMM-Newton satellite (Kaastra et al., 2004; Peterson et al., 2001), however, was that strong limits could be placed on the gas that cools below a few keV. The cooling therefore has to be – at least partly – offset by some mechanism that maintains the gas temperature or reheats gas that cools, and is very likely connected to X-ray holes (also called cavities or bubbles) that are observed as small regions with low flux counts. They are famously observed in the Perseus cluster (see, e.g., Boehringer et al. 1993; Fabian et al. 2000, 2008), but also in other clusters (e.g., Allen et al. 2006; Blanton et al. 2009; Dunn et al. 2010). The bubbles are often seen to coincide with radio jets or relics (e.g., Ensslin et al., 2002) and are now thought to be related to the central AGN and the heating and cooling feedback system in galaxy clusters (e.g., Brüggen et al. 2009; Churazov et al. 2001; Fabian et al. 2006).

2 Mass models from X-ray data

The classic technique to determine galaxy cluster masses in the X-ray regime is the hydrostatic method (e.g., Fabian et al. 1981). This is valid as long as the gas is in hydrostatic equilibrium. For mass analysis studies one chooses therefore primarily relaxed systems. This is a choice that can be made based on fixed criteria, such as regular appearance and a power-law gas density profile. Such a selection minimizes the contribution of non-thermal pressure sources from bulk motions. A further selection can be made based on the comparison of the X-ray mass to lensing masses (e.g., Newman et al. 2009 and references therein) or dynamical mass estimates.

In detail, the mass measurement is done as follows: Firstly, the temperature profile is determined by extracting the X-ray spectra from circular annuli around the cluster centre. These spectra are in fact projected spectra since the correspondings photons were emitted in cylindrical volumes of the cluster. The necessary "deprojection" of the spectra can be accomplished by modelling the projected spectra as the superposition a series of spherical shells, which contribute to the total annulus emission (e.g., the PROJCT model in XSPEC (Arnaud, 1996); see also Russell et al. 2008; Sanders et al. 2007 for a method of direct spectral deprojection). In Fig. 2 the deprojected temperature profile for the galaxy cluster Abell 1835 is shown. Note the cool core and the temperature of $\sim 10\,\mathrm{keV}$ at a radius $r \sim 1$ Mpc.

With the temperature profile in hand, the mass profile of the cluster can be obtained from the hydrostatic equation

$$\frac{1}{\rho_g}\frac{dP}{dr} = -\frac{d\Phi}{dr} = -\frac{GM(r)}{r^2}. \qquad (1)$$

This is usally done in one of two ways: Either (method A) $\rho_g(r)$ and temperature $T(r)$ are somehow obtained from the observations in an analytical form, or (method B) assuming an analytical mass profile $M(r)$, the cluster is divided into spherical shells and the hydrostatic equation is solved from the outside in, in thin spherical shells. For both ways the 3-dimensional gas density needs to be estimated (or deprojected, "onion-peeling method") from the observed X-ray emission using a thermal bremsstrahlung code (such as MEKAL, Kaastra & Mewe 1993; Liedahl et al. 1995). The output is either a numerical estimate for M(r) for method A (e.g., Pointecouteau et al. 2004), or, for method B, a numerical temperature profile $T(r)$ for the "onion shells", which can be compared with the observations to determine the best-fitting mass model (Allen et al., 2001; White et al., 1997). Method B minimizes the need for priors associated with the use of parameterized models for the gas density and/or the temperature profiles. In Fig. 2 the best-fitting numerical temperature profile is indicated with a solid line. For this solution, an analytical Navarro-Frenk-White (Navarro et al., 1997) mass model was used.

3 Testing cold dark matter

One of the most remarkable results of cold dark matter (CDM) simulations of structure formation is that the density profiles of dark matter halos on all resolvable mass scales, from small satellites to the most massive galaxy clusters, can be approximated by a universal profile, the so-called Navarro-Frenk-White (NFW) profile (Navarro et al., 1997)

$$\rho(r) = \frac{\rho_0}{(\frac{r}{r_s})(1 + \frac{r}{r_s})^2}. \qquad (2)$$

Here r_s is the characteristic scale radius of the halo and ρ_0 is a density parameter. NFW also defined the concentration parameter, c, as the ratio of r_{200}, the radius within which the matter density is 200 times the critical density, and the scale radius:

$$c = \frac{r_{200}}{r_s}. \qquad (3)$$

They showed that smaller mass halos are more concentrated than the higher mass halos and interpreted this as reflecting the higher formation redshift of the lower mass systems.

Much numerical work has been devoted to testing these fundamental findings. To explain the mass-concentration relation and its redshift evolution, Bullock et al. (2001) and Eke et al. (2001) introduced simple, but successful models for the formation of dark matter halos. In the Bullock et al. model, for example, only two parameters, K, which determines the initial concentration parameter of collapsing halos and F, the ratio of the initial collapse mass to the final virial mass of the halo at redshift zero, are required to approximately match the simulation predictions. Two- and three-parameter (including redshift evolution) power-laws have been used to characterize the mass-concentration relation (Dolag et al., 2004; Duffy et al., 2008; Shaw et al., 2006).

In Schmidt & Allen (2007) we have analysed the density profiles for 34 of the most massive, dynamically relaxed galaxy clusters known in the redshift range $0 < z < 0.7$ that were observed with the Chandra X-ray Observatory. These clusters are hot ($kT > 5\,\mathrm{keV}$) and X-ray luminous ($L_X > 10^{45}\,\mathrm{erg\,s^{-1}}$). They are among the most promising targets with which to check the predictions of a mass-concentration relation, being both dominated by dark matter (e.g., Allen et al. 2004) and having a size that allows us to resolve well within the scale radius, even at high redshifts. The NFW mass profile leads to acceptable fits in 27 of the 34 clusters and thus provides an overall good fit to the data. We have also tested generalized NFW models with a free power-law exponent of the dark matter density profile for small radii $r \ll r_s$ and find $\alpha = -0.88^{+0.31}_{-0.26}$, in agreement with the prediction $\alpha_{\mathrm{NFW}} = -1.0$ of the NFW model (Eq. 2).

In Fig. 3, a compilation is shown of X-ray determined virial masses and concentration parameters published by us (blue circles) and by Buote et al. (2007, red boxes). The work by Buote et al. also includes a large number of smaller mass group halos and shows the relation down to masses of $\sim 10^{13} M_\odot$. It can be taken from this figure that larger halos have smaller concentration parameters, just as dark matter simulations predict. There is also significant scatter, which is expected from the simulations: The solid line indicates the best-fitting and the 1-σ upper and lower limits found by Duffy et al. (2008) for the concentrations of relaxed halos in their simulations (using the WMAP-5 cosmological parameters, Komatsu et al. 2009). Overall, however, the haloes in the observations appear to be more concentrated than in the

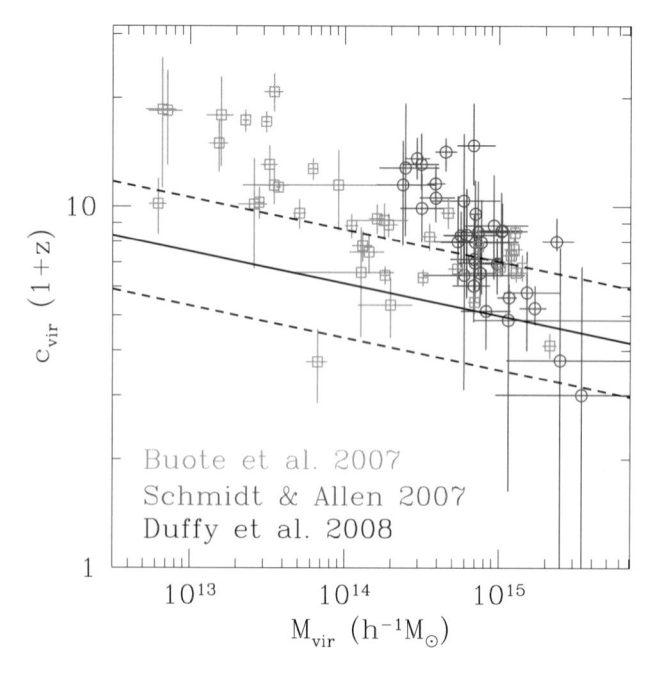

Figure 3: The observed mass-concentration relation for group and cluster haloes based on the work by Buote et al. (2007) and Schmidt & Allen (2007). The solid and dashed lines indicate the location of the best-fitting power-law relation from the simulations by Duffy et al. (2008) and the associated scatter.

simulations. The reason for this is still unknown. In a recent paper Duffy et al. (2010) have shown that baryon physics is probably not to blame for the difference.

For the clusters in our Schmidt & Allen (2007) sample (blue circles in Fig. 3), the mass-concentration relation appears to be significantly steeper (logarithmic slope $\sim -0.4 \pm 0.1$ versus ~ -0.1) than the power-law determined in simulations (the result by Duffy et al. 2008, solid line). We caution, however, that this may be an artifact due to the systematic scatter in the observed data and the limited sample size.

4 Cosmology with gas mass fractions of galaxy clusters

After discussing the application of galaxy cluster masses to dark matter, I will now show that the ratio of the gas mass and the total cluster mass, called the gas mass fraction, in fact is sensitive to the effect of a cosmological constant or dark energy component on the distance scale in the universe.

An important assumption for drawing such conclusions is that the matter content of the largest clusters of galaxies is expected to provide an almost fair sample of the matter content of the Universe (White et al., 1993). The ratio of baryonic-to-

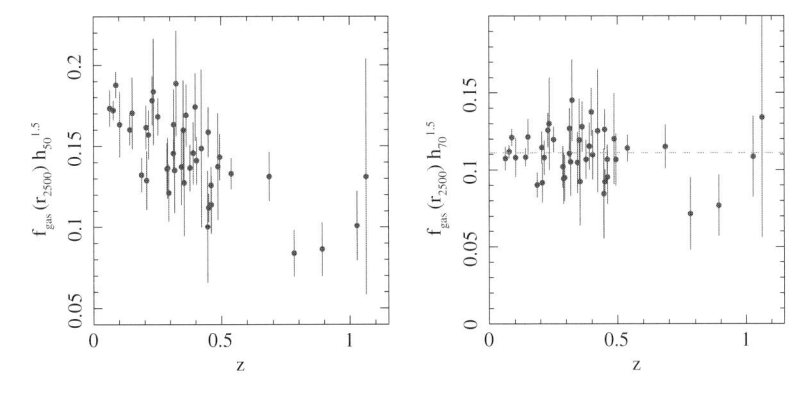

Figure 4: (**Left panel**) Apparent variation of the X-ray gas mass fraction within a radius r_{2500} for the Einstein-de Sitter cosmology. (**Right panel**) The same plot for the ΛCDM cosmology. In this plot the gas mass fraction is consistent with being independent of redshift. h_{50} and h_{70} denote the Hubble constant in units of 50 or 70 km s^{-1} Mpc^{-1}, respectively. Figures from Allen et al. (2008).

total mass in clusters should therefore closely match the ratio of the cosmological parameters $\Omega_{\rm b}/\Omega_{\rm m}$. If $\Omega_{\rm b}$ is obtained independently (either from cosmic microwave background data or from big-bang nucleosynthesis calculations [including the Hubble constant]), one can therefore infer $\Omega_{\rm m}$ by measuring this ratio. Since the baryonic mass content of clusters is dominated by the X-ray emitting gas, this computation can be done by the mass analysis method outlined above (the mass of optically luminous material is about a factor ~ 6 less, e.g., Fukugita et al. 1998).

Importantly, the gas mass fraction computed from the X-ray data depends on the reference cosmology used for the calculation. The dependence on the cosmology can be parametrized by the assumed distance d as follows:

$$f_{\rm gas} = \frac{\rm gas\,mass}{\rm total\,mass} \propto d^{1.5} \tag{4}$$

(Pen, 1997; Sasaki, 1996). We can therefore demand that for the *best-fitting* cosmology the gas mass fraction should be constant with redshift since numerical simulations show that for the largest, dynamically relaxed clusters (beyond the innermost core) $f_{\rm gas}$ should not vary over the observable universe.

This method was carried out by Allen et al. (2008) using Chandra observations of 42 hot, X-ray luminous, dynamically relaxed systems spanning the redshift range $0 < z < 1$. This includes the clusters in Schmidt & Allen (2007) in Sect. 3. Fig. 4 shows the apparent redshift variation of $f_{\rm gas}$ measured at r_{2500} (the radius within which the mean, enclosed density is 2500 times the critical density at the redshifts of the clusters) for an Einstein-de Sitter universe ($\Omega_{\rm m} = 1, \Omega_\Lambda = 0$), and a ΛCDM ($\Omega_{\rm m} = 0.3, \Omega_\Lambda = 0.7$) cosmology. Only the results for ΛCDM show little apparent variation with redshift indicating that the ΛCDM cosmology provides a good fit.

To work out constraints, the $f_{\rm gas}$ data are fitted with a model that takes into account the apparent variation of the $f_{\rm gas}$ values as the cosmological parameters to calculate distances are varied, including the variation of the angular size of r_{2500}

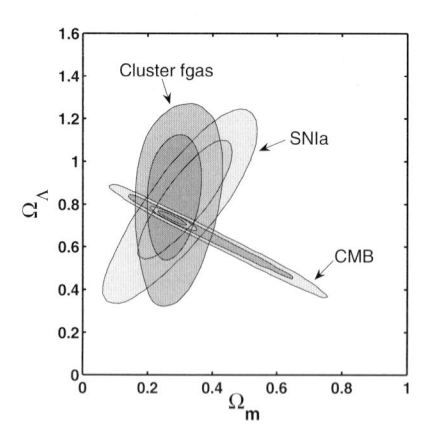

Figure 5: Consistency of the confidence contours (68.4 per cent and 95.4 per cent) in the Ω_m–Ω_Λ plane from the independent techniques X-ray gass mass fraction experiment, Supernovae Ia data and cosmic microwave background (CMB). Plots from Allen et al. (2008).

itself. Various additional factors are also accounted for, such as the depletion of the gas mass fraction at r_{2500} by ~ 17 per cent with respect to the universal mean (a natural consequence of the shock history of the gas, Crain et al. 2007; Eke et al. 1998), a possible moderate evolution of this depletion over the observed redshift range, and the contribution of non-thermal pressure sources due to cosmic ray or magnetic fields (e.g., Nagai et al. 2007). See Allen et al. (2008) for details.

In Fig. 5 the 68 percent and 95.4 per cent confidence regions of this experiment are shown in the Ω_m–Ω_Λ plane (including a prior on the Hubble constant, Freedman et al. 2001). Also shown are the results from Supernova Ia data (Davis et al., 2007) and the corresponding contours from the cosmic microwave background data (from the WMAP–3, CBI, ACBAR and BOOMERanG data sets, see the references in Allen et al. 2008). The results are consistent and show that a region of overlap of all three methods exists (small central contours) that favours a non-vanishing cosmological constant at greater than 99.99 percent confidence. This technique can thus be used in conjunction with these methods (also the Baryon-Acoustic-Oscillations technique, Eisenstein et al. 2005) to tackle even more sophisticated parameterizations of dark energy cosmologies (Allen et al., 2008; Rapetti et al., 2008).

5 Outlook

There are considerable efforts underway to increase the sample of known galaxy clusters in the X-ray regime. The eROSITA instrument (Germany/Russia) aboard the Russian "Spectrum-Roentgen-Gamma" satellite is planned for a launch in 2012 and will carry out an all-sky X-ray survey up to an energy of 10 keV (Predehl et al., 2010). The expected science results include a sample of $>100,000$ active galactic nuclei and more than 50,000 galaxy clusters up to $z \sim 1.3$ (in the spirit of Haiman et al. 2005). Conservative computations show that ≈ 500 new hot and dynamically relaxed clusters could be found, including objects at $z > 1$, and leading to a new quality of cosmology (especially detailed dark energy work) to be done with the gas mass fraction technique (Rapetti et al., 2008).

In the planning phase is the even more ambitious WFXT mission (US, Italy) for which galaxy clusters are also a prime focus (Vikhlinin et al., 2009). Considering the dynamically relaxed galaxy clusters that the already approved eROSITA mission is expected to find at high redshifts $z > 1$, it will be necessary to use an instrument that is sensitive enough to efficiently observe such objects. This will become possible with the next big planned observatory for X-ray studies, the International X-ray Observatory (IXO, Bookbinder 2010; White et al. 2010). This observatory is in fact designed to be a large X-ray space observatory with ~ 3 square meter collection area and a planned image resolution of 5 arcsec – a factor of 20 in collecting area at 1 keV compared to any other previous X-ray observatory. IXO is planned for 2021 and will in all aspects be a revolution for X-ray astronomy.

Acknowledgements

I thank Steve Allen for the collaboration on the galaxy cluster work.

References

Allen, S. W., Ettori, S., & Fabian, A. C. 2001, MNRAS, 324, 877

Allen, S. W., Schmidt, R. W., Ebeling, H., Fabian, A. C., & van Speybroeck, L. 2004, MNRAS, 353, 457

Allen, S. W., Dunn, R. J. H., Fabian, A. C., Taylor, G. B., & Reynolds, C. S. 2006, MNRAS, 372, 21

Allen, S. W., Rapetti, D. A., Schmidt, R. W., Ebeling, H., Morris, R. G., & Fabian, A. C. 2008, MNRAS, 383, 879

Arnaud, K. A. 1996, Astronomical Data Analysis Software and Systems V, 101, 17

Bautz, M. W., et al. 2009, PASJ, 61, 1117

Blanton, E. L., Randall, S. W., Douglass, E. M., Sarazin, C. L., Clarke, T. E., & McNamara, B. R. 2009, ApJ, 697, L95

Boehringer, H., Voges, W., Fabian, A. C., Edge, A. C., & Neumann, D. M. 1993, MNRAS, 264, L25

Böhringer, H., & Werner, N. 2010, A&AR, 18, 127

Bookbinder, J. 2010, arXiv:1003.2847

Brüggen, M., & Scannapieco, E. 2009, MNRAS, 398, 548

Bullock, J. S., Kolatt, T. S., Sigad, Y., Somerville, R. S., Kravtsov, A. V., Klypin, A. A., Primack, J. R., & Dekel, A. 2001, MNRAS, 321, 559

Buote, D. A., Gastaldello, F., Humphrey, P. J., Zappacosta, L., Bullock, J. S., Brighenti, F., & Mathews, W. G. 2007, ApJ, 664, 123

Churazov, E., Brüggen, M., Kaiser, C. R., Böhringer, H., & Forman, W. 2001, ApJ, 554, 261

Crain, R. A., Eke, V. R., Frenk, C. S., Jenkins, A., McCarthy, I. G., Navarro, J. F., & Pearce, F. R. 2007, MNRAS, 377, 41

Crawford, C. S., Allen, S. W., Ebeling, H., Edge, A. C., & Fabian, A. C. 1999, MNRAS, 306, 857

Davis, M., Efstathiou, G., Frenk, C. S., & White, S. D. M. 1985, ApJ, 292, 371

Davis, T. M., et al. 2007, ApJ, 666, 716

Dolag, K., Bartelmann, M., Perrotta, F., Baccigalupi, C., Moscardini, L., Meneghetti, M., & Tormen, G. 2004, A&A, 416, 853

Duffy, A. R., Schaye, J., Kay, S. T., & Dalla Vecchia, C. 2008, MNRAS, 390, L64

Duffy, A. R., Schaye, J., Kay, S. T., Dalla Vecchia, C., Battye, R. A., & Booth, C. M. 2010, arXiv:1001.3447

Dunn, R. J. H., Allen, S. W., Taylor, G. B., Shurkin, K. F., Gentile, G., Fabian, A. C., & Reynolds, C. S. 2010, MNRAS, 343

Eisenstein, D. J., et al. 2005, ApJ, 633, 560

Eke, V. R., Navarro, J. F., & Frenk, C. S. 1998, ApJ, 503, 569

Eke, V. R., Navarro, J. F., & Steinmetz, M. 2001, ApJ, 554, 114

Ensslin, T. A., & Brüggen, M. 2002, MNRAS, 331, 1011

Fabian, A. C., Hu, E. M., Cowie, L. L., & Grindlay, J. 1981, ApJ, 248, 47

Fabian, A. C. 1994, ARA&A, 32, 277

Fabian, A. C., et al. 2000, MNRAS, 318, L65

Fabian, A. C., Sanders, J. S., Taylor, G. B., Allen, S. W., Crawford, C. S., Johnstone, R. M., & Iwasawa, K. 2006, MNRAS, 366, 417

Fabian, A. C., Johnstone, R. M., Sanders, J. S., Conselice, C. J., Crawford, C. S., Gallagher, J. S., III, & Zweibel, E. 2008, Nature, 454, 968

Freedman, W. L., et al. 2001, ApJ, 553, 47

Fukugita, M., Hogan, C. J., & Peebles, P. J. E. 1998, ApJ, 503, 518

George, M. R., Fabian, A. C., Sanders, J. S., Young, A. J., & Russell, H. R. 2009, MNRAS, 395, 657

Haiman, Z., et al. 2005, arXiv:astro-ph/0507013

Kaastra J. S., Mewe R., 1993, Legacy, 3, 16

Kaastra, J. S., et al. 2004, A&A, 413, 415

Komatsu, E., et al. 2009, ApJS, 180, 330

Leccardi, A., & Molendi, S. 2008, A&A, 486, 359

Liedahl, D. A., Osterheld, A. L., & Goldstein, W. H. 1995, ApJ, 438, L115

Markevitch, M., et al. 2000, ApJ, 541, 542

Markevitch, M., Gonzalez, A. H., David, L., Vikhlinin, A., Murray, S., Forman, W., Jones, C., & Tucker, W. 2002, ApJ, 567, L27

Nagai, D., Kravtsov, A. V., & Vikhlinin, A. 2007, ApJ, 668, 1

Navarro, J. F., Frenk, C. S., & White, S. D. M. 1997, ApJ, 490, 493

Newman, A. B., Treu, T., Ellis, R. S., Sand, D. J., Richard, J., Marshall, P. J., Capak, P., & Miyazaki, S. 2009, ApJ, 706, 1078

Pen, U.-L. 1997, New Astronomy, 2, 309

Peterson, J. R., et al. 2001, A&A, 365, L104

Pointecouteau, E., Arnaud, M., Kaastra, J., & de Plaa, J. 2004, A&A, 423, 33

Pratt, G. W., Böhringer, H., Croston, J. H., Arnaud, M., Borgani, S., Finoguenov, A., & Temple, R. F. 2007, A&A, 461, 71

Predehl, P., et al. 2010, arXiv:1001.2502

Rapetti, D., Allen, S. W., & Mantz, A. 2008, MNRAS, 388, 1265

Reiprich, T. H., et al. 2009, A&A, 501, 899

Russell, H. R., Sanders, J. S., & Fabian, A. C. 2008, MNRAS, 390, 1207

Russell, H. R., Sanders, J. S., Fabian, A. C., Baum, S. A., Donahue, M., Edge, A. C., McNamara, B. R., & O'Dea, C. P. 2010, arXiv:1004.1559

Sanders, J. S., & Fabian, A. C. 2007, MNRAS, 381, 1381

Sasaki, S. 1996, PASJ, 48, L119

Schmidt, R. W., & Allen, S. W. 2007, MNRAS, 379, 209

Shaw, L. D., Weller, J., Ostriker, J. P., & Bode, P. 2006, ApJ, 646, 815

Springel, V., Frenk, C. S., & White, S. D. M. 2006, Nature, 440, 1137

Vikhlinin, A., et al. 2009, astro2010: The Astronomy and Astrophysics Decadal Survey, 2010, 305

White, S. D. M., Navarro, J. F., Evrard, A. E., & Frenk, C. S. 1993, Nature, 366, 429

White, D. A., Jones, C., & Forman, W. 1997, MNRAS, 292, 419

White, N. E., Parmar, A., Kunieda, H., Nandra, K., Ohashi, T., & Bookbinder, J. 2010, arXiv:1001.2843

High-fidelity spectroscopy at the highest resolutions

Dainis Dravins

Lund Observatory, Box 43
SE-22100 Lund, Sweden
dainis@astro.lu.se

Abstract

High-fidelity spectroscopy presents challenges for both observations and in designing instruments. High-resolution and high-accuracy spectra are required for verifying hydrodynamic stellar atmospheres and for resolving intergalactic absorption-line structures in quasars. Even with great photon fluxes from large telescopes with matching spectrometers, precise measurements of line profiles and wavelength positions encounter various physical, observational, and instrumental limits. The analysis may be limited by astrophysical and telluric blends, lack of suitable lines, imprecise laboratory wavelengths, or instrumental imperfections. To some extent, such limits can be pushed by forming averages over many similar spectral lines, thus averaging away small random blends and wavelength errors. In situations where theoretical predictions of lineshapes and shifts can be accurately made (e.g., hydrodynamic models of solar-type stars), the consistency between noisy observations and theoretical predictions may be verified; however this is not feasible for, e.g., the complex of intergalactic metal lines in spectra of distant quasars, where the primary data must come from observations. To more fully resolve lineshapes and interpret wavelength shifts in stars and quasars alike, spectral resolutions on order $R = 300\,000$ or more are required; a level that is becoming (but is not yet) available. A grand challenge remains to design efficient spectrometers with resolutions approaching $R = 1\,000\,000$ for the forthcoming generation of extremely large telescopes.

1 Why the highest resolution?

Sophisticated astrophysical models often predict spectral-line shapes with asymmetries and wavelength shifts (e.g., 3-dimensional hydrodynamics in stellar atmospheres; isotopic constituents of interstellar lines; velocity gradients in regions of line formation; or simply the hyperfine structure of the atomic transitions themselves). However, the confrontation with observations is often limited by blends, lack of suitable lines, imprecise laboratory wavelengths, insufficient spectral resolution and instrumental imperfections. Observational limits can be pushed by averaging many similar lines, thus averaging small random blends and wavelength errors, although the resolution cannot be increased beyond that at which the original data are sampled. Instrumental designs are often limited by the optical interface of high-resolution spectrometers to (very) large telescopes with their (very) large image scales.

2 Solar and stellar spectra

Naturally, high-resolution spectroscopy was first explored for the brightest sources such as the Sun and brighter stars. This has also enabled paradigm changes in the analysis of stellar atmospheres in that, contrary to the days of classical one-dimensional homogeneous and static models, it has been realized that it is not possible – not even in principle – to infer detailed stellar properties from analyzing observed line parameters alone, no matter how precisely the spectrum would be measured. Any stellar absorption line is built up by great many contributions from a wide variety of temporally variable inhomogeneities across the stellar surface, whose statistical averaging over time and space produce the line shapes and shifts that can be observed in integrated starlight. While it is possible to compute resulting line profiles (with their asymmetries and wavelength shifts) from hydrodynamic models, the opposite is not feasible because the 3-dimensional structure cannot be uniquely deduced from observed line shapes alone (e.g., Asplund 2005).

While hydrodynamic models may predict detailed properties for a hypothetical spectral line, a confrontation with observations may be unfeasible because stellar lines of the desired species, strength, ionization level, etc., might simply not exist, or be unobservable in practice. In real spectra, lines are frequently blended by stellar rotation, by overlapping telluric lines from the terrestrial atmosphere or else smeared by inadequate instrumental resolution, in practice precluding more detailed studies.

While, in the past, it may have been sufficient to merely resolve the presence of lines, and to determine their strengths, finding line asymmetries implies measurements over smaller fractions of each line-width. Any point on a spectral-line bisector (median) is obtained from intensities at two wavelength positions on either side of the line center. To define not only the bisector slope, but also its curvature requires at least some five points, implying at least ten measurements across the line profile. Given a width of a typical photospheric line of, say, 12 km s^{-1}, an ordinary resolution of $R = 100\,000$ (3 km s^{-1}) can only indicate the general sense of line asymmetry. Both simulated and observed spectra show how the bisectors degrade when the spectral resolutions decrease towards such values (e.g., Dravins & Nordlund 1990b; Allende Prieto et al. 2002; Ramírez et al. 2009). The general appearance of line profiles recorded under different resolutions is shown in Fig. 1.

3 Advancing high-resolution spectroscopy

High spectral resolution alone may not suffice to obtain high-fidelity spectra. The desired line may be blended with stellar or telluric ones; its laboratory wavelength could be uncertain; the data may be noisy, and the line might be distorted due to stellar rotation or oscillations.

High-fidelity spectroscopy at the highest resolutions

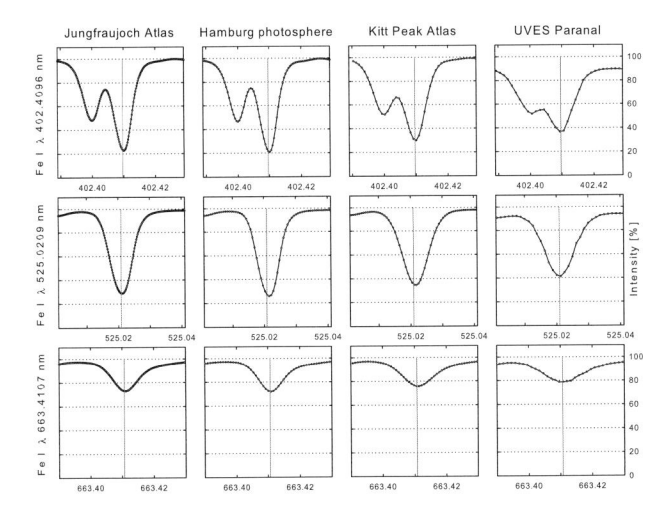

Figure 1: Individual line profiles in solar spectra recorded at different resolutions. Three representative Fe I lines are shown: a deep but somewhat blended line in the violet, a clean line in the green, and a weaker line in the red. Left to right: spectra of solar disk center recorded with a grating spectrometer at $R = \lambda/\Delta\lambda \approx 700\,000$ (Delbouille et al. 1989); solar disk center (Neckel 1999) and integrated sunlight (Kurucz et al. 1984), both with a Fourier-transform spectrometer at $R \approx 350\,000$; and moonlight with a stellar spectrometer at $R \approx 80\,000$ (Bagnulo et al. 2003). Vertical lines mark laboratory wavelengths and thus the 'naively' expected line positions; from Dravins (2008).

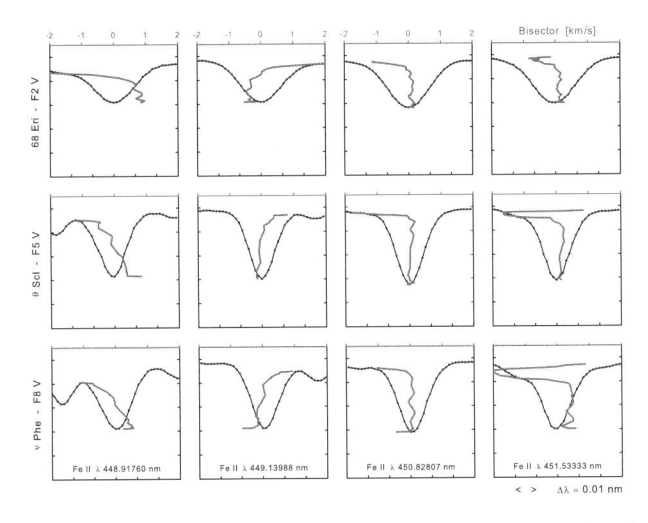

Figure 2: Examples of individual line bisectors (medians), overplotted on the absorption-line profiles, for three representative Fe II lines in $R \approx 80\,000$ spectra (Bagnulo et al. 2003) of different F-type stars: 68 Eri (F2 V), θ Scl (F5 V), and ν Phe (F8 V). The bisector scale (top) is expanded a factor of 10 relative to the wavelength one; from Dravins (2008).

Figure 2 illustrates the difficulty of finding truly 'unblended' lines. Visually, all lines selected appear quite clean but comparing the same line between different stars, it is seen that while bisectors often share some common features, the bisector shapes differ strongly among different lines in the same star (where, under similar conditions of line formation, one would expect them to be similar). Thus, the 'noise' – the cause of bisector deviations from a representative mean – does not originate from photometric errors but instead is largely 'astrophysical' in character, caused by blending lines and similar (Dravins 2008). These spectra represent current high-fidelity spectroscopy, taken from the UVES Paranal Project (Bagnulo et al. 2003), where data of particularly low noise were recorded at resolution $R \approx 80\,000$ with the UVES spectrometer (Dekker et al. 2000) at the ESO VLT Kueyen unit.

3.1 Averages over similar lines

The omnipresence of weak blends makes it impossible to extract reliable asymmetries or shifts from any single line only. However, effects of at least slight blends may be circumvented by forming averages over groups of similar lines which can be expected to have similar physical signatures. Figure 3 shows such average Fe II bisectors for both solar disk center (Delbouille et al. 1989) and for integrated sunlight (Kurucz et al. 1984). The wavelength scale now is absolute, using laboratory data from the FERRUM project (Johansson 2002; Dravins 2008). For solar disk center, 104 lines were found to be clean enough for such averaging, and 93 for integrated sunlight. From such numbers, reasonably well-defined mean bisectors emerge; however, most atomic species have rather fewer lines.

3.2 Spatially resolved stellar spectroscopy

A major milestone in stellar astronomy will be the observation of stars not as point sources, but as extended surface objects. Already, disks of some large stars can be imaged with large telescopes and interferometers, with many more resolvable with future facilities.

3-dimensional models predict spectral changes across stellar disks, differing among various stars and indicating their level of surface 'corrugation' (Dravins & Nordlund 1990a). In stars with 'smooth' surfaces, convective blueshifts decrease from disk center towards the limb, since vertical velocities become perpendicular to the line of sight, and the horizontal ones contributing Doppler shifts appear symmetric. Stars with surface 'hills' and 'valleys' show the opposite, a blueshift increasing towards the limb, where one sees approaching (blueshifted) velocities on the slopes of those 'hills' that are facing the observer. Further effects appear in time variability: on a 'smooth' star, temporal fluctuations are caused by the random evolution of granules. Near the limb of a 'corrugated' star, further variability is added since the swaying stellar surface sometimes hides some granules from direct view.

To realize corresponding observations requires spectrometers with integral-field units and adaptive optics on extremely large telescopes or interferometers. Options for two-dimensional imaging include Fourier transform spectrometers. These pro-

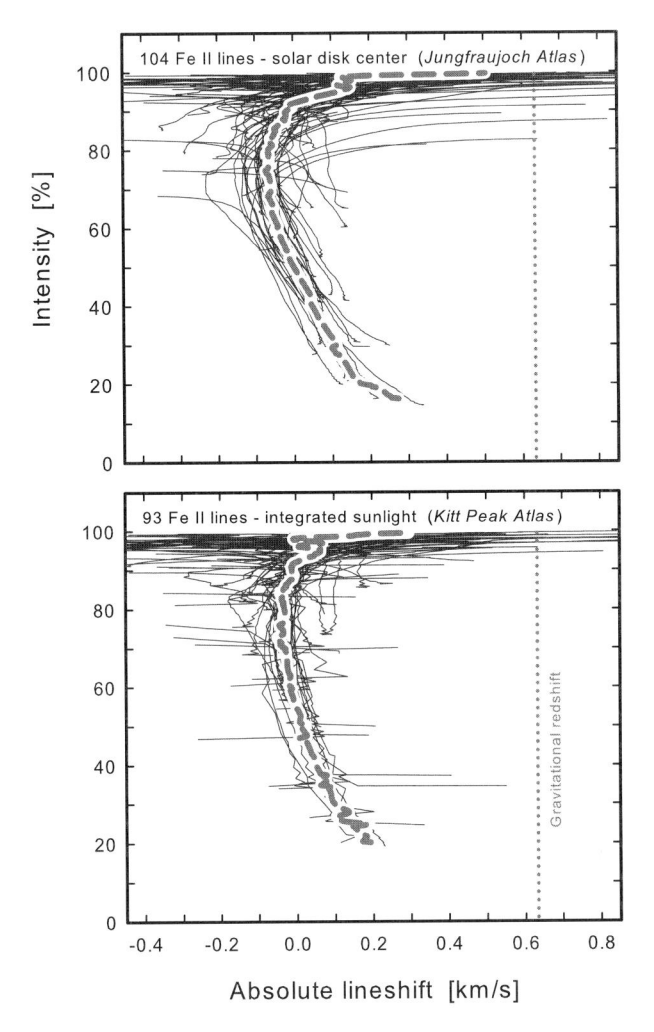

Figure 3: Fe II bisectors in spectra of solar-disk center and of integrated sunlight. Each thin curve is the bisector of one spectral line; the thick dashed is their average. Such averaging over many similar lines (which can physically be expected to have similar intrinsic shapes) is required to reduce effects of weak random blends. The smaller bisector curvature for integrated sunlight reflects both intrinsic line-profile changes across the solar disk, and line broadening due to solar rotation. The vertical scale denotes the intensity in units of the spectral continuum, while the dotted line marks the wavelength expected in a classical 1-dimensional model atmosphere, given by the laboratory wavelength, displaced by the solar gravitational redshift; from Dravins (2008).

vide high spectral resolution for extended objects (e.g., Maillard 2005), although their noise properties have hitherto limited their use in point-source observations

4 Intergalactic wavelength shifts in quasars

Wavelength shifts analogous to those in convective stellar photospheres may be expected also in intergalactic absorption lines from various metals, as seen in quasar spectra. These lines typically form in the intergalactic medium of galaxy clusters that happen to fall in the line of sight towards the more distant quasar. Such intergalactic gas undergoes non-random convective motions qualitatively analogous to those in stellar photospheres (even if the dynamic timescales, now on order 100 million years, are *very* much longer). Active galactic nuclei are normally located near such cluster cores and are sources of energetic cosmic rays, heating the surrounding plasma which becomes buoyantly unstable, thus driving convective motions within clusters of galaxies (e.g., Chandran 2004, 2005; Reynolds et al. 2005; Sharma et al. 2009; Vernaleo & Reynolds 2006).

Corresponding 3-dimensional hydrodynamic models reproduce several features of the observed structures in the intergalactic gas, and should become possible to use as a model input for computing synthetic spectral-line profiles and wavelength shifts. Differential shifts for lines formed within the same cluster, among those with different oscillator strength, ionization level, etc., could then diagnose intergalactic hydrodynamics, even for the distant Universe. For example, lines formed in the deeper gravitational potential closer to the cluster cores would be expected to be slightly more gravitationally redshifted than lines formed further out. Lines of high excitation potential may be expected to predominantly form in hotter gas (where more atoms are in high-excitation states). If these lines are strong, they will predominantly form in the near-side of the clusters, and therefore be seen as more blueshifted (the convective flow of hot gas is mainly radially outward in the cluster, thus approaching and blueshifted on its near-side, while the observable contributions from the receding and redshifted flow on the far side are weakened due to self-absorption in any stronger line). Such shifts must be modeled and segregated from other possible shifts – perhaps due to cosmologically changing 'constants' – if such other shifts are to be reliably deduced.

Even if such lineshift modeling is not yet available, one can try to estimate plausible orders of magnitude. If the analogy to stellar line formation holds, one could expect wavelength shifts due to the different statistical weighting of line contributions from different inhomogeneities on the order of perhaps 1% of the 'turbulent' line broadening. Since predicted gas-flow velocities are on order $50–100\,\mathrm{km\,s^{-1}}$, this implies lineshifts on order $500–1000\,\mathrm{m\,s^{-1}}$. In order to segregate such intrinsic shifts in complex spectra, to map depth structure from multiple line components or perhaps even to map lateral structure from secular time changes, one clearly will need resolving powers of at least 300 000, and ideally even 1 000 000. Further, since the sources are faint while low-noise spectra are required, extremely large telescopes are probably called for.

Telescope	SALT	Keck I	VLT Kueyen	HET	Subaru	LBT
Diameter [m]	10	10	8.2	9.2	8.2	2 × 8.4
Spectrometer	HRS	HIRES	UVES	HRS	HDS	PEPSI
Maximum R	65,000	84,000	110,000	120,000	160,000	320,000
Wavelengths [μm]	0.37–0.89	0.3–1.0	0.3–1.1	0.39–1.1	0.3–1.0	0.38–0.91

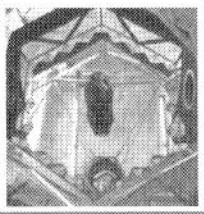

Figure 4: Existing or planned high-resolution spectrometers for the visual range at existing 8–10 m class telescopes.

5 Spectrometers at the largest telescopes

Existing or planned high-resolution stellar spectrometers for the visual at current very large telescopes are summarized in Fig. 4. While some specialized instruments ($R \approx 10^6$) have been used in particular to measure interstellar lines (e.g., the Ultra-High-Resolution Facility at the Anglo-Australian Telescope; Diego et al. 1995), the currently only seriously high-resolution night-time instrument foreseen at any very large telescope is PEPSI (Potsdam Echelle Polarimetric and Spectroscopic Instrument) for the Large Binocular Telescope, planned to reach $R = 320\,000$ (Strassmeier et al. 2003). Among the PEPSI science cases (Strassmeier et al. 2004), diagnosing 3-dimensional stellar hydrodynamics was examined, noting that studies of bisector curvature indeed require such resolutions. Steffen & Strassmeier (2007) further discuss the PEPSI 'deep spectrum' project for the highest quality spectra for any star other than the Sun.

At the largest telescopes, challenges in highest-resolution spectroscopy stem from the difficulty to optically match realistically-sized grating spectrometers to the large image scales, i.e., to squeeze starlight into a narrow spectrometer entrance slit while avoiding unrealistically large optical elements (Spanò et al. 2006). The use of image slicers would limit the number of measurable spectral orders although some remedy could be offered by adaptive optics (Ge et al. 2002; Sacco et al. 2004).

Pasquini et al. (2005) and D'Odorico et al. (2007) discuss future spectrometers. High efficiency requires large optics: the two-grating mosaic for PEPSI makes up a 20×80 cm R4 echelle, while ESPRESSO for the VLT combined focus, and CODEX for the E-ELT aim at twice that: 20×160 cm (Pasquini et al. 2008; Spanò et al. 2008). Nevertheless, resolutions would not reach much above $R = 100\,000$.

5.1 Avoiding telluric absorption?

Even with superlative spectral resolution and photometric precision, spectra are distorted by telluric absorption and emission lines and bands, compromising the spectral fidelity. Actually, in some wavelength regions, the level of contamination makes it doubtful whether it will ever be possible to record truly high-fidelity solar or stellar spectra from the ground.

In some regions, telluric lines from, say, H_2O, are sharp and well confined in wavelength, and then (at least in principle) can be identified and removed in the analysis (e.g., Hadrava 2006). However, an accurate removal requires the terrestrial lines to be spectrally resolved, and not even the 'high' resolution in solar atlases is fully adequate here. Inspection of telluric lines in clean spectral regions of the Kitt Peak Atlas of integrated sunlight (Kurucz et al. 1984) shows that these lines are still unresolved (for an example, see λ 1250.7 nm), instead displaying the characteristic 'ringing' instrumental profile of a Fourier transform spectrometer, even at the resolution $R = 500\,000$.

There are other and more treacherous telluric features, extending over wider wavelength regions. These include diffuse absorptions due to ozone O_3 and the oxygen dimer $[O_2]_2$ that appear throughout the visual. Ozone produces the atmospheric transmission cutoff in the near ultraviolet, where old stellar spectrograms can actually be used to infer past amounts of terrestrial ozone (Griffin 2005).

Only in a few cases has it been possible to compare high-resolution spectra recorded from above and beneath the atmosphere. Sirius was observed with the Goddard High-Resolution Spectrograph on the Hubble Space Telescope in the ultraviolet up to 320 nm, while ground-based observations reach down to 305 nm. The overlapping interval is very instructive in demonstrating how the apparent continuum in ground-based recordings can locally be wrong by some 10% due to diffuse telluric absorptions (Griffin 2005, and private comm.).

Kurucz (2005) computed synthetic telluric spectra, including O_3 and $[O_2]_2$, showing how they affect the highest-resolution solar spectra. The richness of the telluric spectra means that weak lines with depths of perhaps only some percent or less are often superposed onto the flanks or cores of solar ones, making the reduction for their effects awkward or practically impossible (especially since atmospheric extinction is variable in time; Stubbs et al. 2007). In fact, for many spectral regions, this sets a signal-to-noise limit on the order of 100 or worse, irrespective of the photometric precision reached. The affected regions can be inspected in a spectral atlas of telluric absorption ($R \approx 300\,000$), derived from solar data by Hinkle et al. (2003).

Observations from orbit would eliminate telluric absorptions (although geocoronal emission would still be present), but any space mission for high-resolution stellar spectroscopy would probably be complex, considering not only that a large telescope is needed but also that the Doppler shift induced by the spacecraft motion implies continuous wavelength changes.

5.2 Accurate wavelength calibration?

Grating and Fourier transform instruments establish their wavelength scales through different schemes, and each is likely to have its characteristic signatures in its wavelength noise.

Grating spectrometers commonly use an emission-line lamp (thorium, etc.), and thus their wavelength scale depends upon the accuracy of the corresponding laboratory wavelengths and on how well the lamp at the telescope reproduces the laboratory sources. Traditionally, the wavelength noise among individual lines may have been around 100 m s^{-1} but recent calibrations of thorium-argon hollow-cathode lamps have now decreased the internal scatter to some 10 m s^{-1} (Lovis & Pepe 2007), further permitting a selection of subsets of lines for best calibration (Murphy et al. 2007a).

Fourier transform spectrometers have their specific issues. The interferogram is recorded sequentially for its sinusoidal components, between low Fourier frequencies and high ones (how high determines the spectral resolution). However, the realities of detectors and photon noise normally preclude the entire spectrum being measured at once; instead only one piece at a time must be selected by some pre-filter or similar device. The transmission functions of these pre-filters have to be precisely calibrated to enable a correct continuum level to be set. Its placement to better than 1:100 is no trivial task, as illustrated by the efforts to improve the continuum in the Kitt Peak solar atlases by Neckel (1999) and Kurucz (2005).

5.3 Future high-fidelity spectroscopy?

Although several parameters limiting highest-fidelity spectroscopy can be identified, their immediate remedies cannot. Avoiding the terrestrial atmosphere requires ambitious space missions, and improving laboratory wavelengths requires dedicated long-term efforts. Possibly, the least difficult part is to improve instrumental calibrations.

Laboratory devices, in particular the laser frequency comb, permit remarkably precise wavelength determinations, and are also envisioned for future astronomical spectrometers (Araujo-Hauck et al. 2007; Murphy et al. 2007b; Steinmetz et al. 2008). However, it is not necessarily the stability of the calibration device that limits the accuracy in astrophysical spectra because it may in addition require identical light paths for the source and its calibration, and without the calibration light contaminating the source spectrum.

The telescope-spectrometer interface is another issue. Analyses of the UVES spectrometer during asteroid observations reveal a noise of typically 10–50 m s^{-1}, apparently caused by a non-uniform and variable illumination in the image projected by the telescope onto the spectrometer entrance slit (Molaro et al. 2008). Given the chromatic nature of atmospheric dispersion, the wavelength dependence of such shifts could also mimic line-depth dependences since stronger atomic lines often occur at shorter wavelengths. Although a more uniform illumination is provided by image slicers or fiber-optics feeds, their use becomes awkward if one requires both extended spectral coverage and the use of (very) large telescopes. For non-solar work, the former requires multi-order echelle spectrometers, and the latter implies (very) large image scales. Light from all image slices, projected onto the focal plane, would overlap adjacent echelle orders (precluding the recording of extended spectral regions), and an optical fiber would need to have a large diameter to embrace most starlight, requiring entrance apertures that are too wide for adequate spectral resolution (or else cause severe light losses). As already mentioned, possible solutions that avoid the construction of huge instruments could include spectrometers with adaptive optics.

Some instrumentation planning committees appear to concentrate on maximum light efficiency for future instruments. While of course desirable, it is a fallacy to believe that highest optical efficiency would be crucial to scientific discovery: what is required is *adequate* efficiency to reveal novel features in astronomical sources. Still, a grand challenge remains in designing an efficient truly high-resolution ($R \approx 10^6$) and high-fidelity spectrometer for future extremely large telescopes!

Acknowledgements

This work is supported by The Swedish Research Council and The Royal Physiographic Society in Lund. It used data from the UVES Paranal Observatory Project of the European Southern Observatory (ESO DDT Program ID 266.D-5655). It also used solar spectral atlases obtained with the Fourier Transform Spectrometer at the McMath/Pierce Solar Telescope situated on Kitt Peak, Arizona, operated by the National Solar Observatory, a Division of the National Optical Astronomy Observatories.

References

Allende Prieto, C., Asplund, M., García López, R.J., Lambert, D.L.: 2002, ApJ 567, 544

Araujo-Hauck, C., Pasquini, L., Manescau, A., et al.: 2007, ESO Messenger 129, 24

Asplund, M.: 2005, ARA&A 43, 481

Bagnulo, S., Jehin, E., Ledoux, C., et al.: 2003, ESO Messenger 114, 10, http://www.sc.eso.org/santiago/uvespop

Chandran, B.D.G.: 2004, ApJ 616, 169

Chandran, B.D.G.: 2005, ApJ 632, 809

Dekker, H., D'Odorico, S., Kaufer, A., Delabre, B., Kotzlowski, H.: 2000, in: M. Iye, A.F. Moorwood (eds.), *Optical and IR Telescope Instrumentation and Detectors*, SPIE 4008, p. 534

Delbouille, L., Roland, G., Neven, L.: 1989, *Atlas Photométrique du Spectre Solaire de λ 3000 a λ 10000*, Université de Liège, Institut d'Astrophysique, Liège

Diego, F., Fish, A.C., Barlow, M.J., et al.: 1995, MNRAS 272, 323

D'Odorico, V., the CODEX/ESPRESSO team: 2007, Mem. Soc. Astron. Ital. 78, 712

Dravins, D.: 2008, A&A 492, 199

Dravins, D., Nordlund, Å.: 1990a, A&A 228, 184

Dravins, D., Nordlund, Å.: 1990b, A&A 228, 203

Ge, J., Angel, J.R.P., Jacobsen, B., et al.: 2002, PASP 114, 879

Griffin, R.E.: 2005, PASP 117, 885

Hadrava, P.: 2006, A&A 448, 1149

Hinkle, K.H., Wallace, L., Livingston, W.: 2003, BAAS 35, 1260

Johansson, S.: 2002, in: H. Rickman (ed.), *Highlights of Astronomy*, Vol. 12, p. 84

Kurucz, R.L.: 2005, Mem. Soc. Astron. Ital. Suppl. 8, 189

Kurucz, R.L., Furenlid, I., Brault, J.: 1984, *Solar Flux Atlas from 296 to 1300 nm*, National Solar Observatory, New Mexico

Lovis, C., Pepe, F.: 2007, A&A 468, 1115

Maillard, J.-P.: 2005, in: H.U. Käufl et al. (eds.), *High Resolution Infrared Spectroscopy in Astronomy*, p. 528

Molaro, P., Levshakov, S.A., Monai, S., et al.: 2008, A&A 481, 559

Murphy, M.T., Tzanavaris, P., Webb, J.K., Lovis, C.: 2007a, MNRAS 378, 221

Murphy, M.T., Udem, Th., Holzwarth, R., et al.: 2007b, MNRAS 380, 839

Neckel, H.: 1999, Solar Phys. 184, 421

Pasquini, L., Cristiani, S., Dekker, H., et al.: 2005, ESO Messenger 122, 10

Pasquini, L., Avila, G., Dekker, H., et al.: 2008, in: I.S. McLean, M.M. Casali (eds.), *Ground-Based and Airborne Instrumentation for Astronomy II*, SPIE 7014, p. 70141I

Ramírez, I., Allende Prieto, C., Koesterke, L., et al.: 2009, A&A 501, 1087

Reynolds, C.S., McKernan, B., Fabian, A.C., Stone, J.M., Vernaleo, J.C.: 2005, MNRAS 357, 242

Sacco, G.G., Pallavicini, R., Spanò, P., et al.: 2004, in: D.B. Calia, B.L. Ellerbroek, R. Ragazzoni (eds.), *Advancements in Adaptive Optics*, SPIE 5490, p. 398

Sharma, P., Chandran, B.D.G., Quataert, E., Parrish, I.J.: 2009, ApJ 699, 348

Spanò, P., Zerbi, F.M., Norrie, C.J., et al.: 2006, AN 327, 649

Spanò, P., Delabre, B., Dekker, H., Avila, G.: 2008, in: I.S. McLean, M.M. Casali (eds.), *Ground-Based and Airborne Instrumentation for Astronomy II*, SPIE 7014, p. 70140M

Steffen, M., Strassmeier, K.G.: 2007, AN 328, 632

Steinmetz, T., Wilken, T., Araujo-Hauck, C., et al.: 2008, Sci 321, 1335

Strassmeier, K.G., Hofmann, A., Woche, M.F., et al.: 2003, in: S. Fineschi (ed.), *Polarimetry in Astronomy*, SPIE 4843, p. 180

Strassmeier, K.G., Pallavicini, R., Rice, J.B., Andersen, M.I.: 2004, AN 325, 278

Stubbs, C.W., High, F.W., George, M.R., et al.: 2007, PASP 119, 1163

Vernaleo, J.C., Reynolds, C.S.: 2006, ApJ 645, 83

Spectroscopy of Solar Neutrinos

Michael Wurm[1], Franz von Feilitzsch[1], Marianne Göger-Neff[1],
Tobias Lachenmaier[1,2], Timo Lewke[1], Qurin Meindl[1], Randolph Möllenberg[1],
Lothar Oberauer[1], Walter Potzel[1], Marc Tippmann[1], Christoph Traunsteiner[1,2],
and Jürgen Winter[1]

[1]Physik-Department E15, Technische Universität München
James-Franck-Str., D-85748 Garching, Germany
michael.wurm@ph.tum.de

[2]Excellence Cluster Universe, Technische Universität München
Boltzmannstr. 2, D-85748 Garching, Germany

Abstract

In the last years, liquid-scintillator detectors have opened a new window for the observation of low-energetic astrophysical neutrino sources. In 2007, the solar neutrino experiment Borexino began its data-taking in the Gran Sasso underground laboratory. High energy resolution and excellent radioactive background conditions in the detector allow the first-time spectroscopic measurement of solar neutrinos in the sub-MeV energy regime. The experimental results of the Beryllium 7 neutrino flux measurements (Arpesella et al. 2008b) as well as the prospects for the detection of solar Boron 8, pep and CNO neutrinos are presented in the context of the currently discussed ambiguities in solar metallicity. In addition, the potential of the future SNO+ and LENA experiments for high-precision solar neutrino spectroscopy will be outlined.

1 Introduction

Solar neutrinos offer a unique way to study the fusion processes in the center of our Sun. The bulk of the released energy, about 98 %, is converted into photons, that take about 10^5 years to reach the solar surface. Contrariwise, neutrinos released in solar fusion reactions (Sect. 2) leave the Sun unimpeded, taking seconds to arrive at the solar surface and minutes for their travel to the Earth. This free-streaming is based on the fact that neutrinos are only subject to weak nuclear forces, featuring typical interaction cross sections of 10^{-44} cm^2.

The inherently low cross sections pose the main challenge to neutrino detection: In solar neutrino experiments, target masses of several tens or hundreds of tons are required to extract a meaningful signal. A variety of solar neutrino detectors has been devised since the late 1960s (Sect. 3): While in the early stages only integral measurements of the solar neutrino flux were possible, state-of-the-art detectors allow for a time and energy-resolved measurement. The most recent experiment to go into operation is Borexino, a large-volume liquid-scintillator detector allowing the real-time detection of neutrinos in the sub-MeV range (Sect. 4).

The main concern of solar neutrino detection was for a long time to reveal the particle properties of the neutrino themselves. After the establishment of neutrino oscillations, experiments are now arriving at a stage at which information on the Sun can be extracted from the measured neutrino spectrum. Recently, it has been proposed that neutrinos could be used as indicators of the metallicity of the solar core, as helioseismological and spectroscopic evidence is incongruent (Sect. 5). While the available neutrino data is for the moment inconclusive, refinements of the measurements of current experiments might improve the situation. Moreover, future experiments like the SNO+ and LENA detectors will further improve the precision of the solar neutrino measurement. Section 6 will highlight the range of open questions in particle and solar physics that these experiments will be able to answer.

2 The solar neutrino spectrum

The Sun is one of the strongest natural neutrino sources on Earth, producing a flux of 6.6×10^{10} per cm^2 s (Bahcall, Serenelli & Basu 2006). Neutrinos are generated as a byproduct of the fusion processes fueling the solar energy production, mainly by the proton-proton (pp) fusion chain, but to a small extent (1–2 %) also by the catalytic CNO cycle. The net reaction of both processes is

$$4\,p \rightarrow {}^4He + 2\,e^+ + 2\,\nu_e + 24.7\,MeV.$$

Two neutrinos are generated per completed ^{4}He fusion.

Inside the pp chain, most of the neutrinos are generated in the basic fusion reaction of two protons to one deuteron (compare Fig. 1). These so-called pp neutrinos appear at low energies in the solar neutrino spectrum depicted in Fig. 2, the maximum neutrino energy limited to 420 keV. However, side branches of the pp chain allow for the production of neutrinos at higher energies. Depending on the reaction kinematics, the resulting neutrinos can either be distributed over a broad spectral range (in case of a three-particle final state), or are produced at fixed energy (if there are only two particles involved). The corresponding neutrinos are called after their production reaction; the line energies or spectral endpoint, respectively, are listed in Table 1.

The competing CNO cycle fulfills the same net reaction as the pp chain, and as a consequence two neutrinos are generated per ^{4}He fusion. There are several possible catalytic processes that are nevertheless all based on the addition of protons to a ^{12}C nucleus: The corresponding neutrinos are emitted at medium energies (Fig. 2, Table 1).

The predictions for the flux contributions of different solar fusion reactions to the neutrino spectrum are based on the Standard Solar Model (SSM) (Bahcall et al. 2006). The results from spectroscopic and helioseismological observations of

Spectroscopy of Solar Neutrinos

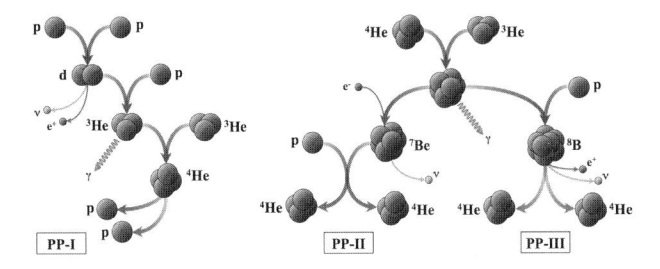

Figure 1: The basic reactions of the pp chain: In the pp-I chain, two protons are fused to a deuteron, releasing a positron and a ν_e, the so-called pp neutrino. While the pp-II chain produces a monoenergetic neutrino by the electron-induced conversion of a ^{7}Be to a ^{7}Li nucleus, in the pp-III chain high-energetic neutrinos are released by the decay of a ^{8}B nucleus.

Table 1: Energy and predicted flux for the solar neutrinos produced along the pp chain and the CNO cycle. The fluxes are derived assuming high (GS) and low (AGS) solar metallicity. Relative uncertainties are given in brackets. The last column shows the relative difference Δ in flux predictions. The fluxes are presented in units of 10^{10} (pp), 10^9 (^{7}Be), 10^8 (pep, ^{13}N, ^{15}O), 10^6 (^{8}B, ^{17}F), 10^3 (hep) cm^{-2}s^{-1} (Peña-Garay & Serenelli 2008).

Source	Energy [MeV]	Neutrino Flux [see caption] BPS08(GS)	BPS08(AGS)	Δ
pp	≤ 0.420	5.97 (0.6%)	6.04 (0.5%)	1.2%
pep	1.442	1.41 (1.1%)	1.45 (1.0%)	2.8%
hep	≤ 18.773	7.90 (15%)	8.22 (15%)	4.1%
^{7}Be	0.862	5.07 (6%)	4.57 (6%)	10%
^{8}B	≤ 14.6	5.94 (11%)	4.72 (11%)	21%
^{13}N	≤ 1.199	2.88 (15%)	1.89 ($^{+14\%}_{-13\%}$)	34%
^{15}O	≤ 1.732	2.15 ($^{+17\%}_{-16\%}$)	1.34 ($^{+16\%}_{-15\%}$)	31%
^{17}F	≤ 1.740	5.82 ($^{+19\%}_{-17\%}$)	3.25 ($^{+16\%}_{-14\%}$)	44%

the Sun are used as input parameters. For a long time, astronomical data was self-consistent, while the measured electron neutrino rates were a factor 2 to 3 lower than the expected values. As will be described in Sect. 3, this inconsistency was finally resolved by the discovery of neutrino oscillations that are responsible for the observed deficit.

Currently, there are two different predictions for the observable solar neutrino flux that are based on the deferring values for solar metallicity derived from optical spectroscopy and helioseismology (Sect. 5) (Peña-Garay & Serenelli 2008). The fluxes are presented in Table 1, including the theoretical uncertainties. The question whether a precision measurement of solar neutrino fluxes might resolve this ambiguity will be discussed in Sect. 5.

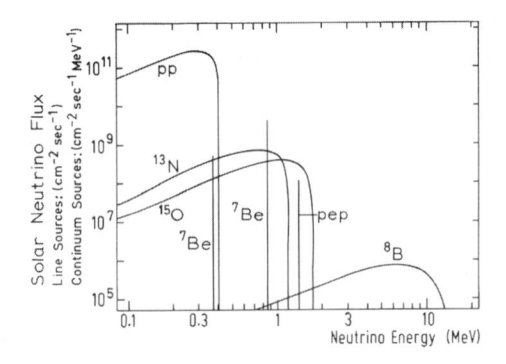

Figure 2: The solar neutrino spectrum. While pp-neutrinos are the dominant contribution to the spectrum at low energies, the spectrum of ^{8}B-neutrinos reaches a maximum energy of 15 MeV. ^{7}Be and pep-neutrinos are monoenergetic and contribute as lines to the spectrum. Also shown are the two dominant neutrino fluxes produced by the CNO cycles (Bahcall et al. 2005).

3 Past and present experiments

In 1968, the Homestake "Chlorine" experiment was the first to measure the neutrinos originating from solar fusion processes, an achievement for which Ray Davis was awarded the Nobel Prize in 2002 (Davis 1994). It was the first of a series of experiments that relied on a neutrino-induced conversion of single nuclei of the element making up the target: In the case of the Homestake experiment, the reaction

$$\nu_e + {}^{37}\text{Cl} \rightarrow {}^{37}\text{Ar} + e^-$$

was taking place in a target volume of 615 t of liquid tetrachlorethylene. The minimum neutrino energy required for the conversion is 814 keV, excluding pp-neutrinos from the reaction. Every month, the produced Argon was removed from the target and measured by its re-decay to Chlorine. The corresponding neutrino flux was about a factor of 3 smaller than the SSM suggested (Davis 1994).

Two follow-up experiments, the European GALLEX experiment (later GNO) (Hampel et al. 1999; Altmann et al. 2005) and the Russian-American SAGE experiment (Abdurashitov et al. 2009) were based on a similar reaction on Gallium and provided an energy threshold low enough to include also the pp neutrinos in the measurement. Again, the measured rate was about half the expectation (Altmann et al. 2005).

These radiochemical experiments had in common that they performed only an integral measurement of the solar neutrino flux, both in time and energy, as the calculated solar neutrino rate is based on counting the overall number of converted nuclei in between two extractions. Opposed to that, since the early 1980's the Japanese Kamiokande and later on Super-Kamiokande experiments were able to detect neutrinos on an event-by-event basis (Suzuki 1995; Fukuda et al. 2001): This realtime detectors used the elastic neutrino-electron scattering as detection reaction in a huge water tank. Photomultipliers mounted to the tank walls registered the Cherenkov light emitted by the recoil electrons, allowing an energy and direction-resolved measurement: The amount of light is proportional to the electron energy, while the light is emitted in a cone centered on the electron trajectory, very much like a supersonic Mach-cone. As electron and incident neutrino direction are closely correlated, the neutrino events could be clearly identified by their tracks pointing back towards the Sun. The measured rate was considerably lower than the predicted one (Fukuda et al. 2001).

By the time the Super-Kamiokande solar data was released, there were already hints from atmospheric neutrino observation that the observed deficit in solar neutrino rates was not due to an incomplete understanding of solar neutrino production but to an intrinsic property of the particles themselves (Fukuda et al. 1998): In the Standard Model of particle physics, every charged lepton, electron e, muon μ, and tauon τ, is accompanied by its own neutrino via weak interactions. The three kinds or "flavors" of neutrinos (ν_e, ν_μ, ν_τ) were thought to be massless particles well separated from each other. However, the new results pointed towards the possibility that the neutrinos had the ability to periodically change flavor and to convert into each other, a phenomenon described as neutrino oscillations (e.g. Kayser 2008).

Up to this point, all solar neutrino experiments had been exclusively or primarily sensitive to electron neutrinos. Oscillations could therefore explain the deficit in the observed event rates. The final proof was obtained in 2002, when two experiments independently confirmed the oscillation hypothesis: The Sudbury Neutrino Observatory (SNO), a water-Cherenkov detector based on heavy water, was able to measure the overall solar neutrino flux by interactions on deuterons, independent of the neutrino flavor. The sum result corresponded to the SSM prediction (Ahmad et al. 2002). At the same time, the liquid-scintillator detector KamLAND discovered the disappearance of electron antineutrinos emitted by the reactors of Japanese nuclear power plants: The observed energy dependence perfectly matched the predictions from oscillation theory (Eguchi et al. 2003).

4 The Borexino experiment

With the establishment of neutrino oscillations, solar neutrino observation returns to the original objective of determining the rates at which different neutrino-producing fusion reactions occur in the solar core. The detection of solar neutrinos by the Homestake experiment proved that the Sun's energy production is fueled by thermonuclear fusion. Now, measuring the relative rates of different fusion processes by their produced neutrino flux will allow to determine solar parameters that influence the fusion rate.

A spectral measurement is mandatory in order to disentangle the contributions of different reactions to the overall neutrino flux. Of the experiments presented in Sect. 3, merely water-Cherenkov detectors are able to provide a spectral measurement of solar neutrinos. However, their sensitivity is limited to the energy range above 5 MeV, in which only ^{8}B and hep neutrinos contribute. Below this threshold, only the integral data of radiochemical experiments was available.

The situation changed in 2007, when the liquid-scintillator experiment Borexino came into operation (Alimonti et al. 2008). As all preceding solar neutrino detectors, it is located at an underground laboratory, in this case the Gran Sasso National Laboratories (LNGS), to shield the detector from the background of cosmic rays. The rock overburden corresponds to about 3 500 meters water equivalent (mwe), and only cosmic muons (and, of course, neutrinos) are able to penetrate this barrier; however, the muon flux is reduced by six orders of magnitude compared to surface level.

Like in water Cherenkov detectors, neutrinos are detected via the elastic scattering on electrons in the target volume. The recoil electrons excite the molecules of the organic scintillator solvent, which in turn de-excitate and emit light. The photons travel through the scintillation volume and are registered by photosensors (photomultiplier tubes, PMTs) that are mounted to the detector walls. The light output of a typical liquid scintillator is about 50 times greater than that of the Cherenkov effect in water, leading to both an increase in the energy resolution and a significantly lower detection threshold. This allows for the first time a spectroscopic measurement of solar neutrinos in the sub-MeV region.

A major obstacle for neutrino detection is the background generated by radioisotopes dissolved in the liquid scintillator and by penetrating radiation from outside the detection volume, primarily gamma rays and cosmic muons. Therefore, the Borexino detector is structured in several concentric shielding layers of increasing radiopurity towards the central detection volume that consists of about 300 tons of liquid scintillator (Fig. 3) (Alimonti et al. 2008). Purification, mainly distillation, reduced the level of radioactive impurities in the organic liquid to 10^{-18} g/g in uranium and thorium, which is about 12 orders of magnitude less than the natural abundance. The target is contained in a 125 μm thick nylon sphere of 4.25 m radius, the Inner Vessel. It floats in about 1000 tons of inactive organic buffer liquid which are in turn contained in a stainless steel sphere of 6.85 m radius, to which the 2212 inward-facing PMTs are mounted. This Inner Detector is surrounded by an external water tank contained in a steel dome of 9 m radius. This Outer Detector provides additional shielding from external radioactivity and is equipped with 208 PMTs to identify cosmic muons crossing the detector by their Cherenkov light. This active muon veto is necessary to reject muons and muon-induced events from the neutrino analysis.

In 2008, the Borexino collaboration released data on the first flux measurement of solar ^{7}Be neutrinos (Arpesella et al. 2008a, 2008b). Figure 4 shows the electron recoil spectrum from about 0.2 to 2 MeV, based on 192 live days of the detector. Spacial reconstruction of the events allows to restrict the analysis to the innermost 100 tons of the target; this fiducial-volume cut removes most of the background signals induced by the decay of radioactive isotopes on the vessel surface and the gamma rays of ^{40}K contaminating the PMT glass of the PMTs. The information of the external veto is used to reject muon events. Coincidence signals from fast decays in the U/Th decay chains are exploited to identify further events from radioimpurities in the liquid. Finally, pulse-shape discrimination allows to remove an otherwise prominent peak caused by the α decay of ^{210}Po at a visible energy of about 500 keV in the spectrum.

Due to the application of these cuts, the Compton-like shoulder of electron recoil events caused by ^{7}Be-neutrinos becomes visible at about 660 keV in the spectrum.

Spectroscopy of Solar Neutrinos 209

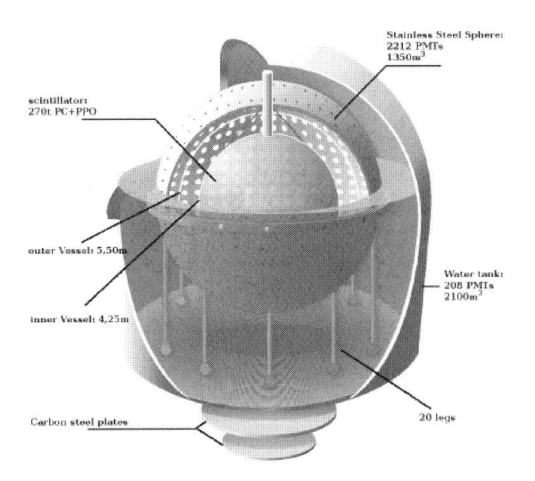

Figure 3: Schematic view of the Borexino detector.

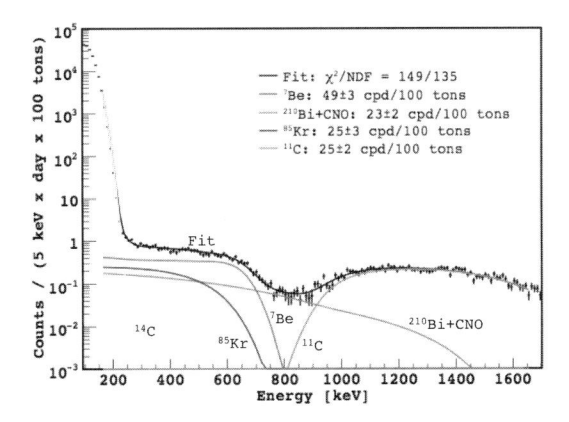

Figure 4: The electron recoil spectrum below 2 MeV based on 192 live days of Borexino data (Arpesella et al. 2008b): The black dots represent the data points, colored lines indicate the fit to the data. Both the spectra of neutrino-induced electron recoils and radioactive sources of background are included. The individual contributions are discussed in the text.

Superimposed are two residual sources of background: The β decays of ^{85}Kr and ^{210}Bi cannot be discriminated by their pulse shape, but separated from the neutrino signal due to their known spectral shape. The ^{7}Be-neutrino event rate obtained by a spectral fit to the data is $49\pm3_{(\text{stat})}\pm4_{(\text{syst})}$ counts per day and 100 tons, corresponding to about 58 % of the value expected from the SSM without taking neutrino oscillations into account. Correcting for the expected conversion of ν_e into $\nu_{\mu,\tau}$, the corresponding total ^{7}Be-neutrino flux is $\Phi(^7\text{Be}) = (5.18 \pm 0.51) \times 10^9 \text{ cm}^{-2}\text{ s}^{-1}$ (Arpesella et al. 2008b).

Table 2: Results of the direct measurements of solar neutrino experiments (Arpesella et al. 2008b; Aharmim et al. 2008), compared to the two SSM predictions based on different values of solar metallicity: BPS08(GS) corresponds to high metallicity, BPS08(AGS) to the novel low metallicity values (Peña-Garay & Serenelli 2008). Values in brackets correspond to the relative 1σ uncertainties. While the ^{7}Be result seems to give a slight indication for BPS08(GS), the measured ^{8}B-neutrino flux lies between the predicted values.

Source	Neutrino Flux [^{7}Be: 10^9/cm^2 s, ^{8}B: 10^6/cm^2 s]		
	Experiment	BPS08(GS)	BPS08(AGS)
^{7}Be	5.18 (10%)	5.07 (6%)	4.57 (6%)
^{8}B	5.54 (9%)	5.94 (11%)	4.72 (11%)

5 Solar neutrinos and solar metallicity

In recent years, a novel analysis of the solar abundances of heavy elements, i. e. the solar metallicity, has introduced inconsistencies in the astrophysical understanding of the Sun: Based on a refined modeling of the shape of the absorption lines in the solar electromagnetic spectrum, Asplund, Grevesse & Sauval (2005) re-determined the metallicity of the solar surface to considerably lower values compared to previous analyses. Where formerly had been an excellent agreement, the new results are dramatically in conflict with helioseismological measurements. An analysis of low-degree acoustic oscillations of the Sun with BISON (Birmingham Solar-Oscillations Network) suggests that the discrepancy is not only present in the convective zone close to the solar surface but extends also to the solar core, and therefore in the production region of solar neutrinos (Basu et al. 2007). As fusion rates depend on the metallicity, an accurate measurement of the corresponding neutrino fluxes might be used to determine the chemical composition of the solar core region.

The current status is presented in Table 2: Up to now, merely the fluxes of ^{7}Be and ^{8}B neutrinos were directly determined by liquid-scintillator and water Cherenkov detectors, respectively (Arpesella et al. 2008b; Aharmim et al. 2008). For comparison, the neutrino flux predictions based on the most recent SSM calculations are presented in which the solar metallicity enters as input parameter: The column BPS08(GS) denotes the fluxes corresponding to the older high values for metallicity, while BPS08(AGS) corresponds to the lower values of the new analysis (Peña-Garay & Serenelli 2008). Values in brackets represent the relative uncertainties (1σ).

Neither the accuracies of the experimental results nor of the SSM predictions are for the moment sufficient to allow a definite conclusion. While the Borexino result for the ^{7}Be-neutrino flux slightly favors the higher metallicity values, the ^{8}B result lies between the two SSM predictions. However, improvement is expected on both sides: An analysis of the error budgets of the SSM neutrino fluxes reveals that especially the precision on the ^{8}B-neutrino flux could be increased by a more accurate determination of the solar iron and carbon abundances and of the nuclear cross section $S_{1,14}$. At the same time, this would greatly decrease the uncertainty for the CNO-neutrino flux. The impact on the currently best-determined ^{7}Be-neutrino flux would be low (Peña-Garay & Serenelli 2008).

From the experimental side, no large improvement is expected for the ^{8}B-neutrino flux measurements in present detectors. While the SNO experiment was terminated, Super-Kamiokande results are for the moment dominated by systematical uncertainties (Fukuda et al. 2001). Nevertheless, a new analysis of the third phase of the experiment from Aug 2006 to Aug 2008 aims on reducing the systematic effects (Yang et al. 2009).

The present uncertainty of the ^{7}Be-neutrino flux as determined in Borexino is about 12 % (Arpesella et al. 2008b). A new analysis based on a considerably larger data set is planned for 2010. Moreover, a calibration campaign using radioactive sources placed inside the detection volume was conducted in early 2009. The results will be used to decrease the systematic uncertainties introduced by the detector response function, i.e. mainly the energy calibration, and by the position reconstruction. The latter is necessary to precisely determine the target mass used for analysis. An overall precision of better than 5 % on the flux measurement is envisaged.

Borexino might also contribute in a further way: In the 2008 analysis shown in Fig. 4, the radioactive background ^{210}Bi and the spectral contribution of the CNO neutrinos is fitted by the same function (Arpesella et al. 2008b). This simplification is valid in the range of the ^{7}Be-neutrino shoulder at 660 keV where the spectral differences are small. However, the endpoint of the ^{210}Bi β-decay is at 1.16 MeV, while the spectra of some of the CNO neutrinos reach up to 1.7 MeV. Also the Compton-like shoulder caused by the monoenergetic pep neutrinos should show up in this regime of the electron recoil spectrum. Unfortunately, the background created by ^{11}C β^+ decays dominates the neutrino signal between 1 and 2 MeV by about an order of magnitude.

Currently, efforts are made to suppress this background using the signature of the production mechanism: ^{11}C is a cosmogenic isotope created by cosmic muons crossing the target volume, knocking out a single neutron from a ^{12}C nucleus present in the organic liquid. The most promising approach is a veto using both the spatial and time information of the parent muon track, the capture of the knock-out neutron on a hydrogen nucleus that is visible in liquid-scintillator, and the decay of the ^{11}C nucleus itself (Galbiati et al. 2005). At the moment, both improvements in the reconstruction software as well as in the DAQ hardware are on-going to reach the rejection efficiency of about 90 to 95 % that is required to resolve the spectral contributions of residual ^{11}C background, CNO neutrinos and pep neutrinos.

6 Potential of future experiments

Even if the level of cosmogenic background in Borexino proves to be too large for an actual measurement of the CNO and pep neutrino fluxes, the best limits on their spectral contributions will significantly be improved. However, the currently upcoming SNO+ experiment will be in a much better position to measure these contributions: Based on the setup of the terminated SNO experiment, the acrylic vessel formerly holding the heavy water target will be refilled with liquid scintillator. The resulting

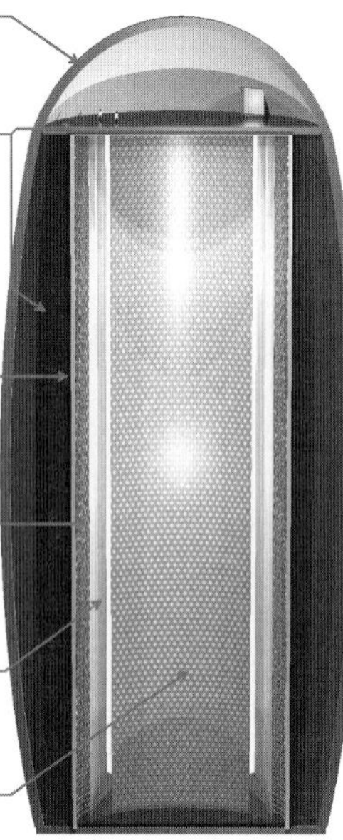
Cavern
height: 115 m, diameter: 50 m
shielding from cosmic rays: ~4,000 m.w
Muon Veto
plastic scintillator panels (on top)
Water Cherenkov Detector
1,500 phototubes
100 kt of water
reduction of fast
neutron background
Steel Cylinder
height: 100 m, diameter: 30 m
70 kt of organic liquid
13,500 phototubes
Buffer
thickness: 2 m
non-scintillating organic liquid
shielding external radioactivity
Nylon Vessel
parting buffer liquid
from liquid scintillator
Target Volume
height: 100 m, diameter: 26 m
50 kt of liquid scintillator

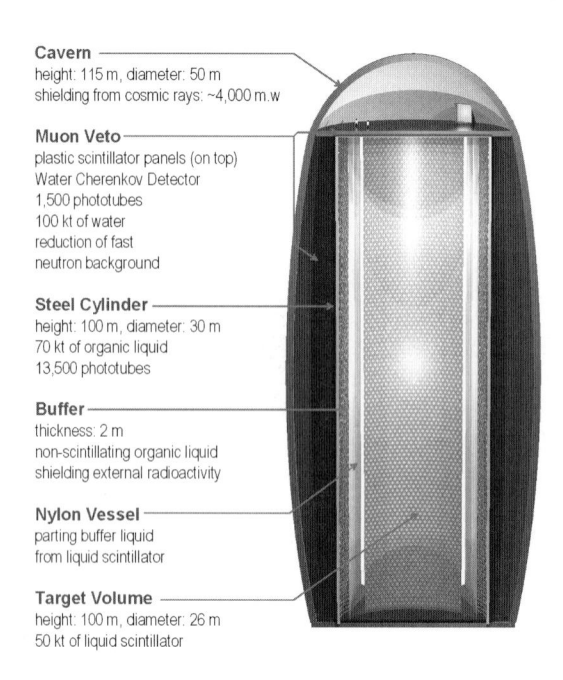

Figure 5: Envisaged layout of the LENA detector.

target mass will be three times larger than in Borexino. More importantly, SNO+ will be located at a depth corresponding to 6000 mwe in the Creighton Mine at Sudbury, the additional shielding reducing the residual muon flux by about 2 orders of magnitude compared to the Borexino site. Therefore, the background of cosmogenic ^{11}C will play a minor role in SNO+, the achievable accuracy mainly depending on the radiopurity of the liquid scintillator and the detector materials.

Further in the future, the LENA (Low Energy Neutrino Astronomy) experiment is planned to go into operation in 2020 (Marrodan Undagoitia et al. 2008). A schematic of the setup is shown in Fig. 5. With a target of 50 kilotons of liquid scintillator, LENA will be an observatory for a variety of terrestrial and astrophysical neutrino sources: About 15000 neutrino events are expected in case of a core-collapse Supernova (SN) at the center of our galaxy, giving insights into the nuclear processes leading to the collapse of the stellar core region and the subsequent neutrino cooling of the proto-neutron star (Wurm et al. 2007b). Even in the absence of a galactic SN, the faint neutrino flux generated by extra-galactic SNe offers a possibility to obtain information both on the star formation rate and the SN neutrino spectrum. While the event rates caused by this diffuse SN neutrino background are too low to be discovered in present-day detectors, the sensitivity of LENA will be sufficient to identify about 10 events per year (Wurm et al. 2007a). Also the chemical composition and heat production of our own planet can be investigated by a precision measurement of geoneutrinos (Hochmuth et al. 2007): These neutrinos are produced by the radioactive elements of the natural Uranium and Thorium decay chains that are embedded in the Earth's crust and mantle.

Regarding the spectroscopy of solar neutrinos, a new era of precision measurements will begin: Based on a fiducial volume of 20 kt, rates of several hundred events per day are expected for CNO, pep and ^{8}B neutrinos in LENA, allowing a precise determination of their relative contributions to the solar neutrino spectrum (Wurm et al. 2007b). As LENA will be placed at greater depth than Borexino (albeit not as deep as SNO+), cosmogenic background levels will be significantly lower. Most remarkable is probably the daily rate of 10 000 ^{7}Be-neutrino events per day: These statistics would allow to search for modulations in the neutrino flux on a per mill level. Such modulations could be induced by the interaction of the propagating neutrinos with solar or terrestrial matter (Blennow, Ohlsson & Snellman 2005), or by periodical changes in the neutrino production rate itself. The range of interest reaches from periods of tens of minutes typical for helioseismic g-modes to the scale of decades correlated to the solar cycle (Krauss 1990).

7 Conclusions

In the last 20 years, the spectroscopy of solar neutrinos has greatly enhanced our knowledge of neutrino particle parameters, but has also confirmed the elaborate models describing the solar energy production by fusion in the pp-chain and the CNO cycle. While the currently available data on solar neutrino fluxes is not accurate enough to resolve the conflict in the determination of solar metallicity, a considerable improvement in the near future can be expected.

Up to now, only ^{7}Be and ^{8}B neutrino fluxes have spectroscopically been measured. However, there is a good chance that the contributions of pep and CNO neutrinos will be discovered either in the currently running Borexino detector or in the upcoming SNO+ experiment. In 10 years from now, a large liquid-scintillator based neutrino observatory like LENA might allow precision measurements of the solar neutrino spectrum, possibly revealing time variations in the solar energy production. For sure, the next years will deepen both our understanding of neutrinos as well as of their sources.

Acknowledgements

This work was supported by the Deutsche Forschungsgesellschaft, the Munich Cluster of Excellence "Universe", and the Maier-Leibnitz-Laboratorium in Garching. We would like to thank the Borexino collaboration for the common work.

References

Abdurashitov, J.N., Gavrin, V.N., Gorbachev, V.V., et al.: 2009, Phys. Rev. C 80, 015807

Aharmim, B., Ahmed, S.N., Amsbauh, J.F., et al.: 2008, Phys. Rev. Lett. 101, 111301

Ahmad, Q.R., Allen, R.C., Andersen, T.C., et al.: 2002, Phys. Rev. Lett. 89, 011301

Alimonti, G., Arpesella, C., Back, H.O., et al.: 2008, NIMPA 600, 568

Altmann, M., Balata, M., Belli, P., et al.: 2005, Phys. Lett. B 616, 174

Arpesella, C., Back, H.O., Balata, M., et al.: 2008b, Phys. Rev. Lett. 101, 091302

Arpesella, C., Bellini, G., Benziger, J., et al: 2008a, Phys. Lett. B 658, 101

Asplund, M., Grevesse, N., Sauval, A.J.: 2005, in: T.G. Barnes III, F.N. Bash (eds.), *Cosmic Abundances as Records of Stellar Evolution and Nucleosynthesis*, ASPC 336, p. 25

Bahcall, J.N., Serenelli, A.M., Basu, S.: 2006, ApJS 165, 400

Basu, S., Chaplin, W.J., Elsworth, Y., New, R., Serenelli, A.M., Verner, G.A.: 2007, ApJ 655, 660

Blennow, M., Ohlsson, T., Snellman, H.: 2005, NuPhS 143, 578

Davis, R.: 1994, PrPNP 32, 13

Eguchi, K., Enomoto, S., Furuno, K., et al.: 2003, Phys. Rev. Lett. 90, 021802

Fukuda, S., Fukuda, Y., Ishitsuka, M., et al.: 2001, Phys. Rev. Lett. 86, 5651

Fukuda, S., Hayakawa, T., Ichihara, E., et al.: 1998, Phys. Rev. Lett. 81, 1562

Galbiati, C., Pocar, A., Franco, D., Ianni, A., Cadonati, L., Schönert, S.: 2005, Phys. Rev. C 71, 055805

Hampel, W., Handt, J., Heusser, G., et al.: 1999, Phys. Lett. B 447, 127

Hochmuth, K.A, Feilitzsch, F., Fields, B.D., Marrodan Undagoitia, T., Oberauer, L., Potzel, W., Raffelt, G.G., Wurm, M.: 2007, APh 27, 21

Kayser, B.: 2008, Phys. Lett. B 667, 163

Krauss, L.M.: 1990, Nat 348, 403

Marrodan Undagoitia, T., Feilitzsch, F., Göger-Neff, M., Oberauer, L., Potzel, W., Ulrich, A., Winter, J., Wurm, M.: 2008, JPhCS 120, 052018

Peña-Garay, C., Serenelli, A.M.: 2008, astro-ph/0811.2424

Suzuki, Y.: 1995, NuPhS 38, 54

Wurm, M., Feilitzsch, F., Göger-Neff, M., Hochmuth, K.A., Marrodan Undagoitia, T., Oberauer, L., Potzel, W.: 2007, Phys. Rev. D 75, 023007

Wurm, M., Feilitzsch, F., Göger-Neff M., Marrodan Undagoitia, T., Oberauer, L., Potzel, W.: 2007, in: J.R. Wilkes 9ed.), *Next Generation Nucleon Decay and Neutrino Detectors: NNN06*, AIPC 944, p. 82

Yang, B.S., et al.: 2009, hep-ex/0909.5469

Open clusters and the galactic disk

Siegfried Röser[1], Nina V. Kharchenko[1,2,3], Anatoly E. Piskunov[1,2,4],
Elena Schilbach[1], Ralf-Dieter Scholz[2], and Hans Zinnecker[2]

[1] Astronomisches Rechen-Institut
Zentrum für Astronomie der Universität Heidelberg
Mönchhofstr. 12-14, 69120 Heidelberg, Germany
roeser@ari.uni-heidelberg.de

[2] Astrophysikalisches Institut Potsdam
An der Sternwarte 16, D-14482 Potsdam, Germany

[3] Main Astronomical Observatory
27 Academica Zabolotnogo Str., 03680 Kiev, Ukraine

[4] Institute of Astronomy of the Russian Acad. Sci.
48 Pyatnitskaya Str., 109017 Moscow, Russia

Abstract

It is textbook knowledge that open clusters are conspicuous members of the thin disk of our Galaxy, but their role as contributors to the stellar population of the disk was regarded as minor. Starting from a homogenous stellar sky survey, the ASCC-2.5, we revisited the population of open clusters in the solar neighbourhood from scratch. In the course of this enterprise we detected 130 formerly unknown open clusters, constructed volume- and magnitude-limited samples of clusters, re-determined distances, motions, sizes, ages, luminosities and masses of 650 open clusters. We derived the present-day luminosity and mass functions of open clusters (not the stellar mass function in open clusters), the cluster initial mass function CIMF and the formation rate of open clusters. We find that open clusters contributed around 40 percent to the stellar content of the disk during the history of our Galaxy. Hence, open clusters are important building blocks of the Galactic disk.

1 Introduction

Open clusters constitute an important part of a process transforming gas and dust into stars. They are observed as the most prominent parts in the regions of active star formation, or as tracers of the ceased star formation process in the general Galactic field. However, the role they are playing in this process has still not been fully understood. In spite of their prominence, there had been indications that classical open clusters contribute only 10% or even less input (Miller & Scalo, 1978; Piskunov et al., 2006; Wielen, 1971) to the total stellar population of the Galactic disk. This discordance can be explained either by an early decay of a considerable fraction of newly formed star clusters (see e.g. Kroupa et al., 2001; Lamers et al., 2005; Tutukov, 1978) or by an insufficient knowledge of cluster formation statistics. In this context, one should note that the most important items of cluster formation like the distribution of cluster masses at birth (i.e., the initial mass function of star clusters) and the cluster formation rate were still poorly known a decade ago.

In principle, basic parameters like distance, motion, age, and metallicity can be determined for an open cluster more accurately than for a single field star. Indeed, they are better tracers of large scale structures of the Galactic disk population than field stars. Nevertheless, the most comprehensive studies of the Galactic cluster population are about 20 years old (Janes et al., 1988; Lyngå, 1982). They were based on the best data available at that time, the Lund Catalogue of Open Cluster Data (Lyngå , 1987, hereafter, the Lund Catalogue) and its subset of clusters with 3-colour photometry (Janes & Adler, 1982). Although these studies represent an important step in our understanding of the general properties of the cluster population, they suffer from incompleteness of the cluster samples and from inhomogeneity of the cluster parameters.

About 1200 clusters were known in the Lund catalogue by 1988. Only 400 of them had accurate, but heterogeneous UBV photometry, and photometric distances, reddening and age values. Although for almost all clusters apparent diameters were given in the Lund catalogue (estimated by eye from sky charts or defined by the size of detector field of view), only about 100 clusters were studied in a systematic way on the basis of star counts (Danilov & Seleznev, 1994).

Currently, the on-line list of open cluster data by Dias et al. (2002, DLAM hereafter), which can be considered as a continuation of the Lund Catalogue, contains by a factor of 1.5 more clusters than its predecessor, but the degree of completeness of this list is still unknown. Since the cluster data in the DLAM list are taken from literature, the set of the derived parameters differ from cluster to cluster. Also, the parameters themselves are based on heterogeneous observations and different methods of cluster definition and of parameter determination. Whenever using these data for cluster population studies, one is confronted with problems caused by uncertain cluster statistics and data heterogeneity.

In the following sections we describe our approach to construct a representative sample of open clusters in the solar neighbourhood, derive a homogeneous set of cluster parameters, especially age, mass and luminosity which are basic for estimating the role of open clusters as building blocks in the evolution of the Galactic disk. A previous review on this topic was given by Zinnecker et al. (2009) at IAU Symposium 254.

2 A volume- and magnitude-limited sample of open clusters

The first goal in this cooperation which started in 2003 was the construction of a sample of Galactic open clusters whose properties and biases are known (as good as possible). In this approach we started from the very beginning, namely from a magnitude-limited catalog (sky survey), the All-Sky Compiled Catalogue of 2.5 million stars (ASCC-2.5[1]). The latest version of ASCC-2.5 (Kharchenko & Röser, 2009) gives absolute proper motions in the ICRS, Johnson B, V and 2MASS J, H, K_s as well as spectral types and radial velocities if available. The catalog is to 90% complete down to $V = 11.5$. ASCC-2.5 was used to identify known open clusters and compact associations from the Lund Catalogue (Lyngå, 1987), the Dias et al. (2002) on-line data collection, and the Ruprecht et al. (1981) list of associations. Cluster membership of stars was determined based on kinematic and photometric criteria. In the ASCC-2.5 we found 520 of about 1700 known clusters (Kharchenko et al., 2005a), and discovered 130 new open clusters (Kharchenko et al., 2005b).

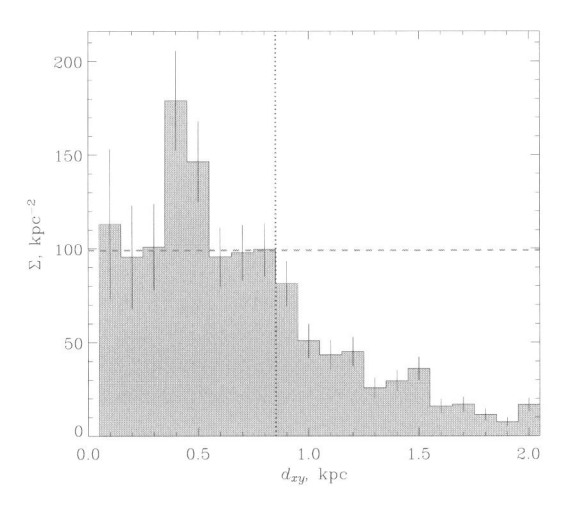

Figure 1: Distribution of the surface density Σ of open clusters versus distance d_{xy} from the Sun projected onto the Galactic plane. The dotted line indicates the completeness limit, the dashed horizontal line corresponds to the average density of "field clusters". The peak at $d_{xy} = 0.4$ to 0.5 is due to a population of young clusters mainly connected to Gould's belt. The bars are Poisson errors derived from cluster counts. From Piskunov et al. (2006) where a detailed analysis can be found.

From this full sample of 650 clusters in the ASCC-2.5 we extract 2 important sub-samples: a volume-limited sample of 256 open clusters complete to a distance of 850 pc from the Sun, see Piskunov et al. (2006) and Fig. 1, and a magnitude-limited sample complete down to apparent integrated magnitude $I_V = 8$, with 440 clusters above this completeness limit. For details of the construction of the magnitude-limit-

---
[1] ftp://cdsarc.u-strasbg.fr/pub/cats/I/280B

ed sample see Piskunov et al. (2008b). Finally, we must keep in mind the evolutionary status of open clusters included in our sample. Since cluster membership is based on the proper motion data mainly obtained in the optical spectral range, we consider our sample as representative of optical clusters or "classical" open clusters. Embedded objects are not included in this sample since their members usually are fainter (in the visual) than the limiting magnitude of the ASCC-2.5. Therefore, we assume the beginning of the transparency phase after the removal of the bulk of the placental matter to be a starting point of the evolution of a classical open cluster. The corresponding age t_0 is defined by the lowest age of our clusters, that is, about 4 Myr.

2.1 Distribution in the Galactic disk

Having reliably determined membership of stars in open clusters, the distances from the Sun together with the interstellar extinction to the cluster was derived (Piskunov et al., 2006). The symmetry plane of the clusters' distribution is determined to be at $Z_0 = -22 \pm 4$ pc, and the scale height of open clusters is only 56 ± 3 pc. The total surface density (Fig. 1) and volume density in the symmetry plane are $\Sigma = 114$ kpc^{-2} and $D(Z_0) = 1015$ kpc^{-3}, respectively. We estimate the total number of open clusters in the Galactic disk to be of order of 10^5 at present.

3 Astrophysical parameters of open clusters

3.1 Age distribution of open clusters

Ages of open clusters in the Galaxy can be determined from the fitting of theoretical isochrones to the loci of member stars in the CMD. It should be kept in mind that this makes the basic difference to the age determination of open clusters in other galaxies. Our method of age determination is described in detail in Kharchenko et al. (2005a). Ages were determined for 506 out of 520 clusters, of which 196 are first estimates. Our values for cluster ages are in good agreement with earlier results by Loktin et al. (2001) and Loktin (2004).

In the upper panel of Fig. 2 the distributions of clusters versus age are shown for the total sample as well as for those within the completeness area. Since young clusters contain, in general, more luminous stars, they can be observed at larger distances (beyond 850 pc) and their proportion in the total sample is somewhat higher than that of older clusters. Hence the total sample is biased towards young clusters. Such a bias has a strong impact onto the determination of cluster formation rate and lifetime. Therefore we determined these parameters from the volume-limited sample.

In the bottom panel of Fig. 2 we show the same distributions together with results from Wielen (1971) based on the Becker & Fenkart (1971) sample. The pronounced deficiency of older clusters in the latter sample is the reason for the smaller mean lifetime derived by Wielen (1971) of 231 Myr compared to the 327±25 Myr for our volume-limited sample.

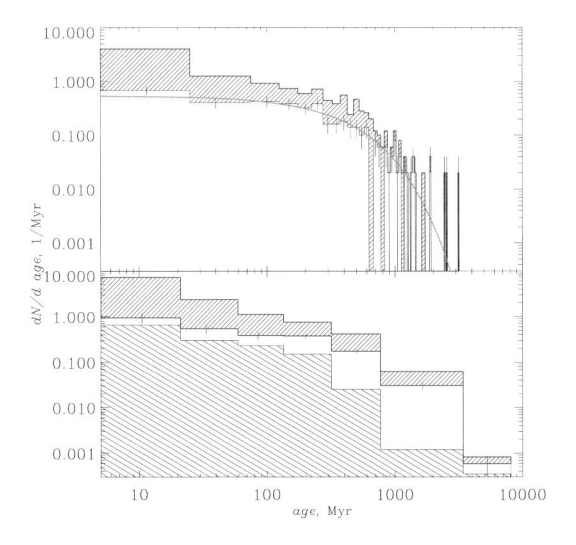

Figure 2: Distribution of open clusters versus age. For an easier comparison, the distributions for the different samples are not normalised to unit area. *Upper panel*: our total sample is shown as the hatched histogram. The sample of field clusters within the completeness area is marked as the filled histogram, and the solid curve marks the fitted age distribution. The age step is 50 Myr. *Lower panel*: the same distributions as in the upper panel together with the age sample used by Wielen (1971). The data from Wielen (1971) are shown as the backhatched histogram in the foreground. The vertical bars correspond to Poissonian errors derived from cluster counts. The binning is chosen same as in Wielen (1971). From Piskunov et al. (2006).

3.2 Integrated magnitudes and colours of open clusters

In the past decades quite a number of authors determined the integrated colours and magnitudes of MW open clusters. Among them we mention Gray (1965), Piskunov (1974), Sagar et al. (1983), Spassova & Baev (1985), Pandey et al. (1989), Battinelli et al. (1994), and Lata et al. (2002). In many cases, however, the underlying data are strongly inhomogeneous since the photometric observations of different clusters were obtained with different instruments and detectors, and the data reduction was carried out with different methods by different authors. Frequently, the integrated magnitudes and colours were only "by-products" of studies aiming primarily at constructing photometric sequences (e.g., sets of photometric standards, or cluster CMDs), where the data completeness is not essential.

In Fig. 3 we show the colour-absolute magnitude diagrams for 650 clusters in the solar neighbourhood. At first sight, it looks rather surprising that, in the optical, most of the open clusters in the solar neighbourhood appear rather blue even if they are not as young as, e.g., open clusters in galaxies with active star formation. Piskunov et al. (2009) discuss this finding and show that this blue colour must be expected in clusters with masses less than about $10^4 \, M_\odot$ where due to the discreteness of the IMF red giants do pop up only for a short period during a cluster's lifetime. Hence,

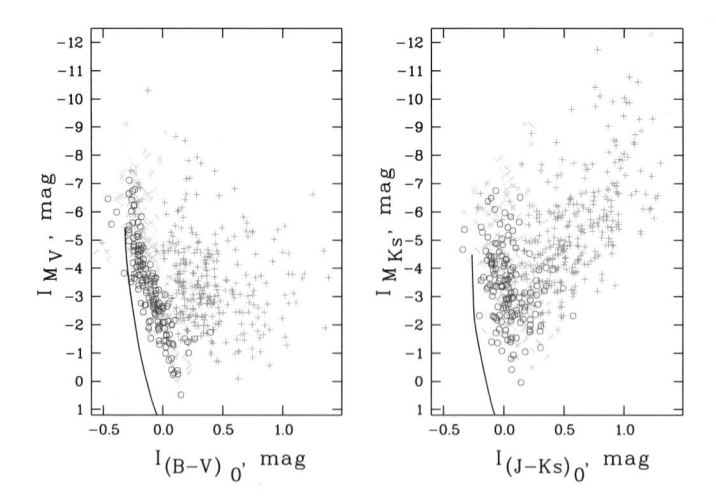

Figure 3: Integrated magnitudes and colours of 650 Galactic open clusters. *Left panel*: the colour-magnitude diagram I_{M_V} vs. $I_{(B-V)_0}$. *Right panel*: the colour-magnitude diagram $I_{M_{K_s}}$ vs. $I_{(J-K_s)_0}$. Clusters with only MS-members are marked by open circles, clusters with evolved members which did not yet cross the Hertzsprung gap are marked by crosses, clusters containing red giants are marked by plusses. It is remarkable that in the optical most of the brightest clusters have blue colours, whereas in the NIR they have red colours.

the usually adopted SSP models are not suited to reproduce the colours of the open clusters in the solar neighbourhood.

3.3 The luminosity function of Galactic open clusters

Even in the close vicinity of the Sun, the only previous attempt to construct the luminosity function of open clusters (van den Bergh & Lafontaine, 1984) is based on a sample of 142 clusters that is to 2/3 complete within 400 pc.

Figure 4 (from Piskunov et al., 2008b) shows the present-day luminosity function (CPDLF) constructed from the magnitude-limited sample of 440 clusters. At brighter magnitudes the CPDLF follows a power law with an exponent α in $\mathrm{d}N/\mathrm{d}L_V \propto L^{-\alpha}$ which comes out as $\alpha = 2.02 \pm 0.02$. This is comparable to the slope in extragalactic clusters (see e.g. Larsen, 2002). Notice, that for Galactic star clusters the CPDLF can be observed much deeper than for clusters in other galaxies. For the Large Magellanic Cloud (LMC), the faint limit is reached already at about $I_{M_V} = -5$, and it is much brighter in more distant galaxies (Larsen, 2002). As a consequence of going deeper we find a turnover in the CPDLF between $I_{M_V} = -3$ and -2, and an apparent decrease at fainter magnitudes. This turnover is a real phenomenon, since the luminosity function is obtained from the distribution of clusters within the completeness limit.

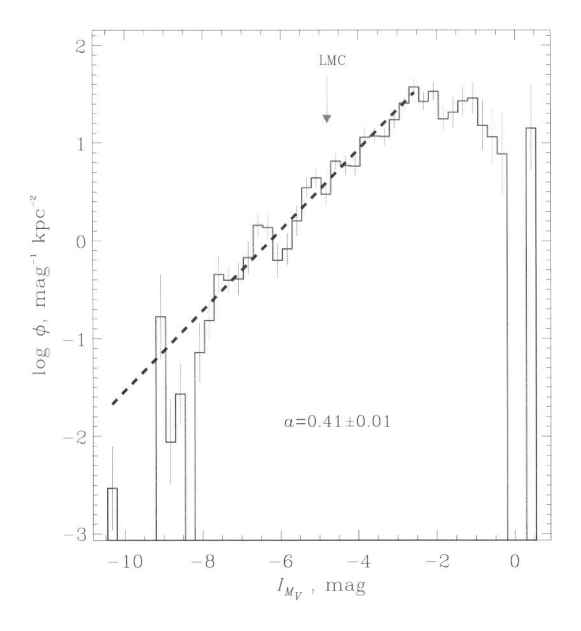

Figure 4: Luminosity function of Galactic open clusters based on 440 local clusters brighter than the completeness limit I_V of the sample. The bars are Poisson errors, the dashed line shows a linear fit for the brighter part of the histogram ($I_{M_V} < -2.5$) where a is the corresponding slope. The arrow indicates the limit of integrated absolute magnitudes reached for open clusters in the LMC (see Larsen, 2002). Figure from Piskunov et al. (2008b).

3.4 Masses and the mass function of Galactic open clusters

Mass is one of the fundamental parameters of star clusters which, from an observational point of view, is difficult to determine.

There are at least three independent methods for estimating cluster masses, each with advantages and disadvantages. The simplest and most straightforward way is to count cluster members and to sum up their masses. This requires a complete census of cluster members (down to the lowest masses). The real situation is, however, far from being ideal: incompleteness comes from either the limiting magnitude or the limited field of view or both. The extrapolation of the observed mass spectrum to "unseen" cluster members by choosing some inappropriate IMF can lead to unjustified and unpredictable modifications of the observed cluster mass, i.e. to biases. Nevertheless, due to its simplicity, the method is currently widely used (see Danilov & Seleznev, 1994; Lamers et al., 2005; Tadross et al., 2002).

Another method is based on the virial theorem: the mass of a cluster is determined from the stellar velocity dispersion. It does not require the observation and membership determination of all cluster stars. The application of the method is, however, limited to sufficiently massive stellar systems (globular clusters and dwarf spheroidals) with dispersions of internal motions large enough to be measurable. For open clusters, the typical velocity dispersion is about or less than 1 km s^{-1}, so,

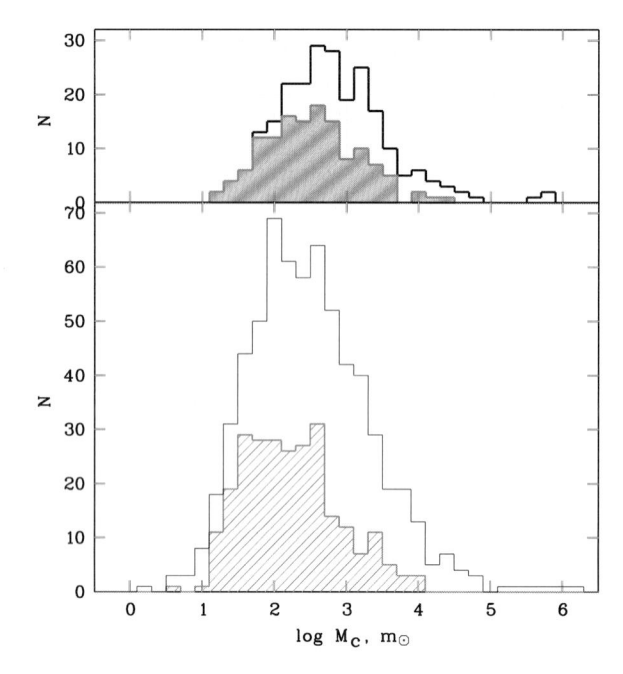

Figure 5: Distributions of tidal masses of open clusters. The upper panel shows masses calculated from measured radii for 236 clusters. The lower panel displays the masses calculated from calibrated radii of all 650 clusters of our sample. The open histograms correspond to total distributions, the filled/hatched ones are distributions of corresponding clusters residing in the area where our sample is complete. Figure from Piskunov et al. (2008a).

only for a few clusters with the most accurate proper motions and/or radial velocities, attempts have been undertaken to derive "virial masses" (see the references in Piskunov et al., 2007).

The third method uses the interpretation of the tidal interaction of a cluster with the parent galaxy, and requires knowledge of the tidal radius of a cluster. It gives so-called 'tidal masses', and goes back to the fundamental paper by King (1962). Considering globular clusters which, in general, have elliptical orbits, King (1962) differentiates between the tidal and the limiting radius of a cluster. For open clusters revolving at approximately circular orbits, one can expect the observed tidal radius to be approximately equal to the limiting one, although a probable deviation of the cluster shape from sphericity may have some impact on the computed cluster mass. Nevertheless, this method gives a mass estimate of a cluster (Raboud & Mermilliod, 1998a,b) that is independent of the results of the two methods mentioned above. As tidal masses grow with r_t cubed, their precision is strongly influenced by the uncertainties of the r_t determination. For details of the application of King's model the reader is referred to Piskunov et al. (2007) and Piskunov et al. (2008a).

For 236 clusters of our sample we could determine core and tidal radii directly from the fitting of King profiles to the density profile of cluster members. The distribution of the corresponding cluster masses is shown in the upper part of Fig. 5, where the filled histogram shows the distribution within the completeness area. Most of the clusters have tidal masses between 50 and 5000 $M_\odot$, and for about half of the clusters, the masses were obtained with a relative error better than 50%. To obtain a mass estimate for all the 650 clusters we calibrated the semi-major axis $\mathcal{A}$ (of the observed stellar density distribution) of the clusters using the tidal radii of the subsample of 236 clusters. The resulting mass distribution is shown in the lower part of Fig. 5. Although the masses are of moderate accuracy the large cluster sample should lead to reliable statistical evaluation.

The corresponding present-day mass function CPDMF is shown as the upper curve in Fig. 6. Notice that in Fig. 6 we show the logarithmic mass function in the form $\eta_t = \mathrm{d}N/\mathrm{d}\log M_c$ while in the further discussion (and to compare with other authors) we refer to the mass function $\mathrm{d}N/\mathrm{d}M_c$. On its high-mass part ($\log M_c > 3.3$) the CPDMF follows a power law with an exponent α in $\mathrm{d}N/\mathrm{d}M_c \propto M_c^{-\alpha}$ which comes out as $\alpha = 2.17 \pm 0.08$. de Grijs et al. (2003) derived masses for open clusters in four galaxies and found a typical value of the CPDMF slope $\alpha \approx 2$ within a mass range (10^3–10^6 $M_\odot$). This is in remarkable agreement with the slope found for our CPDMF. Notice, however, that the masses in de Grijs et al. (2003) are based on the mass-luminosity relation used to convert the observed photometric or spectroscopic data into masses. Nevertheless, this can be interpreted as indirect evidence of the coincidence of the relative mass scales of Galactic and extragalactic clusters.

In Fig. 6 we also show the cluster mass function with different upper limits for age. The youngest clusters with $\log t \leqslant 6.9$ (49 in our sample) are considered to represent the open cluster initial mass function (CIMF). It is shown as the lowest curve (with solid dots). The CIMF has a power-law shape with $\alpha = 1.66$ between $\log M_c \approx 3.3$ and the cut-off at about $\log M_c \approx 5$. The low-mass part ($\log M_c \leqslant 3.3$) can also be fitted by a power-law with $\alpha = 0.82 \pm 0.14$. With time, the slope of the high-mass part increases, for clusters with $\log t \leqslant 8.5$ we find $\alpha = 2.13$, and for $\log t \leqslant 9.5$, the CPDMF, α increases to 2.17. So, at every age the cluster mass function shows the same basic features, i.e., a quasi-linear high-mass portion, and a non-linear portion at lower masses. The low-mass portion changes from an approximately flat distribution at $\log t = 6.9$ to a clearly non-linear behaviour displaying a broad maximum with a peak at about 100 $M_\odot$ for the CPDMF and a decline towards lower masses.

With the same completeness arguments as for the CPDLF we infer that this maximum of the CPDMF and the decline towards lower masses is real and not due to an observational bias. Moreover, the maximum in the CIMF at about 1000 $M_\odot$ and the decline shortward is also real, as the detection probability of young clusters of, say, a few 100 $M_\odot$ is larger than that of old clusters, because the former contain brighter stars.

The steepening of the time-integrated mass function is a direct consequence of the mass-loss of clusters due to dynamical evolution together with the cut-off at the high-mass end of the CIMF. When the clusters grow older the high-mass cut-off shifts towards lower masses. Hence, the number of clusters in the higher mass bins

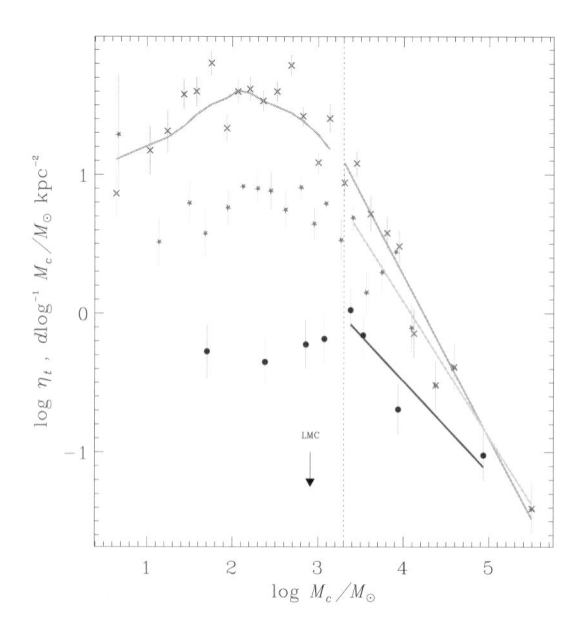

Figure 6: Evolution of the mass function $\eta_t = \mathrm{d}N/\mathrm{d}\log M_c$ of Galactic open clusters. Different symbols mark samples with different upper limits of cluster ages. The filled circles are for clusters with $\log t \leqslant 6.9$ (CIMF), stars show the CMF for $\log t \leqslant 7.9$, and crosses indicate the CPDMF based on all 440 clusters ($\log t \leqslant 9.5$). The bars are Poisson errors. The straight lines are the corresponding fits to linear parts of the mass functions at masses greater than $\log M_c = 3.3$ indicated by the vertical dotted line. The curve is the smoothed CPDMF at $\log M_c < 3.3$. The arrow indicates the lower mass limit reached for open clusters in the LMC. Figure from Piskunov et al. (2009).

increases more slowly or does not increase at all. This leads to a steepening as well as a shift of the maximum to lower masses.

According to present belief, the classical gas-free open clusters stem from a population of clusters which are surrounded by the remnants of the molecular cloud in which their stars have formed. Lada & Lada (2003) have compiled a catalogue of about 100 embedded clusters within 2.4 kpc from the Sun. The sample contains some optical objects and is partly overlapping with our data. Using models of the luminosity function, Lada & Lada (2003) scaled the IR counts within the areas studied, estimated cluster masses, and constructed an embedded cluster mass function (ECMF) shown in Fig. 7. Typically, the clusters are distributed over a mass range from 50 to 1000 $M_\odot$ and follow a power law of the form $\mathrm{d}N/\mathrm{d}M_c \propto M_c^{-\alpha}$, where $\alpha \approx 1.7$ to 2.0. There are striking similarities between the ECMF and the CIMF. Both follow a power law with about the same exponent $\alpha \approx 1.7$, which hints that both groups come from a universal parent distribution. Also, both show cut-offs. The high-mass cut-off of the ECMF coincides remarkably well with the low-mass cut-off of the CIMF. Lada & Lada (2003) also determined the embedded cluster formation rate to be about 2–4 kpc^{-2} Myr^{-1} which is about 10 times larger than our open cluster formation rate of 0.4 kpc^{-2} Myr^{-1} determined from the CIMF in Fig. 6. The latter low rate led Lada & Lada (2003) to the conclusion that about 10% of the embedded clusters do survive to become classical open clusters. Hence it is not surprising that the CIMF in Fig. 6 has a break at about 1000 M$_\odot$. On the other hand, one could ask why in Fig. 7 the embedded counterparts of the classical open cluster more massive than about 1000 M$_\odot$ are absent. In fact only a few of them have been detected in a recent work by Ascenso (2008). An answer to this question may have already been given by Kroupa & Boily (2002). They found that (initially embedded) clusters that form in total $10^3 < N < 10^5$ stars (so-called type II clusters) lose their gas within a dynamical time as a result of the photoionizing flux from O stars. Sparser (type I) clusters get rid of their residual gas on a time-scale longer or comparable to the nominal crossing time and thus evolve approximately adiabatically. For Kroupa & Boily (2002) this effect works on the transformation of the mass function of embedded clusters (ECMF) to the 'initial' mass function of bound gas-free star clusters (CIMF). They estimate that the resulting CIMF has, for a featureless power-law ECMF, a turnover near $10^{4.5}$ M$_\odot$ and a peak near 10^3 M$_\odot$. This explains both the absence of high-mass clusters in the ECMF and the low number of low mass clusters in the CIMF. The latter being related to 'infant mortality'.

4 Evolution of open clusters and their contribution to the disk population

The driving forces in the modification of the cluster mass function with time lies in the evolution of the individual clusters during their life-time: mass-loss both from stellar evolution of massive stars and from dynamical evolution affecting preferentially low-mass stars.

This mass-loss of clusters is determined from comparing the average mass of the newly formed, youngest clusters $M_c \simeq 4.5 \times 10^3$ M$_\odot$ with the average cluster mass from the whole sample ($M_c \simeq 700$ M$_\odot$) (Piskunov et al., 2008b). The typical mass-loss occurring in open clusters during their evolution amounts to about 3–14 M$_\odot$ Myr^{-1}. In the earliest phase of the cluster evolution this mass loss primarily occurs from stellar evolution of massive stars and even from the expulsion of massive stars from the cluster. Schilbach & Röser (2008) have traced back the trajectories of so-called 'field' O-stars and found that the overwhelming majority had their origin in young open clusters. They found that the mass-loss rate from ejected O-stars alone amounts to about 1.5 M$_\odot$ Myr^{-1} during the first 6 Myr of a cluster life. To get this number, a typical O-star mass of 20 M$_\odot$ was assumed.

Provided that the cluster formation history has not changed dramatically in the solar neighbourhood during the evolution of the Galactic disk, we estimate the contribution of mass from open clusters to the thin disk of the Galaxy, or, to be more precise, the fraction of mass in the thin disk from stars that have spent part of their lifetime being members of classical open clusters.

With an assumed lifetime of the thin disk of 10 Gyr, an average mass of open clusters from the CIMF of 4500 M$_\odot$ and a cluster formation rate of 0.4 kpc^{-2} Myr^{-1}

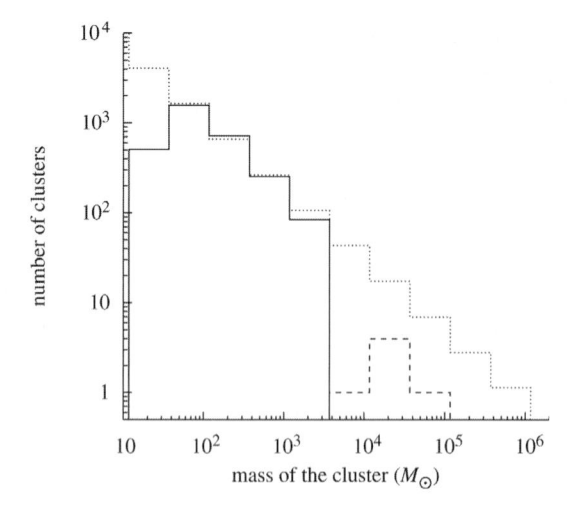

Figure 7: The embedded-cluster mass function (ECMF) for the Galaxy. The scaled ECMF for clusters from Lada & Lada (2003) compilation of cluster masses within 2.5 kpc is plotted as the solid line. The predicted ECMF for all masses and a spectral index of $\alpha = 1.7$ is shown as the dotted line. Massive embedded clusters from the list of Ascenso (2008) are represented by the dashed line. Adapted from Ascenso (2008) and Lada (2010).

we estimate this contribution to be

$$\Sigma = 18 \, M_\odot \, \text{pc}^{-2} \,.$$

This has to be compared to the present total surface density *in form of stars* of the Galactic disk in the solar neighbourhood that, according to Holmberg & Flynn (2004) is $35 \pm 6 \, M_\odot \, \text{pc}^{-2}$. As part of this mass is re-processed (mass-loss from massive stars) one finds (Just, 2009) that the amount of mass in stars ever formed in the thin disk must have been $48 \pm 6 \, M_\odot \, \text{pc}^{-2}$ to explain the present-day Holmberg & Flynn value. With these numbers about 37% of the observed surface density of the thin disk comes from open clusters.

This is considerably higher than the previous estimates for the input of open clusters to the observed stellar population of the Galactic disk that is quoted as about 10% (see Miller & Scalo, 1978; Piskunov et al., 2006) or even less than 10% (Wielen, 1971).

5 Summary and outlook

Summing all this up, it is fair to say that this work on the population of open clusters in the solar neighbourhood that started from an all-sky survey has found that open clusters are larger, more massive, live longer and contribute more to the thin stellar disk of the Galaxy than was believed a decade before.

Although much progress in our knowledge of the statistical properties of the open cluster population has been made in the past decade, there is still a long way to go. The sample discussed here allows us to draw general conclusions, but in some cases the derived parameters are erroneous because of a severe undersampling of the cluster membership due to the bright magnitude limit of the ASCC-2.5 survey. New questions have appeared, such as: is the local cluster population representative for the whole disk? how to discern open clusters from compact associations? are these separate populations or can the latter be seen as the high-mass end of the cluster population? What exactly do we mean by "the mass of a cluster"? Tidal, virial and star-counted masses may not necessarily measure the same mass. To get more insight into this problem, one must carefully define what a "cluster member" is. This becomes especially important near the boundary of a cluster and consequently influences our understanding of "mass".

Progress on some of these topics may be expected from an exploitation of the new deep survey catalogue PPMXL (Röser et al., 2010), and, on a somewhat longer timescale from Gaia.

Acknowledgements

The work on which this highlight talk is based upon was supported by DFG grants 436 RUS 113/757/0-1 and 436 RUS 113/757/0-2, and RFBR grants 03-02-04028, 06-02-16379 and 07-02-91566.

References

Ascenso, J.: 2008, PhD, unpublished Dissertation, University of Porto, Portugal

Battinelli, P., Brandimarti, A., Capuzzo-Dolcetta, R.: 1994, A&AS 104, 379

Becker, W., Fenkart, R.: 1971, A&AS 4, 241

van den Bergh, S., Lafontaine, A.: 1984, AJ 89, 1822

Danilov, V.M., Seleznev, A.F.: 1994, Astronomical and Astrophysical Transactions 6, 85

Dias, W.S., Alessi, B.S., Moitinho, A., Lépine, J.R.D.: 2002, A&A 389, 871

Gray, D.F.: 1965, AJ 70, 362

de Grijs, R., Anders, P., Bastian, N., Lynds, R., Lamers, H.J.G.L.M., O'Neil, E.J.: 2003, MNRAS 343, 1285

Holmberg, J., Flynn, C.: 2004, MNRAS 352, 440

Janes, K., Adler, D.: 1982, ApJS 49, 425

Janes, K.A., Tilley, C., Lyngå, G.: 1988, AJ 95, 771

Just, A.: 2009, private communication

Kharchenko, N.V., Röser, S.: 2009, VizieR Online Data Catalog 1280, 0, I/280B

Kharchenko, N.V., Piskunov, A.E., Röser, S., Schilbach, E., Scholz, R.-D.: 2005a, A&A 438, 1163

Kharchenko, N.V., Piskunov, A.E., Röser, S., Schilbach, E., Scholz, R.-D.: 2005b, A&A 440, 403

King, I.: 1962, AJ 67, 471

Kroupa, P., Boily, C.M.: 2002, MNRAS 336, 1188

Kroupa, P., Aarseth, S., Hurley, J.: 2001, MNRAS 321, 699

Lada, C.J.: 2010, Phil. Trans. R. Soc. London, Ser. A 368, 713

Lada, C.J., Lada, E.A.: 2003, ARA&A 41, 57

Lamers, H.J.G.L.M., Gieles, M., Bastian, N., Baumgardt, H., Kharchenko, N.V., Portegies Zwart, S.: 2005, A&A 441, 117

Larsen, S.S.: 2002, AJ 124, 1393

Lata, S., Pandey, A.K., Sagar, R., Mohan, V.: 2002, A&A 388, 158

Loktin, A.V.: 2004, private communication

Loktin, A.V., Gerasimenko, T.P., Malysheva, L.K.: 2001, A&AT 20, 607

Lyngå, G.: 1982, A&A 109, 213

Lyngå, G. 1987, Catalogue of open clusters data, Fifth edition, CDS, Strasbourg (VII/92)

Miller, G.E., Scalo, J.M.: 1978, PASP 90, 506

Pandey, A.K., Bhatt, B.C., Mahra, H.S., Sagar, R.: 1989, MNRAS 236, 263

Piskunov, A.E.: 1974, Nauchnye Informatsii 33, 101

Piskunov, A.E., Kharchenko, N.V., Röser, S., Schilbach, E., Scholz, R.-D.: 2006, A&A 445, 545

Piskunov, A.E., Schilbach, E., Kharchenko, N.V., Röser, S., Scholz, R.-D.: 2007, A&A 468, 151

Piskunov, A.E., Schilbach, E., Kharchenko, N.V., Röser, S., Scholz, R.-D.: 2008a, A&A 477, 165

Piskunov, A.E., Kharchenko, N.V., Schilbach, E., Röser, S., Scholz, R.-D., Zinnecker, H.: 2008b, A&A 487, 557

Piskunov, A.E., Kharchenko, N.V., Schilbach, E., Röser, S., Scholz, R.-D., Zinnecker, H.: 2009, A&A 507, L5

Raboud, D., Mermilliod, J.-C.: 1998a, A&A 329, 101

Raboud, D., Mermilliod, J.-C.: 1998b, A&A 333, 897

Röser, S., Demleitner, M., Schilbach, E.: 2010, astro-ph/1003.5852

Ruprecht, J., Balazs, B.A., White, R.E.: 1981, *Catalogue of Star Clusters and Associations*, Akademiai Kiado, Budapest

Sagar, R., Joshi, U.C., Sinvhal, S.D.: 1983, Bulletin of the Astronomical Society of India 11, 44

Schilbach, E., Röser, S.: 2008, A&A 489, 105

Spassova, N.M., Baev, P.V.: 1985, Ap&SS 112, 111

Tadross, A.L., Werner, P., Osman, A., Marie, M.: 2002, New A 7, 553

Tutukov, A.V.: 1978, A&A 70, 57

Wielen, R.: 1971, A&A 13, 309

Zinnecker, H., Piskunov, A.E., Kharchenko, N.V., Röser, S., Schilbach, E., Scholz, R.-D.: 2009, in: J. Andersen, J. Bland-Hawthorn, B. Nordström (eds.), *The Galaxy Disk in Cosmological Context*, IAU Symp. 254, p. 221

VLT-CRIRES: "Good Vibrations"
Rotational-vibrational molecular spectroscopy in astronomy

Hans Ulrich Käufl

European Southern Observatory
Karl Schwarzschild Str. 2
D-85748 Garching bei München, Germany
hukaufl@eso.org

Abstract

Near-Infrared high spectral and spatial resolution spectroscopy offers new and innovative observing opportunities for astronomy. The "traditional" benefits of IR-astronomy – strongly reduced extinction and availability of adaptive optics – more than offset for many applications the compared to CCD-based astronomy strongly reduced sensitivity. Especially in high resolution spectroscopy interferences by telluric lines can be minimized. Moreover for abundance studies many important atomic lines can be accessed in the NIR. A novel spectral feature available for quantitative spectroscopy are the molecular rotational-vibrational transitions which allow for fundamentally new studies of condensed objects and atmospheres. This is also an important complement to radio-astronomy, especially with ALMA, where molecules are generally only observed in the vibrational ground state. Rot-vib transitions also allow high precision abundance measurements – including isotopic ratios – fundamental to understand the thermo-nuclear processes in stars beyond the main sequence. Quantitative modeling of atmospheres has progressed such that the unambiguous interpretation of IR-spectra is now well established. In combination with adaptive optics spectro-astrometry is even more powerful and with VLT-CRIRES a spatial resolution of better than one milli-arcsecond has been demonstrated. Some highlights and recent results will be presented: our solar system, extrasolar planets, star- and planet formation, stellar evolution and the formation of galactic bulges.

1 Introduction

The cryogenic pre-dispersed infrared Echelle spectrograph CRIRES provides a nominal resolving power $\nu/\Delta\nu$ approximately 10^5 for a nominal slit width of 0.2 arcsec between 58 000 and 310 000 GHz (aka 950–5000 nm). The CRIRES installation at the Nasmyth focus A of the 8-m VLT UT1 (Antu) marks the completion of the original instrumentation plan for the VLT. A curvature sensing adaptive optics system feed is used to minimize slit losses and to provide 0.2 arcsec spatial resolution along the slit. A mosaic of four Aladdin InSb-arrays packaged on custom-fabricated ceramic boards has been developed. It provides for an effective 4096×512 pixel focal plane array to maximize the free spectral range covered in each exposure. Insertion of gas cells is possible in order to measure radial velocities with high precision. Measurement of circular and linear polarization in Zeeman sensitive lines for magnetic Doppler imaging is foreseen but not yet fully implemented. A cryogenic Wollaston prism on a kinematic mount is already incorporated. The retarder devices will be located close to the Unit Telescope focal plane. Here we briefly recall the major design features of CRIRES and describe the commissioning of the instrument including a report of extensive testing and a preview of astronomical results.

2 Why CRIRES?

The near-infrared spectral domain at high resolution is one of the few uncharted territories in astronomy. "High" in this context means, that the relevant features are resolved. The frequency resolution of CRIRES corresponds to a Doppler velocity of 3 km/s. For many solar system objects this is not sufficient. E.g., Fig. 3 in Lellouch et al. (2010) shows saturated CO lines in the atmosphere of the Neptune's enigmatic moon Triton which are still underresolved by at least a factor of 3.5). Although for the interstellar medium this may not be enough, but a compromise had to be found.

The fundamentally new spectral feature which can be accessed with CRIRES are rotational-vibrational transitions in the electronic ground state. A molecule is a set of n coupled harmonic oscillators with a characteristic frequency ω_i, generally referred to as *normal* vibrations. A molecule is also a spinning top. In first approximation the energy levels of a molecule in the electronic groundstate are given by

$$E(\nu_1 \ldots \nu_n, j) = \sum_{i=1}^{n} \hbar \omega_i (\nu_i + \frac{1}{2}) + \frac{\Theta}{2} \hbar^2 j(j+1). \qquad (1)$$

The rotational moment of inertia, Θ, is a function of all quantum numbers of the molecule, ν_i, j and k, the equivalent to the "magnetic quantum number" in atomic physics, i.e. the projection of the symmetry axis on the vector of angular momentum. Centrifugal stretching affects the oscillator frequencies ω_i with j and the oscillations have some level of unharmonicity. It should be noted, that the energy difference or vibrational levels is of order of 0.1 eV while the spacing of rotational levels is of order of 0.01 eV or less. In local thermal equilibrium each energy level $E(\nu_1 \ldots \nu_i, j, k)$ is populated according to the Boltzmann statistics. Interestingly the distribution of the rotational levels is decoupled from that of the vibrational levels. Due to frequent collisions the population of the rotational levels is almost always a good proxy for the gas temperature, while the vibrational levels in the presence of strong radiation fields (e.g. in a stellar atmosphere) can be orders of magnitude out of LTE (see, e.g., Käufl et al. 1984 for an example). For radiative transitions the normal rules for dipole

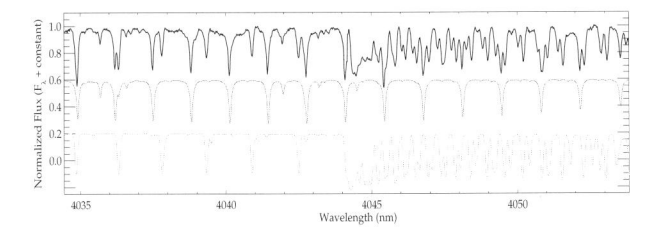

Figure 1: Example of molecular spectra: the panel shows three spectra taken with CRIRES in the region 74 000–74 350 GHz (4035–4054 nm). The top trace (black) shows the spectrum of the semi-regular variable star MS-Vel taken without telluric correction. The second trace in the middle (red) shows an absorption spectrum taken with the CRIRES calibration gas cell (N_2O). This spectrum shows a comb of nearly equally spaced lines as expected from Eq. (1). Comparing the gas-cell spectrum with that of MS-Vel, it is obvious that the dominant telluric absorber in this spectral range is indeed N_2O. The bottom spectrum (green) is model spectrum (PHOENIX, courtesy P. Hauschildt). The spectral feature in the star between 4044 nm and 4054 nm is due to an overtone ($|\Delta\nu| = 2$) transition of circumstellar SiO. At these temperatures j values beyond $100\hbar$ are populated and due to the anharmonicity of the vibration the lines are no longer equally spaced. This gives rise to the "inverted shark fin" feature at 4044 nm, usually referred to as a *band head*. The panel shows only one of the four CRIRES detectors. For more details see Seifahrt & Käufl (2008).

radiation apply. For a molecule in LTE many levels are populated and one species typically produces even in the most simple cases at least 10–20 lines, but a few 100 are more typical. A molecular spectrum hence carries a cornucopia of information and high redundancy. Examples of molecular spectra are shown in Fig. 1.

Another interesting aspect of molecular spectra is, that isotopic shifts are of the order of %. This can be easily understood for the case of 2-particle systems where the isotopic shifts scale with the reduced mass μ,

$$\mu = \frac{m_1 m_2}{m_1 + m_2}. \qquad (2)$$

If m_2 is the electron mass than the nuclear mass m_1 is four orders of magnitudes larger and the change of mass by adding neutrons to the atomic nucleus has little influence. On the other hand, for example for SiO, changing from $^{28}Si^{16}O$ to $^{29}Si^{16}O$ amounts to a shift of $\Delta\mu \approx 1.3\%$. For an example of the isotopic shift see e.g. the low resolution spectra shown in Aringer et al. (1999).

The small fraction of spectrum shown in Fig. 1 samples some 50 different j values from SiO in the circumstellar envelope and a comparison of the line strengths allows to constrain the thermal population of the rotational sub-states, hence the determination of the rotational temperature T_{rot}. The same measurement can than be done for the rare isotopes, and in this case one is probing deeper into the atmosphere. In combination with a detailed atmospheric model altitude resolved temperature measurements in stellar atmospheres become possible.

Another interesting aspect is, that the observation of ro-tational-vibrational transitions in the near-IR samples many different j values simultaneously with nearly identical geometry. In radio astronomy on the other hand pure rotational transitions are observed and for cold gas the transition frequencies, are proportional to j. So e.g. the $(2\rightarrow 1)$ transition of a diatomic molecule has the double frequency of the $(1\rightarrow 0)$ transition, hence for a given telescope dish, the beam diameter changes by a factor of two, which often makes modeling a bit cumbersome (if one does not have to use different equipment altogether).

Last, not least it should be mentioned that the near-IR also allows to observe the "usual suspects", atomic lines, but in some cases exceptionally better than in the optical (see, e.g., Ryde et al. 2009, 2010). More information as to background and importance of high spectral resolution infrared spectroscopy can be found in Wahlgren et al. (2010).

3 Instrument lay-out and main characteristics

3.1 Optical design

The instrument has been designed as a pre-dispersed spectrograph, to allow for also for long slit observations. To avoid an unreasonably large collimated beam entailing a much bigger cryostat and difficult (or impossible) to procure optics a rather narrow slit (nominally 0.2") was chosen. To reduce slit losses CRIRES was equipped with a specialized version (Paufique et al. 2006) of the general purpose curvature sensing adaptive optics system of the ESO-VLT (MACAO, Arsenault et al. 2004).

An overview of the instrument optics is given in Fig. 2. A full report on the instrument design can be found in Käufl et al. (2004) and an in detail report on commissioning and first results is given in Käufl et al. (2008).

The CRIRES detector mosaic consists of four detectors which have been hybridized to custom fabricated ceramics. The arrays are rotated such that always the best two of the four quadrants are illuminated by the spectrum. The overall characteristics of this detector assembly have already been reported elsewhere (cf. Dorn et al. 2004). While this was a rather economic approach in achieving a mega-pixel focal plane, it entails, that the readout direction is for two chips perpendicular and for two parallel to the dispersion, which puts an extra burden on flat-fielding and data processing.

3.2 Performance characteristics

The adaptive optics system closes the loop under normal conditions for a natural guide star R-magnitude of 16–17. As the CRIRES nominal slit width (0.2 arcsec) is less than the diffraction limit of an 8 m telescope[1] slit losses are usually acceptable if not negligible.

---

[1] 1.2 λ/D corresponds to 0.14 arcsec at $\nu \approx 65\,200$ GHz ($\lambda \approx 4.6$ μm), the location of the CO fundamental band and the lowest frequency CRIRES is normally used.

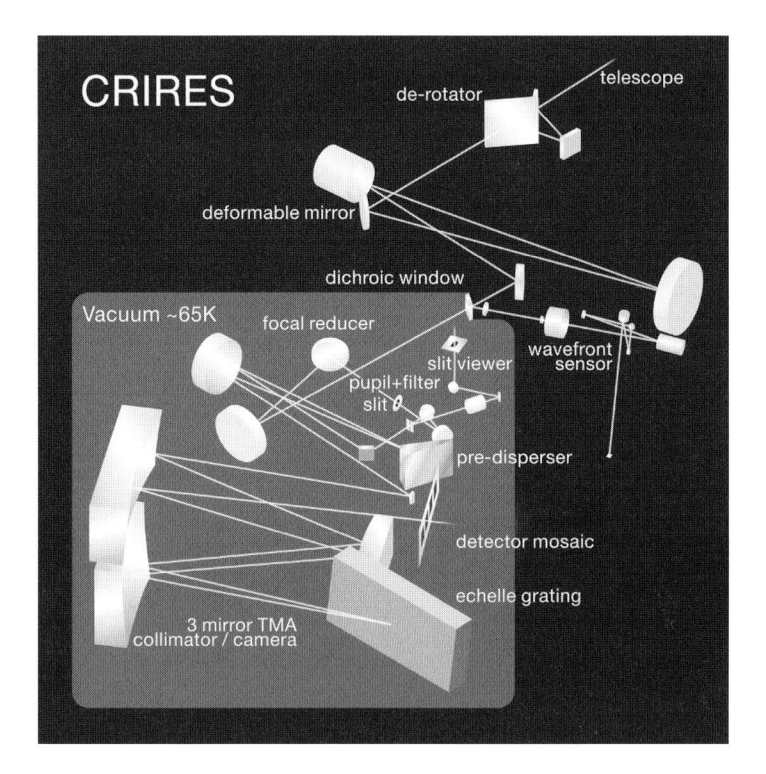

Figure 2: CRIRES optical design: the VLT Nasmyth focus (f# 15) is close to the first mirror of the de-rotator assembly. A calibration unit (krypton arc-lamp, IR-HeNe-laser, halogen lamp and an infrared glower in combination with an integration sphere) serves for flat fielding and spectral calibration. For the near-IR a ThAr spectrum can be imaged onto the slit using IR-fibers. In addition, gas-cells can be moved into the beam for calibration and for search for very small radial velocity changes similar to an iodine-cell in optical spectrograph. In future, in addition to the gas-cells retarders in a motorised mount can be placed to use CRIRES for spectro-polarimetry. The de-rotator is followed by the curvature sensing adaptive optics system with the deformable mirror on a kinematic gimbal mount. The entrance window to the cryostat is a dichroic, separating the visible light with high efficiency for the AO wavefront control. The entire cryogenic optical bench (highlighted by the grey background) is cooled with three Closed Cycle Coolers to ≈ 65 K. The pre-slit optics of CRIRES consists of an all-reflective re-imager with a cold-pupil stop, reducing the f-ratio to f# 7.5. Close to the cold pupil a Wollaston prism (MgF_2) can be inserted, eventually with a linear polarizer compensating instrumental polarization. The slit-viewer (J, H, and K filters, 1024^2 InSb Aladdin detector) gives a pixel scale of 0.05 arcsec/pixel and an un-vignetted field-of-view of $\approx 20 \times 40$ arcsec^2. The main slit is continuously adjustable up to several arcsec with a closed loop encoder controlling the slit separation. The pre-disperser has a collimated beam diameter of 100 mm and uses a ZnSe prism in retro-reflection. The collimator mirror can be slightly tilted with a piezo. The instrument is equipped with optical metrology. Optical fibers at the level of main and intermediate slit illuminated with the ThAr lamp and a bandpass filter produce reference marks which allow with the piezo to set grating and prism with a reproducibility of a fraction of a pixel. Order selection in the pre-dispersed spectrum is provided by a second intermediate slit located close to the small folding mirror. The main collimator, a three-mirror anastigmat (TMA), produces a 200 mm beam which illuminates a R2 échelle grating (31.6 gr/mm). The spectrum is imaged with a mosaic of four 1024^2 InSb (Aladdin III) detectors. From this mosaic, a strip with 512×4096 pixels is being used.

Table 1: CRIRES main characteristics in a nut-shell. More details are given in the resources shown in Appendix A.

Spectral coverage	ν: 58 000–310 000 GHz (λ: 950–5200 nm)
Spectral resolution	$\nu/\Delta\nu \approx 10^5$ or 3 km/s (2 pixel Nyquist sampling)
Spatial resolution	0.2 arcsec, 2 pixel Nyquist sampling of 40 arcsecslit
Array detector mosaic	$4 \times 1024 \times 512$ Aladdin III InSb arrays, instantaneous ν (λ) coverage $\geq 2.0\%$
Infrared slit viewer	Aladdin III, with J, H, and K filters, scale 0.05 arcsec/pix
Precision stability and calibration	≈ 75 m/s, i.e. 1/20 th of a pixel or 5 mas tracking error (goal)
Dark current	Int. background and detector ≤ 0.05 e$^-$/s
Sensitivity	Read-out or background noise limited, for details see the exposure time calculator
Ghosts, straylight	Specification $\leq 10^{-3}$, achieved $\leq 10^{-4}$, except for $\lambda < 1500$ nm: a white light ghost, displaced from the spectrum

The best frequency stability achieved so far is reported by Bean et al. (2010) (≈ 5 m/s) using a special NH$_3$ gas-cell in absorption in the light path. As compared to the I$_2$ gas-cell used in the optical, the pressure and column density of the absorber is much better controlled, hence the method is quite robust to artifacts. Seifarth & Käufl (2008) have reported a stability of ≈ 10 m/s using telluric atmospheric absorption lines as a local frequency standard.

The very first scientific paper of CRIRES using science verification data reported a S/N of 330 for a specially selected metal poor star (Nissen et al. 2007). Meanwhile the highest S/N ratio ever reported is ≈ 1000 (Villanueva & Mumma, priv. communication).

3.3 Precision absorption spectroscopy

With some experience in the use of CRIRES, it was found, that in certain spectral regions, where there is strong atmospheric absorption, the intensity of a continuum source should go down to zero, indeed residual flux remained in the cores of the lines. This was well studied around the ν_3 Q-branch[2] of CH$_4$.

With a precision measurement the in-dispersion instrumental profile was measured. It can well be represented by a Voigt function (Fig. 4). A convolution of the

---
[2] A Q-branch is formed by superposition of lines for which $\Delta j = 0$ and $\Delta k = \pm 1$ which nearly all have the same transition frequency.

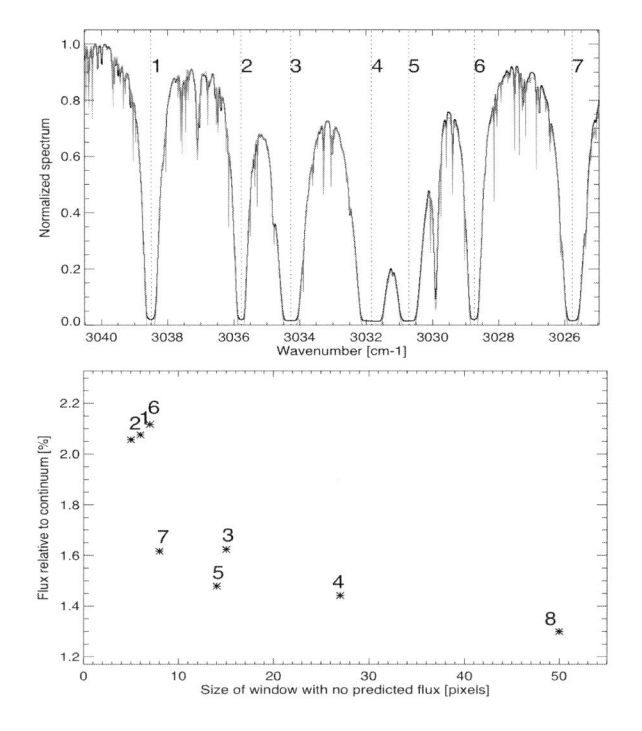

Figure 3: Example of straylight in absorption spectra. The *top panel* shows telluric CH$_4$ taken with CRIRES at ≈90 900 GHz in the ν_3 vibration band (C-H-stretch-mode). The black trace is the measured spectrum, the red one is a model spectrum. There are seven distinct regions, where the flux should go down to zero. The *bottom panel* shows the residuals for the 7 cases plotted again the width of the region where the atmosphere should be totally opaque.

theoretical telluric spectra in this wavelength domain with this function reproduces the residuals shown in Fig. 3 extremely well. Such an understanding of the response of the instrument is of high importance for high S/N observations and for precision modeling. While the HeNe laser happens to have a line conveniently located in the 90 000 GHz range a similar monochromatic source for the K-band, ≈150 000 GHz, is missing. The goal is to make a table available, which gives the instrument profile parametrisation as a function of frequency, so that this modeling can be employed for all CRIRES observations (Villanueva & Käufl, in preparation).

It should be noted, that the lack of understanding of the in-dispersion straylight is also important to settle the issue, if in the near-IR there is a continuum between the OH recombination lines. Detailed understanding is important to assess the sensitivity of the next generation of low and medium resolution spectrographs for faint object work in the context of extremely large telescopes.

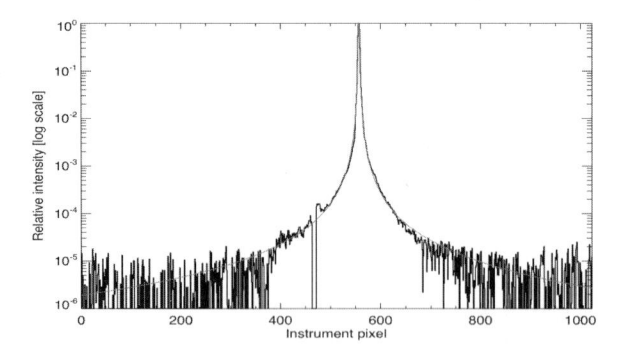

Figure 4: Deep exposure using the IR-HeNe laser. The spectrum (black) is the in-dispersion line profile of CRIRES measured at $\nu \approx 88\,402.43$ GHz (3391.2 nm) in logarithmic scale for a slit width of 0.4 arcsec. It can be represented at least over 4.5 dex with a Voigt profile (red trace).

The quantity of the residual flux remained a source of uncertainty, especially with the discovery of a set of previously unknown weak absorption bands for rare isotopes.

3.4 Calibration

The optical concept (pre-dispersion) combined with cryogenically movable grating and prism mounts produces variability of the CRIRES through-put. The Aladdin detectors in use are non-linear. In addition there is a systematic pixel to pixel intrinsic response modulation which is introduced by the CMOS read-out. This makes CRIRES flatfielding both challenging and highly important (or rewarding).

Frequency calibration in the near infrared down to $\nu \approx 120\,000$ GHz (up to $\lambda \approx 2500$ nm) uses a ThAr lamp. Kerber et al. (2008) re-measured the ThAr spectrum to sufficient precision. As this lamp is too faint to illuminate the entrance slit via the integrating sphere, a fiber system is being used to illuminate the entrance slit.

For the thermal IR, $\nu \leq 100\,000$ GHz (or $\lambda \geq 3000$ nm), a N_2O gas cell is being used. This will be later complemented by OCS. A sufficient number of these transitions has been measured with frequency chains relative to a caesium clock[3] i.e. to the time standard. Therefore, CRIRES is indeed frequency calibrated, traceable to the time standard.[4]

---

[3] http://www.nist.gov/physlab/data/wavenum/spectra.cfm
[4] In order not to loose precision it should be avoided to convert ν into λ as the c has only 8 significant digits.

4 Some science highlights

As CRIRES is actually an instrument which gives access to uncharted territory, a selection of highlights may be biased, incomplete and soon obsolete (see Appendix A how to access the official bibliography for all ESO installations).

For the solar system, with the exception of the rapidly rotating giant gas planets, the spectral features are quite narrow (≤ 1 km/s). Thus even the spectral resolution of CRIRES is not yet sufficient. As an example how high spectral resolution is the key to sensitivity, reference is made to cometary research. Villanueva et al. (2009), eg. report a D/H survey in comets or Boehnhard et al. (2008), have analyzed the gas coma of comet 8P/Tuttle to constrain the existence of the "inner Oort cloud". A study of the composition and the structure of the tenuous atmospheres of trans-Neptunian objects such as Triton or Pluto was reported by Lellouch et al. (2009) and Lellouch et al. (2010).

Constraining stellar evolution and planet formation is a focus point of IR astronomy. Pontoppidan et al. (2008) have used *spectroastrometry*, that is to use the cross-dispersion profile of the feature free stellar continuum to deconvolve in spatial direction the spectrum at certain transitions. They used the fundamental band of CO at 63 300 GHz to map the inner disk around YSOs to determine, if the gas is co-rotating with the dust. The spatial resolution achieved given by the formal fitting error is less than one milli-arcsec.[5]

Planet detection with the radial velocity (RV) method at this point is an industry. The *holy grail* would be to detect an Earth-like planet in the habitable zone around a G-star. This may be beyond the scope of this method. The *easiest*, as technically feasible with existing telescopes and instrumentation – at least when allowing for some extrapolation toward the next generation of large telescopes – is, to search for *Earths* in the habitable zone around M-dwarfs. Results with CRIRES so far are encouraging, but as of today, there was no detection yet reported. But CRIRES and the VLT have destroyed one object! Huelamo et al. (2008), not only found that the RV signal in TW Hya detected in the optical range (Setiawan et al. 2008) is not present with CRIRES, and indeed can be explained better by stellar activity. Interestingly due to various more or less trivial effects, the IR-range appears to be more "forgiving" when it comes to distinguish between stellar activity and a true RV signal.

Some work on abundance studies has already been mentioned here (Ryde et al. 2009, 2010) which is focused on the formation of the Milky Way bulge. Another interesting example how IR rotational vibrational spectroscopy helps for a better understanding of nucleosynthesis, is the case of fluorine, which in principle is much easier to destroy than to create. Fluorine has no suitable transitions in the visible spectral range. In the infrared, however, in the K-band around 130 000 GHz ($\lambda \approx 2300$ nm), the HF molecule has its fundamental band, readily available for ground based astronomy. Uttenthaler et al. (2008) have studied HF in oxygen-rich stars in the galactic bulge to establish fluorine production on the asymptotic giant branch. Abia et al. (2009) searched for fluorine in carbon-rich AGB-stars, and claim to have solved the missing fluorine problem there.

---

[5]To achieve such a spatial resolution, typically ≤ 0.2 AU in the observed systems in interferometry, would need a baseline of at least 1000 m!

Accessible to CRIRES there are a variety of important metal lines from damped Lyman-α systems along the line of sight to bright quasars. Still the true extragalactic and cosmologic applications of CRIRES have not come yet.

5 Conclusions and outlook

Since its commissioning in mid-2006 at the VLT, CRIRES has proven to be a reliable and useful complement of the VLT instrumentation suite. While CRIRES is prepared to do also spectro-polarimetry in lines for Zeeman-Doppler imaging and other magnetic field studies, this could not yet be implemented. Still moving to the infrared for measuring magnetic fields has at least three significant advantages: i) low mass stars and brown dwarfs show some high activity, while their SEDs peak in the infrared; ii) the ratio Zeeman splitting to trivial line broadening effects (e.g turbulence) scales with $1/\nu$, hence optical observations are less sensitive; iii) the depth of the lines, associated with active regions becomes more favorable for lower frequencies as the spectra become more Jeans-like (for a detailed discussion see Käufl et al. 2003). Implementation of the spectro-polarimetry, however, had to be delayed, as the focus of work after commissioning was to improve the open-shutter time, as many observation programs were overhead driven.

Even though CRIRES is a unique instrument at an 8-10m class telescope and no competing instrument is on the horizon elsewhere, still at this point the process of drafting an upgrade plan has started; such an upgrade could entail in order of the optical path: i) add a long-path length gas-cell ($\approx$1.5 m path length are possible) for very high precision radial velocity work; ii) add $\lambda/2$ and $\lambda/4$ rotating plates plus a polarimetric calibration unit to the calibration slide; iii) upgrade of the adaptive optics real-time computer to improve the slit efficiency and Strehl in case of phase of low coherence time; iv) change the optomechanics of the wave front sensor to speed-up target acquisition; v) rebuild the mount for the Wollaston, to be able to implement limited cross dispersion using grisms; vi) rebuild the entrance slit assembly which limits flat field precision and over-all reproducibility; vii) exchange the focal plane detector assembly to improve sensitivity and cosmetics.

At this point this is an – incomplete – wishlist. Other ideas, e.g. adding a frequency comb for spectral calibration or a fiber image scrambler exist as well. Whatever will be implemented at the end will depend on a detailed trade-off between cost, available person power and scientific productivity. The dream would be, to have a limited cross dispersion and detect very high red-shift γ-ray-burst afterglows in rapid response mode.

Acknowledgements

Having CRIRES as a fully operational and scientifically productive instrument conforming to the overall standards of the ESO VLT was possible only with the enthusiastic support of the project by the members of the CRIRES team, both at ESO headquarters Garching and at the VLT-observatory. The availability of spectral templates based on the PHOENIX code, thanks to P. Hauschildt, helped a lot throughout the commissioning and science verification. P. van Hoof's website is a great resource for precision spectroscopy.

References

Abia, C., et al.: 2009, ApJ 694, 971

Aringer, B., et al.: A&A 342, 799

Arsenault, R., et al.: 2004, in: D.B. Calia, B.L. Ellerbroek, R. Ragazzoni (eds.), *Advancements in Adaptive Optics*, SPIE 5490, p. 47

Bean, J.L., et al.: 2010, ApJ 713, 410

Boehnhardt, H., et al.: 2008, ApJ 683, L71

Dorn, R.J., et al.: 2004, in: J.D. Garnett, J.W. Beletic (eds.), *Optical and Infrared Detectors for Astronomy*, SPIE 5499, p. 510

Huelamo, N., et al.: 2008, A&A 489, 9

Käufl, H.U., et al.: 1984, A&A 136, 319

Käufl, H.U., et al.: 2003, in: S. Fineschi (ed.), *Polarimetry in Astronomy*, SPIE 4843, p. 223

Käufl, H.U. et al.: 2004, in: A.F.M. Moorwood, I. Masanori (eds.), *Ground-Based Instrumentation for Astronomy*, SPIE 5492, p. 1218

Käufl, H.U., et al.: 2008, in: I.S. McLean, M.M. Casali (eds.), *Ground-based and Airborne Instrumentation for Astronomy II*, SPIE 7014, p. 29

Kerber, F., et al.: 2008, ApJS 178, 374

Lellouch, E., et al.: 2009 A&A 495, L17

Lellouch, E., et al.: 2010, A&A 512, L8

Nissen, P.E., et al.: 2007, A&A 469, 319

Paufique, J., et al.: 2006, in: B.L. Ellerbroek, C.D. Bonaccini (eds.), *Advances in Adaptive Optics II*, SPIE 6272, p. 36

Pontoppidan, K.M., et al.: 2008, ApJ 684, 1323

Ryde, N., et al.: 2009, A&A 496, 701

Ryde, N., et al.: 2010, A&A 509, A20

Seifahrt, A., Käufl, H.U.: 2008, A&A 491, 929

Setiawan, J., et al.: 2008, Nature 451, 38

Uttenthaler, S., et al.: 2008 ApJ 682, 509

Villanueva, G.L. et al.: 2009, ApJ **690** L5

Wahlgren, G., et al.: 2010, *IR and Sub-mm Spectroscopy: A New Tool for Studying Stellar Evolution*, IAU XXVII General Assembly, in press

A Useful links in the context of CRIRES

As CRIRES is still in an ongoing phase of improvements, for the actual description of the instrument, reference is made to the official user's manual

http://www.eso.org/sci/facilities/paranal/instruments/crires/ .

Similarly, the performance is best assessed by using the official exposure time calculator

http://www.eso.org/observing/etc/ .

CRIRES is a new and unique instrument, hence the bibliographic data base changes rapidly. It can be accessed conveniently via the ESO library

http://archive.eso.org/wdb/wdb/eso/publications/query .

CRIRES will open the near-IR for high precision measurements. Having a reliable source of atomic frequencies available online is a great help. Extensive use was made from the compilation

http://www.pa.uky.edu/~peter/atomic/index.html

maintained by P. v. Hoof, and hosted by the Department of Physics and Astronomy University of Kentucky.

Index of Contributors

Baumgardt, Holger	133	Potzel, Walter	203
Carilli, Chris	167	Röser, Siegfried	215
Carmona, A.	155	Schilbach, Elena	215
Daddi, Emanuele	167	Schmidt, Robert	179
Dravins, Dainis	191	Scholz, Ralf-Dieter	215
Fang, Min	155	Schuh, Sonja	29
Frebel, Anna	53	Sicilia-Aguilar, Aurora	155
Göger-Neff, Marianne	203	Thomas, Jens	143
Haehnelt, Martin	117	Tippmann, Marc	203
Henning, Thomas	155	Traunsteiner, Christoph	203
Käufl, Hans Ulrich	229	van Boekel, Roy	155
Kharchenko, Nina V.	215	von Feilitzsch, Franz	203
Kudritzki, Rolf-Peter	1	Walter, Fabian	167
Lachenmaier, Tobias	203	Wang, Wei	155
Lewke, Timo	203	Wilhelm, Klaus	81
Meindl, Qurin	203	Winter, Jürgen	203
Möllenberger, Randolph	203	Wurm, Michael	203
Oberauer, Lothar	203	Wyse, Rosemary	99
Piskunov, Anatoly E.	215	Zinnecker, Hans	215

General Table of Contents

Volume 1 (1988): Cosmic Chemistry

Geiss, J.: Composition in Halley's Comet:
    Clues to Origin and History of Cometary Matter 1/1

Palme, H.: Chemical Abundances in Meteorites 1/28

Gehren, T.: Chemical Abundances in Stars 1/52

Omont, A.: Chemistry of Circumstellar Shells 1/102

Herbst, E.: Interstellar Molecular Formation Processes 1/114

Edmunds, M.G.: Chemical Abundances in Galaxies 1/139

Arnould, M.: An Overview of the Theory of Nucleosynthesis 1/155

Schwenn, R.: Chemical Composition and Ionisation States of the
    Solar Wind – Plasma as Characteristics of Solar Phenomena 1/179

Kratz, K.-L.: Nucear Physics Constraints to Bring the Astrophysical
    R-Process to the "Waiting Point" 1/184

Henkel, R., Sedlmayr, E., Gail, H.-P.: Nonequilibrium Chemistry
    in Circumstellar Shells 1/231

Ungerechts, H.: Molecular Clouds in the Milky Way: the Columbia-Chile
    CO Survey and Detailed Studies with the KOSMA 3 m Telescope 1/210

Stutzki, J.: Molecular Millimeter and Submillimeter Observations 1/221

Volume 2 (1989)

Rees, M.J.: Is There a Massive Black Hole in Every Galaxy?
    (19th Karl Schwarzschild Lecture 1989) 2/1

Patermann, C.: European and Other International Cooperation
    in Large-Scale Astronomical Projects 2/13

Lamers, H.J.G.L.M.: A Decade of Stellar Research with IUE 2/24

Schoenfelder, V.: Astrophysics with GRO 2/47

Lemke, D., Kessler, M.: The Infrared Space Observatory ISO 2/53

Jahreiß, H.: HIPPARCOS after Launch!?
    The Preparation of the Input Catalogue 2/72

Ip, W.H.: The Cassini/Huygens Mission 2/86

Beckers, J.M.: Plan for High Resolution Imaging with the VLT 2/90

Rimmele, Th., von der Luehe, O.: A Correlation Tracker
    for Solar Fine Scale Studies 2/105

Schuecker, P., Horstmann, H., Seitter, W.C., Ott, H.-A., Duemmler, R.,
    Tucholke, H.-J., Teuber, D., Meijer, J., Cunow, B.:
    The Muenster Redshift Project (MRSP) 2/109

Kraan-Korteweg, R.C.: Galaxies in the Galactic Plane 2/119

Meisenheimer, K.: Synchrotron Light from Extragalactic Radio Jets
    and Hot Spots 2/129

Staubert, R.: Very High Energy X-Rays from Supernova 1987A 2/141

Hanuschik, R.W.: Optical Spectrophotometry
    of the Supernova 1987A in the LMC 2/148
Weinberger, R.: Planetary Nebulae in Late Evolutionary Stages 2/167
Pauliny-Toth, I.I.K., Alberdi, A., Zensus, J A., Cohen, M.H.:
    Structural Variations in the Quasar 2134+004 .... 2/177
Chini, R.: Submillimeter Observations
    of Galactic and Extragalactic Objects 2/180
Kroll, R.: Atmospheric Variations in Chemically Peculiar Stars 2/194
Maitzen, H.M.: Chemically Peculiar Stars of the Upper Main Sequence 2/205
Beisser, K.: Dynamics and Structures of Cometary Dust Tails 2/221
Teuber, D.: Automated Data Analysis 2/229
Grosbol, P.: MIDAS 2/242
Stix, M.: The Sun's Differential Rotation 2/248
Buchert, T.: Lighting up Pancakes –
    Towards a Theory of Galaxy-formation 2/267
Yorke, H.W.: The Simulation of Hydrodynamic Processes
    with Large Computers 2/283
Langer, N.: Evolution of Massive Stars
    (First Ludwig Biermann Award Lecture 1989) 2/306
Baade, R.: Multi-dimensional Radiation Transfer
    in the Expanding Envelopes of Binary Systems 2/324
Duschl, W.J.: Accretion Disks in Close Binarys 2/333

Volume 3 (1990): Accretion and Winds

Meyer, F.: Some New Elements in Accretion Disk Theory 3/1
King, A.R.: Mass Transfer and Evolution in Close Binaries 3/14
Kley, W.: Radiation Hydrodynamics of the Boundary Layer
    of Accretion Disks in Cataclysmic Variables 3/21
Hessman, F.V.: Curious Observations of Cataclysmic Variables 3/32
Schwope, A.D.: Accretion in AM Herculis Stars 3/44
Hasinger, G.: X-ray Diagnostics of Accretion Disks 3/60
Rebetzky, A., Herold, H., Kraus, U., Nollert, H.-P., Ruder, H.:
    Accretion Phenomena at Neutron Stars 3/74
Schmitt, D.: A Torus-Dynamo for Magnetic Fields
    in Galaxies and Accretion Disks 3/86
Owocki, S.P.: Winds from Hot Stars 3/98
Pauldrach, A.W.A., Puls, J.: Radiation Driven Winds
    of Hot Luminous Stars. Applications of Stationary Wind Models 3/124
Puls, J., Pauldrach, A.W.A.: Theory of Radiatively Driven Winds
    of Hot Stars: II. Some Aspects of Radiative Transfer 3/140
Gail, H.-P.: Winds of Late Type Stars 3/156

General Table of Contents

Hamann, W.-R., Wessolowski, U., Schmutz, W., Schwarz, E.,
    Duennebeil, G., Koesterke, L., Baum, E., Leuenhagen, U.:
    Analyses of Wolf-Rayet Stars 3/174

Schroeder, K.-P.: The Transition of Supergiant CS Matter from
    Cool Winds to Coronae – New Insights with X AUR Binary Systems 3/187

Dominik, C.: Dust Driven Mass Lost in the HRD 3/199

Montmerle, T.: The Close Circumstellar Environment
    of Young Stellar Objects 3/209

Camenzind, M.: Magnetized Disk-Winds
    and the Origin of Bipolar Outflows 3/234

Staude, H.J., Neckel, Th.: Bipolar Nebulae Driven by the Winds
    of Young Stars 3/266

Stahl, O.: Winds of Luminous Blue Variables 3/286

Jenkner, H.: The Hubble Space Telescope Before Launch:
    A Personal Perspective 3/297

Christensen-Dalsgaard, J.: Helioseismic Measurements
    of the Solar Internal Rotation 3/313

Deiss, B.M.: Fluctuations of the Interstellar Medium 3/350

Dorfi, E.A.: Acceleration of Cosmic Rays in Supernova Remnants 3/361

Volume 4 (1991)

Parker, E.N.: Convection, Spontaneous Discontinuities,
    and Stellar Winds and X-Ray Emission
    (20th Karl Schwarzschild Lecture 1990) 4/1

Schrijver, C.J.: The Sun as a Prototype
    in the Study of Stellar Magnetic Activity 4/18

Steffen, M., Freytag, B.: Hydrodynamics of the Solar Photosphere:
    Model Calculations and Spectroscopic Observations 4/43

Wittmann, A.D.: Solar Spectroscopy with a 100×100 Diode Array 4/61

Staude, J.: Solar Research at Potsdam:
    Papers on the Structure and Dynamics of Sunspots 4/69

Fleck, B.: Time-Resolved Stokes V Polarimetry
    of Small Scale Magnetic Structures on the Sun 4/90

Glatzel, W.: Instabilities in Astrophysical Shear Flows 4/104

Schmidt, W.: Simultaneous Observations with a Tunable Filter
    and the Echelle Spectrograph of the Vacuum Tower Telescope
    at Teneriffe 4/117

Fahr, H.J.: Aspects of the Present Heliospheric Research 4/126

Marsch, E.: Turbulence in the Solar Wind 4/145

Gruen, E.: Dust Rings Around Planets 4/157

Hoffmann, M.: Asteroid-Asteroid Interactions – Dynamically Irrelevant? 4/165

Aschenbach, B.: First Results from the X-Ray Astronomy Mission ROSAT 4/173

Wicenec, A.: TYCHO/HIPPARCOS A Successful Mission! 4/188

Spruit, H.C.: Shock Waves in Accretion Disks 4/197
Solanki, S.K.: Magnetic Field Measurements on Cool Stars 4/208
Hanuschik, R.W.: The Expanding Envelope of Supernova 1987A
　in the Large Magellanic Cloud
　(2nd Ludwig Biermann Award Lecture 1990) 4/233
Krause, F., Wielebinski, R.: Dynamos in Galaxies 4/260

Volume 5 (1992): Variabilities in Stars and Galaxies

Wolf, B.: Luminous Blue Variables; Quiescent and Eruptive States 5/1
Gautschy, A.: On Pulsations of Luminous Stars 5/16
Richter, G.A.: Cataclysmic Variables – Selected Problems 5/26
Luthardt, R.: Symbiotic Stars 5/38
Andreae, J.: Abundances of Classical Novae 5/58
Starrfield, S.: Recent Advances in Studies of the Nova Outburst 5/73
Pringle, J.E.: Accretion Disc Phenomena 5/97
Landstreet, J.D.: The Variability of Magnetic Stars 5/105
Baade, D.: Observational Aspects of Stellar Seismology 5/125
Dziembowski, W.: Testing Stellar Evolution Theory
　with Oscillation Frequency Data 5/143
Spurzem, R.: Evolution of Stars and Gas in Galactic Nuclei 5/161
Gerhard, O.E.: Gas Motions in the Inner Galaxy
　and the Dynamics of the Galactic Bulge Region 5/174
Schmitt, J.H.M.M.: Stellar X-Ray Variability
　as Observed with the ROSAT XRT 5/188
Notni, P.: M82 – The Bipolar Galaxy 5/200
Quirrenbach, A.: Variability and VLBI Observations
　of Extragalactic Radio Surces 5/214
Kollatschny, W.: Emission Line Variability in AGN's 5/229
Ulrich, M.-H.: The Continuum of Quasars and Active Galactic Nuclei,
　and Its Time Variability 5/247
Bartelmann, M.: Gravitational Lensing by Large-Scale Structures 5/259

Volume 6 (1993): Stellar Evolution and Interstellar Matter

Hoyle, F.: The Synthesis of the Light Elements
　(21st Karl Schwarzschild Lecture 1992) 6/1
Heiles, C.: A Personal Perspective of the Diffuse Interstellar Gas
　and Particularly the Wim 6/19
Dettmar, R.-J.: Diffuse Ionized Gas and the Disk-Halo Connection
　in Spiral Galaxies 6/33
Williams, D.A.: The Chemical Composition of the Interstellar Gas 6/49
Mauersberger, R., Henkel, C.: Dense Gas in Galactic Nuclei 6/69
Krabbe, A.: Near Infrared Imaging Spectroscopy of Galactic Nuclei 6/103

Dorschner, J.: Subject and Agent of Galactic Evolution 6/117

Markiewicz, W.J.: Coagulation of Interstellar Grains in a
  Turbulent Pre-Solar Nebula: Models and Laboratory Experiments 6/149

Goeres, A.: The Formation of PAHs in C-Type Star Environments 6/165

Koeppen, J.: The Chemical History of the Interstellar Medium 6/179

Zinnecker, H., McCaughrean, M.J., Rayner, J.T., Wilking, B.A.,
  Moneti, A.: Near Infrared Images of Star-Forming Regions 6/191

Stutzki, R.: The Small Scale Structure of Molecular Clouds 6/209

Bodenheimer, P.: Theory of Protostars 6/233

Kunze, R.: On the Impact of Massive Stars on their Environment –
  the Photoevaporation by H II Regions 6/257

Puls, J., Pauldrach, A.W.A., Kudritzki, R.-P., Owocki, S.P., Najarro, F.:
  Radiation Driven Winds of Hot Stars – some Remarks on Stationary
  Models and Spectrum Synthesis in Time-Dependent Simulations
  (3rd Ludwig Biermann Award Lecture 1992) 6/271

Volume 7 (1994)

Wilson, R.N.: Karl Schwarzschild and Telscope Optics
  (22nd Karl Schwarzschild Lecture 1993) 7/1

Lucy, L.B.: Astronomical Inverse Problems 7/31

Moffat, A.F.J.: Turbulence in Outflows from Hot Stars 7/51

Leitherer, C.: Massive Stars in Starburst Galaxies
  and the Origin of Galactic Superwinds 7/73

Mueller, E., Janka, H.-T.:
  Multi-Dimensional Simulations of Neutrino-Driven Supernovae 7/103

Hasinger, G.: Supersoft X-Ray Sources 7/129

Herbstmeier, U., Kerp, J., Moritz, P.:
  X-Ray Diagnostics of Interstellar Clouds 7/151

Luks, T.: Structure and Kinematics of the Magellanic Clouds 7/171

Burkert, A.: On the Formation of Elliptical Galaxies
  (4th Ludwig Biermann Award Lecture 1993) 7/191

Spiekermann, G., Seitter, W.C., Boschan, P., Cunow, B., Duemmler, R.,
  Naumann, M., Ott, H.-A., Schuecker, P., Ungruhe, R.:
  Cosmology with a Million Low Resolution Redshifts:
  The Muenster Redshift Project MRSP 7/207

Wegner, G.: Motions and Spatial Distributions of Galaxies 7/235

White, S.D.M.: Large-Scale Structure 7/255

Volume 8 (1995): Cosmic Magnetic Fields

Trümper, J.E.: X-Rays from Neutron Stars
  (23rd Karl Schwarzschild Lecture 1994) 8/1

Schuessler, M.: Solar Magnetic Fields 8/11

Keller, Ch.U.: Properties of Solar Magnetic Fields from Speckle Polarimetry
   (5th Ludwig Biermann Award Lecture 1994) 8/27
Schmitt, D., Degenhardt, U.:
   Equilibrium and Stability of Quiescent Prominences 8/61
Steiner, O., Grossmann-Doerth, U., Knoelker, M., Schuessler, M.:
   Simulation oif the Interaction of Convective Flow
   with Magnetic Elements in the Solar Atmosphere 8/81
Fischer, O.: Polarization by Interstellar Dust –
   Modelling and Interpretation of Polarization Maps 8/103
Schwope, A.D.: Accretion and Magnetism – AM Herculis Stars 8/125
Schmidt, G.D.: White Dwarfs as Magnetic Stars 8/147
Richtler, T.: Globular Cluster Systems of Elliptical Galaxies 8/163
Wielebinski, R.: Galactic and Extragalactic Magnetic Fields 8/185
Camenzind, M.: Magnetic Fields and the Physics of Active Galactic Nuclei 8/201
Dietrich, M.:
   Broad Emission-Line Variability Studies of Active Galactic Nuclei 8/235
Böhringer, H.: Hot, X-Ray Emitting Plasma, Radio Halos,
   and Magnetic Fields in Clusters of Galaxies 8/259
Hopp, U., Kuhn, B.:
   How Empty are the Voids? Results of an Optical Survey 8/277
Raedler, K.-H.: Cosmic Dynamos 8/295
Hesse, M.: Three-Dimensional Magnetic Reconnection
   in Space- and Astrophysical Plasmas and its Consequences
   for Particle Acceleration 8/323
Kiessling, M.K.-H.: Condensation in Gravitating Systems as Pase Transition 8/349

Volume 9 (1996): Positions, Motions, and Cosmic Evolution

van de Hulst, H.:
   Scaling Laws in Multiple Light Scattering under very Small Angles
   (24th Karl Schwarzschild Lecture 1995) 9/1
Mannheim, K.: Gamma Rays from Compact Objects
   (6th Ludwig Biermann Award Lecture 1995) 9/17
Schoenfelder, V.:
   Highlight Results from the Compton Gamma-Ray Observatory 9/49
Turon, C.: HIPPARCOS, a new Start
   for many Astronomical and Astrophysical Topics 9/69
Bastian, U., Schilbach, E.:
   GAIA, the successor of HIPPARCOS in the 21st century 9/87
Baade, D.: The Operations Model for the Very Large Telescope 9/95
Baars, J.W.M., Martin, R.N.: The Heinrich Hertz Telescope –
   A New Instrument for Submillimeter-wavelength Astronomy 9/111
Gouguenheim, L., Bottinelli, L., Theureau, G., Paturel, G., Teerikorpi, P.:
   The Extragalactic Distance Scale and the Hubble Constant:
   Controversies and Misconceptions 9/127

Tammann, G.A.: Why is there still Controversy on the Hubble Constant? 9/139

Mann, I.: Dust in Interplanetary Space:
a Component of Small Bodies in the Solar System 9/173

Fichtner, H.: Production of Energetic Particles at the Heliospheric Shock –
Implications for the Global Structure of the Heliosphere 9/191

Schroeder, K.-P., Eggleton, P.P.: Calibrating Late Stellar Evolution
by means of zeta AUR Systems – Blue Loop Luminosity
as a Critical Test for Core-Overshooting 9/221

Zensus, J.A., Krichbaum, T.P., Lobanov, P.A.:
Jets in High-Luminosity Compact Radio Sources 9/221

Gilmore, G.: Positions, Motions, and Evolution
of the Oldest Stellar Populations 9/263

Samland, M., Hensler, G.: Modelling the Evolution of Galaxies 9/277

Kallrath, J.: Fields of Activity for Astronomers and Astrophysicists
in Industry – Survey and Experience in Chemical Industry – 9/307

Volume 10 (1997): Gravitation

Thorne, K.S.: Gravitational Radiation – a New Window Onto the Universe
(25th Karl Schwarzschild Lecture 1996) 10/1

Grebel, E.K.: Star Formation Histories of Local Group Dwarf Galaxies
(7th Ludwig Biermann Award Lecture 1996 (i)) 10/29

Bartelmann, M.L.: On Arcs in X-Ray Clusters
(7th Ludwig Biermann Award Lecture 1996 (ii)) 10/61

Ehlers, J.: 80 Years of General Relativity 10/91

Lamb, D.Q.: The Distance Scale To Gamma-Ray Bursts 10/101

Meszaros, P.: Gamma-Ray Burst Models 10/127

Schulte-Ladbeck, R.: Massive Stars – Near and Far 10/135

Geller, M.J.: The Great Wall and Beyond –
Surveys of the Universe to $z < 0.1$ 10/159

Rees, M.J.: Black Holes in Galactic Nuclei 10/179

Mueller, J., Soffel, M.: Experimental Gravity and Lunar Laser Ranging 10/191

Ruffert, M., Janka, H.-Th.: Merging Neutron Stars 10/201

Werner, K., Dreizler, S., Heber, U., Kappelmann, N., Kruk, J., Rauch, T.,
Wolff, B.: Ultraviolet Spectroscopy of Hot Compact Stars 10/219

Roeser, H.-J., Meisenheimer, K., Neumann, M., Conway, R.G., Davis, R.J.,
Perley, R.A.: The Jet of the Quasar 3C 273/ at High Resolution 10/253

Lemke, D.: ISO: The First 10 Months of the Mission 10/263

Fleck, B.: First Results from SOHO 10/273

Thommes, E., Meisenheimer, K., Fockenbrock, R., Hippelein, H.,
Roeser, H.-J.: Search for Primeval Galaxies
with the Calar Alto Deep Imaging Survey (CADIS) 10/297

Neuhaeuser, R.: The New Pre-main Sequence Population
    South of the Taurus Molecular Clouds 10/323

Volume 11 (1998): Stars and Galaxies

Taylor, J.H. jr.: Binary Pulsars and General Relativity
    (26th Karl Schwarzschild Lecture 1997 – *not published*) 11/1
Napiwotzki, R.: From Central Stars of Planetary Nebulae to White Dwarfs
    (9th Ludwig Biermann Award Lecture 1997) 11/3
Dvorak, R.: On the Dynamics of Bodies in Our Planetary System 11/29
Langer, N., Heger, A., García-Segura, G.: Massive Stars:
    the Pre-Supernova Evolution of Internal and Circumstellar Structure 11/57
Ferguson, H.C.: The Hubble Deep Field 11/83
Staveley-Smith, L., Sungeun Kim, Putman, M., Stanimirović, S.:
    Neutral Hydrogen in the Magellanic System 11/117
Arnaboldi, M., Capaccioli, M.: Extragalactic Planetary Nebulae
    as Mass Tracers in the Outer Halos of Early-type Galaxies 11/129
Dorfi, E.A., Häfner, S.: AGB Stars and Mass Loss 11/147
Kerber, F.: Planetary Nebulae:
    the Normal, the Strange, and Sakurai's Object 11/161
Kaufer, A.: Variable Circumstellar Structure of Luminous Hot Stars:
    the Impact of Spectroscopic Long-term Campaigns 11/177
Strassmeier, K.G.: Stellar Variability as a Tool in Astrophysics.
    A Joint Research Initiative in Austria 11/197
Mauersberger, R., Bronfman, L.: Molecular Gas in the Inner Milky Way 11/209
Zeilinger, W.W.: Elliptical Galaxies 11/229
Falcke, H.: Jets in Active Galaxies: New Results from HST and VLA 11/245
Schuecker, P., Seitter, W.C.: The Deceleration of Cosmic Expansion 11/267
Vrielmann, S.: Eclipse Mapping of Accretion Disks 11/285
Schmid, H.M.: Raman Scattering
    and the Geometric Structure of Symbiotic Stars 11/297
Schmidtobreick, L., Schlosser, W., Koczet, P., Wiemann, S., Jütte, M.:
    The Milky Way in the UV 11/317
Albrecht, R.: From the Hubble Space Telescope
    to the Next Generation Space Telescope 11/331
Heck, A.: Electronic Publishing in its Context
    and in a Professional Perspective 11/337

Volume 12 (1999):
Astronomical Instruments and Methods at the Turn of the 21st Century

Strittmatter, P.A.: Steps to the Large Binocular Telescope – and Beyond
    (27th Karl Schwarzschild Lecture 1998) 12/1
Neuhäuser, R.: The Spatial Distribution and Origin
    of the Widely Dispersed ROSAT T Tauri Stars
    (10th Ludwig Biermann Award Lecture 1998) 12/27

Huber, C.E.: Space Research at the Threshold of the 21st Century –
    Aims and Technologies 12/47
Downes, D.: High-Resolution Millimeter and Submillimeter Astronomy:
    Recent Results and Future Directions 12/69
Röser, S.: DIVA – Beyond HIPPARCOS and Towards GAIA 12/97
Krabbe, A., Röser, H.P.:
    SOFIA – Astronomy and Technology in the 21st Century 12/107
Fort, B.P.: Lensing by Large-Scale Structures 12/131
Wambsganss, J.: Gravitational Lensing as a Universal Astrophysical Tool ... 12/149
Mannheim, K.: Frontiers in High-Energy Astroparticle Physics 12/167
Basri, G.B.: Brown Dwarfs: The First Three Years 12/187
Heithausen, A., Stutzki, J., Bensch, F., Falgarone, E., Panis, J.-F.:
    Results from the IRAM Key Project:
    "Small Scale Structure of Pre-Star-forming Regions" 12/201
Duschl, W.J.: The Galactic Center 12/221
Wisotzki, L.: The Evolution of the QSO Luminosity Function
    between $z = 0$ and $z = 3$ 12/231
Dreizler, S.: Spectroscopy of Hot Hydrogen Deficient White Dwarfs 12/255
Moehler, S.: Hot Stars in Globular Clusters 12/281
Theis, Ch.: Modeling Encounters of Galaxies: The Case of NGC 4449 .... 12/309

Volume 13 (2000): New Astrophysical Horizons

Ostriker, J.P.: Historical Reflections
    on the Role of Numerical Modeling in Astrophysics
    (28th Karl Schwarzschild Lecture 1999) 13/1
Kissler-Patig, M.: Extragalactic Globular Cluster Systems:
    A new Perspective on Galaxy Formation and Evolution
    (11th Ludwig Biermann Award Lecture 1999) 13/13
Sigwarth, M.: Dynamics of Solar Magnetic Fields –
    A Spectroscopic Investigation 13/45
Tilgner, A.: Models of Experimental Fluid Dynamos 13/71
Eislöffel, J.: Morphology and Kinematics of Jets from Young Stars 13/81
Englmaier, P.: Gas Streams and Spiral Structure in the Milky Way 13/97
Schmitt, J.H.M.M.:
    Stellar X-Ray Astronomy: Perspectives for the New Millenium 13/115
Klose, S.: Gamma Ray Bursts in the 1990's –
    a Multi-wavelength Scientific Adventure 13/129
Gänsicke, B.T.: Evolution of White Dwarfs in Cataclysmic Variables 13/151
Koo, D.: Exploring Distant Galaxy Evolution: Highlights with Keck 13/173
Fritze-von Alvensleben, U.:
    The Evolution of Galaxies on Cosmological Timescales 13/189
Ziegler, B.L.: Evolution of Early-type Galaxies in Clusters 13/211

Menten, K., Bertoldi, F.:
   Extragalactic (Sub)millimeter Astronomy – Today and Tomorrow 13/229
Davies, J.I.: In Search of the Low Surface Brightness Universe 13/245
Chini, R.: The Hexapod Telescope – A Never-ending Story 13/257

Volume 14 (2001): Dynamic Stability and Instabilities in the Universe

Penrose, R.: The Schwarzschild Singularity:
   One Clue to Resolving the Quantum Measurement Paradox
   (29th Karl Schwarzschild Lecture 2000) 14/1
Falcke, H.: The Silent Majority –
   Jets and Radio Cores from Low-Luminosity Black Holes
   (12th Ludwig Biermann Award Lecture 2000) 14/15
Richter, P. H.: Chaos in Cosmos 14/53
Duncan, M.J., Levison, H., Dones, L., Thommes, E.:
   Chaos, Comets, and the Kuiper Belt 14/93
Kokubo, E.: Planetary Accretion: From Planitesimals to Protoplanets 14/117
Priest, E. R.: Surprises from Our Sun 14/133
Liebscher, D.-E.: Large-scale Structure – Witness of Evolution 14/161
Woitke, P.: Dust Induced Structure Formation 14/185
Heidt, J., Appenzeller, I., Bender, R., Böhm, A., Drory, N., Fricke, K. J.,
   Gabasch, A., Hopp, U., Jäger, K., Kümmel, M., Mehlert, D.,
   Möllenhoff, C., Moorwood, A., Nicklas, H., Noll, S., Saglia, R.,
   Seifert, W., Seitz, S., Stahl, O., Sutorius, E., Szeifert, Th.,
   Wagner, S. J., and Ziegler, B.: The FORS Deep Field 14/209
Grebel, E. K.: A Map of the Northern Sky:
   The Sloan Digital Sky Survey in Its First Year 14/223
Glatzel, W.:
   Mechanism and Result of Dynamical Instabilities in Hot Stars 14/245
Weis, K.: LBV Nebulae: The Mass Lost from the Most Massive Stars 14/261
Baumgardt, H.: Dynamical Evolution of Star Clusters 14/283
Bomans, D. J.: Warm and Hot Diffuse Gas in Dwarf Galaxies 14/297

Volume 15 (2002): JENAM 2001 – Five Days of Creation: Astronomy with Large Telescopes from Ground and Space

Kodaira, K.: Macro- and Microscopic Views of Nearby Galaxies
   (30th Karl Schwarzschild Lecture 2001) 15/1
Komossa, S.: X-ray Evidence for Supermassive Black Holes
   at the Centers of Nearby, Non-Active Galaxies
   (13th Ludwig Biermann Award Lecture 2001) 15/27
Richstone, D. O.: Supermassive Black Holes 15/57
Hasinger, G.: The Distant Universe Seen with Chandra and XMM-Newton 15/71
Danzmann, K. and Rüdiger, A.:
   Seeing the Universe in the Light of Gravitational Waves 15/93
Gandorfer, A.: Observations of Weak Polarisation Signals from the Sun 15/113

Mazeh, T. and Zucker, S.: A Statistical Analysis of the Extrasolar Planets
    and the Low-Mass Secondaries 15/133
Hegmann, M.: Radiative Transfer in Turbulent Molecular Clouds 15/151
Alves, J. F.: Seeing the Light through the Dark:
    the Initial Conditions to Star Formation 15/165
Maiolino, R.: Obscured Active Galactic Nuclei 15/179
Britzen, S.: Cosmological Evolution of AGN – A Radioastronomer's View 15/199
Thomas, D., Maraston, C., and Bender, R.: The Epoch(s)
    of Early-Type Galaxy Formation in Clusters and in the Field 15/219
Popescu, C. C. and Tuffs, R. J.: Modelling the Spectral Energy Distribution
    of Galaxies from the Ultraviolet to Submillimeter 15/239
Elbaz, D.: Nature of the Cosmic Infrared Background
    and Cosmic Star Formation History: Are Galaxies Shy? 15/259

Volume 16 (2003): The Cosmic Circuit of Matter

Townes, C. H.: The Behavior of Stars Observed by Infrared Interferometry
    (31th Karl Schwarzschild Lecture 2002) 16/1
Klessen, R. S.: Star Formation in Turbulent Interstellar Gas
    (14th Ludwig Biermann Award Lecture 2002) 16/23
Hanslmeier, A.: Dynamics of Small Scale Motions in the Solar Photosphere 16/55
Franco, J., Kurtz, S., García-Segura, G.:
    The Interstellar Medium and Star Formation: The Impact of Massive Stars ... 16/85
Helling, Ch.: Circuit of Dust in Substellar Objects 16/115
Pauldrach, A. W. A.: Hot Stars: Old-Fashioned or Trendy? 16/133
Kerschbaum, F., Olofsson, H., Posch, Th., González Delgado, D., Bergman, P.,
    Mutschke, H., Jäger, C., Dorschner, J., Schöier, F.:
    Gas and Dust Mass Loss of O-rich AGB-stars 16/171
Christlieb, N.: Finding the Most Metal-poor Stars of the Galactic Halo
    with the Hamburg/ESO Objective-prism Survey 16/191
Hüttemeister, S.: A Tale of Bars and Starbursts:
    Dense Gas in the Central Regions of Galaxies 16/207
Schröder, K.-P.: Tip-AGB Mass-Loss on the Galactic Scale 16/227
Klaas, U.: The Dusty Sight of Galaxies:
    ISOPHOT Surveys of Normal Galaxies, ULIRGS, and Quasars 16/243
Truran, J. W.: Abundance Evolution with Cosmic Time 16/261
Böhringer, H.: Matter and Energy in Clusters of Galaxies as Probes
    for Galaxy and Large-Scale Structure Formation in the Universe 16/275

Volume 17 (2004): The Sun and Planetary Systems – Paradigms for the Universe

Boehm-Vitense, E.: What Hyades F Stars tell us about Heating Mechanisms
    in the outer Stellar Atmospheres
    (32th Karl Schwarzschild Lecture 2003) 17/1
Bellot Rubio, L. R.: Sunspots as seen in Polarized Light
    (15th Ludwig Biermann Award Lecture 2003) 17/21

Stix, M.: Helioseismology 17/51
Vögler, A. Simulating Radiative Magneto-convection in the Solar Photosphere 17/69
Peter, H.: Structure and Dynamics of the Low Corona of the Sun 17/87
Krüger, H.: Jupiter's Dust Disk – An Astrophysical Laboratory 17/111
Wuchterl, G.: Planet Formation – Is the Solar System misleading? 17/129
Poppe, T.: Experimental Studies on the Dusty History of the Solar System 17/169
Ness, J.-U.: High-resolution X-ray Plasma Diagnostics of Stellar Coronae
   in the XMM-Newton and Chandra Era 17/189
Fellhauer, M.: ω Cen – an Ultra Compact Dwarf Galaxy? 17/209
Leibundgut, B.: Cosmology with Supernovae 17/221
Beckers, J. M.: Interferometric Imaging in Astronomy: A Personal Retrospective ... 17/239
Stenflo, J. O.: The New World of Scattering Physics
   Seen by High-precision Imaging Polarimetry 17/269

Volume 18 (2005): From Cosmological Structures to the Milky Way

Giacconi, R.: The Dawn of X-Ray Astronomy
   (33rd Karl Schwarzschild Lecture 2004) 18/1
Herwig, F.: The Second Stars
   (16th Ludwig Biermann Award Lecture 2004) 18/21
Kraan-Korteweg, R.: Cosmological Structures behind the Milky Way 18/49
Schuecker, P.: New Cosmology with Clusters of Galaxies 18/77
Böhm, A., Ziegler, B. L.:
   The Evolution of Field Spiral Galaxies over the Past 8 Gyrs 18/109
Palouš, J.: Galaxy Collisions, Gas Striping and Star Formation in the Evolution
   of Galaxies 18/129
Ferrari, C.: Star Formation in Merging Galaxy Clusters 18/153
Recchi, S., Hensler, G.:
   Continuous Star Formation in Blue Compact Dwarf Galaxies 18/171
Brunthaler, A.: The Proper Motion and Geometric Distance of M33 .... 18/187
Schödel, R., Eckart, A., Straubmeier, C., Pott, J.-U.:
   NIR Observations of the Galactic Center 18/203
Ehlerová, S.: Structures in the Interstellar Medium 18/213
Joergens, V.: Origins of Brown Dwarfs 18/225

Volume 19 (2006): The Many Facets of the Universe – Revelations by New Instruments.

Tammann, G. A.: The Ups and Downs of the Hubble Constant
   (34th Karl Schwarzschild Lecture 2005) 19/1
Richter, P.: High-Velocity Clouds and the Local Intergalactic Medium
   (17th Ludwig Biermann Award Lecture 2005) 19/31
Baschek, B.: Physics of stellar atmospheres – new aspects of old problems
   (Talk in honor of Albrecht Unsöld's 100th anniversary) 19/61

Olofsson, H.: The circumstellar environment of asymptotic giant branch stars 19/75
Hirschi, R. et al.: Stellar evolution of massive stars at very low metallicities 19/101
Röpke, F. K.: Multi-dimensional numerical simulations of
   type Ia supernova explosions 19/127
Heitsch, F.: The Formation of Turbulent Molecular Clouds: A Modeler's View 19/157
Herbst, E.: Astrochemistry and Star Formation: Successes and Challenges 19/167
Kley, W.: Protoplanetary Disks and embedded Planets 19/195
Horneck, G.: Search for life in the Universe –
   What can we learn from our own Biosphere? 19/215
Guenther, E. W.: GQ Lup and its companion 19/237
Posch, T., et al.: Progress and Perspectives in Solid State Astrophysics –
   From ISO to Herschel 19/251
Brüggen, M., Beck, R. & Falcke, H.:
   German LOFAR - A New Era in Radio Astronomy 19/277
Stutzki, J.: SOFIA: The Stratospheric Observatory for Infrared Astronomy 19/293
Sargent, A., Bock, D.: Astronomy with CARMA – Raising Our Sites 19/315

Volume 20 (2008): Cosmic Matter.

Kippenhahn, R.: Als die Computer die Astronomie eroberten
   (35th Karl Schwarzschild Lecture 2007) 20/1
Beuther, H.: Massive Star Formation: The Power of Interferometry
   (18th Ludwig Biermann Award Lecture 2007 (i)) 20/15
Reiners, A.: At the Bottom of the Main Sequence
   Activity and Magnetic Fields Beyond the Threshold to Complete Convection
   (18th Ludwig Biermann Award Lecture 2007 (ii)) 20/40
Klypin, A., Ceverino, D., and Tinker, J.: Structure Formation in the
   Expanding Universe: Dark and Bright Sides 20/64
Bartelmann, M.: From COBE to Planck 20/92
Bœhm, C.: Thirty Years of Research in Cosmology, Particle Physics
   and Astrophysics and How Many More to Discover Dark Matter? 20/107
Kokkotas, K. D.: Gravitational Wave Astronomy 20/140
Horns, D.: High-(Energy)-Lights – The Very High Energy Gamma-Ray Sky 20/167
Hörandel, J. R.: Astronomy with Ultra High-Energy Particles 20/198
Mastropietro, C. and Burkert, A.: Hydrodynamical Simulations of the Bullet Cluster 20/228
Kramer, M.: Pulsar Timing – From Astrophysics to Fundamental Physics 20/255
Meisenheimer, K.: The Assembly of Present-Day Galaxies
   as Witnessed by Deep Surveys 20/279
Bromm, V.: The First Stars 20/307
Przybilla, N.: Massive Stars as Tracers for Stellar and Galactochemical Evolution .. 20/323
Scholz, A.: Formation and Evolution of Brown Dwarfs 20/357
Spiering, C.: Status and Perspectives of Astroparticle Physics in Europe 20/375

Volume 21 (2009): Formation and Evolution of Cosmic Structures.

Sunyaev, R. and Chluba, J.: Signals From the Epoch of Cosmological Recombination
(36th Karl Schwarzschild Lecture 2008) 21/1

Koch, A.: Complexity in small-scale dwarf spheroidal galaxies
(19th Ludwig Biermann Award Lecture 2008) 21/39

Sanders, R.H.: The current status of MOND 21/71

Khochfar, S.: Modeling the High-z Universe: Probing Galaxy Formation 21/87

Meynet, G., Ekström, S., Georgy, C., Chiappini, C., and Maeder, A.:
Evolution of Massive Stars along the Cosmic History 21/97

Beckman, J. E.: Cosmic Evolution of Stellar Disc Truncations: $0 \leq z \leq 1$ 21/127

Alves, J. and Trimble, V.: Star Formation from Spitzer (Lyman)
to Spitzer (Space Telescope) and Beyond 21/141

Trimble, V.: Catastrophism versus Uniformitarianism
in the History of Star Formation 21/147

Elmegreen, B. G.: Lyman Spitzer, Jr. and the Physics of Star Formation 21/157

Parmentier, G.: What cluster gas expulsion can tell us about star formation,
cluster environment and galaxy evolution 21/183

Hilker, M.: The high mass end of extragalactic globular clusters 21/199

Crida, A.: Solar System Formation 21/215

Solanki, S. K. and Marsch, E.: Solar Space Missions: present and future 21/229

Urry, M.: Women in (European) Astronomy 21/249

Volume 22 (2010): Deciphering the Universe through Spectroscopy.

Kudritzki, R.-P.: Dissecting Galaxies with Quantitative Spectroscopy
of the Brightest Stars in the Universe
(37th Karl Schwarzschild Lecture 2009) 22/1

Schuh, S.: Pulsations and Planets: The Asteroseismology-Extrasolar-Planet
Connection
(20th Ludwig Biermann Award Lecture 2009) 22/29

Frebel, A.: Stellar archaeology: Exploring the Universe with Metal-poor stars
(20th Ludwig Biermann Award Lecture 2009) 22/53

Wilhelm, K.: Quantitative Solar Spectroscopy 22/81

Wyse, R.: Metallicity and Kinematical Clues
To the Formation of the Local Group 22/99

Haehnelt, M.: Probing Dark Matter, Galaxies and the Expansion History
of the Universe with Lyα in Absorption and Emission 22/117

Baumgardt, H.: Hypervelocity Stars in the Galactic Halo 22/133

Thomas, J.: Schwarzschild Modelling of Elliptical Galaxies
and their Black Holes 22/143

van Boekel, R., Fang, M., Wang, W., Carmona, A., Sicilia-Aguilar, A., and
Henning, Th.: Star and Protoplanetary Disk Properties in Orion's Suburbs .. 22/155

Walter, F., Carilli, C., and Daddi, E.: Molecular Gas at High Redshift 22/167

Schmidt, R.: X-ray Spectroscopy and Mass Analysis of Galaxy Clusters 22/179

General Table of Contents

Dravins, D.: High-fidelity Spectroscopy at the Highest Resolution 22/191

Wurm, M., von Feilitzsch, F., Göger-Neff, M., Lachenmaier, T., Lewke, T., Meindl, Q., Möllenberger, R., Oberauer, L., Potzel, W., Tippmann, M., Traunsteiner, C., and Winter, J.: Spectroscopy of Solar Neutrinos 22/203

Röser, S., Kharchenko, N.V., Piskunov, A.E., Schilbach, E., Scholz, R.-D., and Zinnecker, H.: Open Clusters and the Galactic Disk 22/215

Käufl, H. U.: Rotational-Vibrational Molecular Spectroscopy in Astronomy 22/229

General Index of Contributors

Alberdi, A.	2/177	Ceverino, D.	20/64
Albrecht, R.	11/331	Chiappini, C.	21/97
Alves, J. F.	15/165, 21/141	Chini, R.	2/180, 13/257
Andreae, J.	5/58	Chluba, J.	21/1
Appenzeller, I.	14/209	Christensen-Dalsgaard, J.	3/313
Arnaboldi, M.	11/129	Christlieb, N.	16/191
Arnould, M.	1/155	Cohen, M.H.	2/177
Aschenbach, B.	4/173	Conway, R.G.	10/253
Baade, D.	5/125, 9/95	Crida, A.	21/215
Baade, R.	2/324	Cunow, B.	2/109, 7/207
Baars, J.W.M.	9/111	Daddi, E.	22/167
Bartelmann, M.L.	5/259, 10/61	Danzmann, K.	15/93
Bartelmann, M.	20/92	Davies, J.I.	13/245
Baschek, B.	19/61	Davis, R.J.	10/253
Basri, G.B.	12/187	Degenhardt, U.	8/61
Bastian, U.	9/87	Deiss, B.M.	3/350
Baum, E.	3/174	Dettmar, R.-J.	6/33
Baumgardt, H.	14/283, 22/133	Dietrich, M.	8/235
Beck, R.	19/277	Dominik, C.	3/199
Beckman, J. E.	21/127	Dones, L.	14/93
Beckers, J.M.	2/90, 17/239	Dorfi, E.A.	3/361, 11/147
Beisser, K.	2/221	Dorschner, J.	6/117, 16/171
Bellot Rubio, L.R.	17/21	Downes, D.	12/69
Bender, R.	14/209, 15/219	Dravins, D.	22/191
Bensch, F.	12/201	Dreizler, S.	10/219, 12/255
Bergman, P.	16/171	Drory, N.	14/209
Bertoldi, F.	13/229	Duemmler, R.	2/109, 7/207
Beuther, H.	20/15	Duennebeil, G.	3/174
Bodenheimer, P.	6/233	Duncan, M.J.	14/93
Bock, D.	19/315	Duschl, W.J.	2/333, 12/221
Böhm, A.	14/209, 18/109	Dvorak, R.	11/29
Bœhm, C.	20/107	Dziembowski, W.	5/143
Boehm-Vitense, E.	17/1	Eckart, A.	18/203
Böhringer, H.	8/259, 16/275	Edmunds, M.G.	1/139
Bomans, D.J.	14/297	Eggleton, P.P.	9/221
Boschan, P.	7/207	Ehlerová, S.	18/213
Bottinelli, L.	9/127	Ehlers, J.	10/91
Britzen, S.	15/199	Eislöffel, J.	13/81
Bromm, V.	20/307	Ekström, S.	21/97
Bronfman, L.	11/209	Elbaz, D.	15/259
Brüggen, M.	19/277	Elmegreen, B. G.	21/157
Brunthaler, A.	18/187	Englmaier, P.	13/97
Buchert, T.	2/267	Fahr, H.J.	4/126
Burkert, A.	7/191, 20/228	Falcke, H.	11/245, 14/15, 19/277
Camenzind, M.	3/234, 8/201	Falgarone, E.	12/201
Capaccioli, M.	11/129	Fang, M.	22/155
Carilli, C.	22/167	Fellhauer, M.	17/209
Carmona, A.	22/155	Ferguson, H.C.	11/83

Ferrari, C.	18/153	Henkel, R.	1/231
Fichtner, H.	9/191	Henning, T.	22/155
Fischer, O.	8/103	Hensler, G.	9/277, 18/171
Fleck, B.	4/90, 10/273	Herbst, E.	1/114, 19/167
Fockenbrock, R.	10/297	Herbstmeier, U.	7/151
Fort, B.P.	12/131	Herold, H.	3/74
Franco, J.	16/85	Herwig, F.	18/21
Frebel, A.	22/53	Hesse, M.	8/323
Freytag, B.	4/43	Hessman, F.V.	3/32
Fricke, K.J.	14/209	Hilker, M.	21/199
Fritze-von Alvensleben, U.	13/189	Hippelein, H.	10/297
Fröhlich, C.	19/101	Hirschi, R.	19/101
Gabasch, A.	14/209	Hoffmann, M.	4/165
Gandorfer, A.	15/113	Hörandel, J. R.	20/198
Gänsicke, B.T.	13/151	Hopp, U.	8/277, 14/209
Gail, H.-P.	1/231, 3/156	Horneck, G.	19/215
García-Segura, G.	11/57, 16/85	Horns, D.	20/167
Gautschy, A.	5/16	Horstmann, H.	2/109
Gehren, T.	1/52	Hoyle, F.	6/1
Geiss, J.	1/1	Huber, C.E.	12/47
Geller, M.J.	10/159	Hüttemeister, S.	16/207
Georgy, C.	21/97	Ip, W.H.	2/86
Gerhard, O.E.	5/174	Jäger, C.	16/171
Giacconi, R.	18/1	Jäger, K.	11/317, 14/209
Gilmore, G.	9/263	Jahreiß H.	2/72
Glatzel, W.	4/104, 14/245	Janka, H.-T.	7/103, 10/201
Goeres, A.	6/165	Jenkner, H.	3/297
Göger-Neff, M.	22/203	Joergens, V.	18/225
González Delgado, D.	16/171	Jütte, M.	11/317
Gouguenheim, L.	9/127	Kallrath, J.	9/307
Grebel, E.K.	10/29, 14/223	Kappelmann, N.	10/219
Grosbol, P.	2/242	Kaufer, A.	11/177
Grossmann-Doerth, U.	8/81	Käufl, H. U.	22/229
Gruen, E.	4/157	Keller, Ch.U.	8/27
Guenther, E. W.	19/237	Kerber, F.	11/161
Haehnelt, M.	22/117	Kerp, J.	7/151
Häfner, S.	11/147	Kerschbaum, F.	16/171, 19/251
Hamann, W.-R.	3/174	Kharchenko, N. V.	22/215
Hanslmeier, A.	16/55	Kessler, M.	2/53
Hanuschik, R.W.	2/148, 4/233	Khochfar, S.	21/87
Hasinger, G.	3/60, 7/129, 15/71	Kiessling, M.K.-H.	8/349
Heber, U.	10/219	King, A.R.	3/14
Heck, A.	11/337	Kippenhahn, R.	20/1
Heger, A.	11/57	Kissler-Patig, M.	13/13
Hegmann, M.	15/151	Klaas, U.	16/243
Heidt, J.	14/209	Klessen, R. S.	16/23
Heiles, C.	6/19	Kley, W.	3/21, 19/195
Heithausen, A.	12/201	Klose, S.	13/129
Heitsch, F.	19/157	Klypin, A.	20/64
Helling, Ch.	16/115	Knoelker, M.	8/81
Henkel, C.	6/69	Koch, A.	21/39

Koczet, P.	11/317	Mastropietro, C.	20/228
Kodaira, K.	15/1	Mauersberger, R.	6/69, 11/209
Koeppen, J.	6/179	Mazeh, T.	15/133
Koesterke, L.	3/174	McCaughrean, M.J.	6/191
Kokkotas, K. D.	20/140	Meijer, J.	2/109
Kokubo, E.	14/117	Meindl, Q.	22/203
Kollatschny, W.	5/229	Meisenheimer, K.	
Komossa, S.	15/27		2/129, 10/253, 10/297, 20/279
Koo, D.	13/173	Mehlert, D.	14/209
Kraan-Korteweg, R.C.	2/119, 18/49	Menten, K.	13/229
Krabbe, A.	6/103, 12/107	Meszaros, P.	10/127
Kramer, M.	20/255	Meyer, F.	3/1
Kratz, K.-L.	1/184	Meynet, G.	21/97
Kraus, U.	3/74	Moehler, S.	12/281
Krause, F.	4/260	Möllenberger, R.	22/203
Krichbaum, T.P.	9/221	Möllenhoff, C.	14/209
Kroll, R.	2/194	Moffat, A.F.J.	7/51
Krüger, H.	17/111	Moneti, A.	6/191
Kruk, J.	10/219	Montmerle, T.	3/209
Kudritzki, R.-P.	6/271, 22/1	Moorwood, A.	14/209
Kuhn, B.	8/277	Moritz, P.	7/151
Kümmel, M.	14/209	Mueller, E.	7/103
Kunze, R.	6/257	Mueller, J.	10/191
Kurtz, S.	16/85	Mutschke, H.	16/171, 19/251
Lachenmaier, T.	22/203	Najarro, F.	6/271
Lamb, D.Q.	10/101	Napiwotzki, R.	11/3
Lamers, H.J.G.L.M.	2/24	Naumann, M.	7/207
Landstreet, J.D.	5/105	Neckel, Th.	3/266
Langer, N.	2/306, 11/57	Ness, J.-U.	17/189
Lebzelter, T.	19/251	Neuhäuser, R.	10/323, 12/27
Leibundgut, B.	17/221	Neumann, M.	10/253
Leitherer, C.	7/73	Nicklas, H.	14/209
Lemke, D.	2/53, 10/263	Noll, S.	14/209
Leuenhagen, U.	3/174	Nollert, H.-P.	3/74
Levison, H.	14/93	Notni, P.	5/200
Lewke, T.	22/203	Oberauer, L.	22/203
Liebendörfer, M.	19/101	Olofsson, H.	16/171, 19/75
Liebscher, D.-E.	14/161	Omont, A.	1/102
Lobanov, P.A.	9/221	Ostriker, J.P.	13/1
Lucy, L.B.	7/31	Ott, H.-A.	2/109, 7/207
Luks, T.	7/171	Owocki, S.P.	3/98, 6/271
Luthardt, R.	5/38	Palme, H.	1/28
Maeder, A.	21/97	Palouš, J.	18/129
Maiolino, R.	15/179	Panis, J.-F.	12/201
Maitzen, H.M.	2/205	Parker, E.N.	4/1
Mann, I.	9/173	Parmentier, G.	21/183
Mannheim, K.	9/17, 12/167	Patermann, C.	2/13
Maraston, C.	15/219	Paturel, G.	9/127
Markiewicz, W.J.	6/149	Pauldrach, A.W.A.	
Marsch, E.	4/145, 21/229		3/124, 3/140, 6/271, 16/133
Martin, R.N.	9/111	Pauliny-Toth, I.I.K.	2/177

Penrose, R.	14/1	Schöier, F.	16/171
Perley, R.A.	10/253	Scholz, A.	20/357
Peter, H.	17/87	Scholz, R.-D.	22/215
Piskunov, A. E.	22/215	Schrijver, C.J.	4/18
Popescu, C. C.	15/239	Schroeder, K.-P.	3/187, 9/2210, 16/227
Poppe, T.	17/169	Schuecker, P.	2/109, 7/207, 11/267, 18/77
Posch, Th.	16/171, 19/251	Schuessler, M.	8/11, 8/81
Pott, J.-U.	18/203	Schuh, S.	22/29
Potzel, W.	22/203	Schulte-Ladbeck, R.	10/135
Priest, E.R.	14/133	Schwarz, E.	3/174
Pringle, J.E.	5/97	Schwenn, R.	1/179
Przybilla, N.	20/323	Schwope, A.D.	3/44, 8/125
Puls, J.	3/124, 3/140, 6/271	Sedlmayr, E.	1/231
Putman, M.	11/117	Seifert, W.	14/209
Quirrenbach, A.	5/214	Seitter, W.C.	2/109, 7/207, 11/267
Raedler, K.-H.	8/295	Seitz, S.	14/209
Rauch, T.	10/219	Sicilia-Aguilar, A.	22/155
Rayner, J.T.	6/191	Sigwarth, M.	13/45
Rebetzky, A.	3/74	Soffel, M.	10/191
Recchi, S.	18/171	Solanki, S.K.	4/208, 21/229
Rees, M.J.	2/1, 10/179	Spiekermann, G.	7/207
Reiners, A.	20/40	Spiering, C.	20/375
Richstone, D. O.	15/57	Spruit, H.C.	4/197
Richter, G.A.	5/26	Spurzem, R.	5/161
Richter, P.	19/31	Stahl, O.	3/286, 14/209
Richter, P.H.	14/53	Stanimirović, S.	11/117
Richtler, T.	8/163	Starrfield, S.	5/73
Rimmele, Th.	2/105	Staubert, R.	2/141
Röpke, F. K.	19/127	Staude, H.J.	3/266
Roeser, H.-J.	10/253, 10/297	Staude, J.	4/69
Röser, H.P.	12/107	Staveley-Smith, L.	11/117
Röser, S.	12/97, 22/215	Steffen, M.	4/43
Ruder, H.	3/74	Steiner, O.	8/81
Rüdiger, A.	15/93	Stenflo, J.O.	17/269
Ruffert, M.	10/201	Stix, M.	2/248, 17/51
Saglia, R.	14/209	Strassmeier, K.G.	11/197
Samland, M.	9/277	Straubmeier, C.	18/203
Sanders, R.H.	21/71	Strittmatter, P.A.	12/1
Sargent, A.	19/315	Stutzki, J.	1/221, 6/209, 12/201, 19/293
Schilbach, E.	9/87, 22/215	Sungeun K.	11/117
Schlosser, W.	11/317	Sunyaev, R.	21/1
Schmid, H.M.	11/297	Sutorius, E.	14/209
Schmidt, G.D.	8/147	Szeifert, T.	14/209
Schmidt, R.	22/179	Tammann, G.A.	9/139, 19/1
Schmidt, W.	4/117	Teerikorpi, P.	9/127
Schmidtobreick, L.	11/317	Teuber, D.	2/109, 2/229
Schmitt, D.	3/86, 8/61	Theis, Ch.	12/309
Schmitt, J.H.M.M.	5/188, 13/115	Theureau, G.	9/127
Schmutz, W.	3/174	Thielemann, F.-K.	19/101
Schoenfelder, V.	2/47, 9/49	Thomas, D.	15/219
Schödel, R.	18/203	Thomas, J.	22/143

Thommes, E.	10/297	Weinberger, R.	2/167
Thorne, K.S.	10/1	Weis, K.	14/261
Tilgner, A.	13/71	Werner, K.	10/219
Tinker, J.	20/64	Wessolowski, U.	3/174
Tippmann, M.	22/203	White, S.D.M.	7/255
Townes, Ch. H.	16/1	Wicenec, A.	4/188
Traunsteiner, C.	22/203	Wielebinski, R.	4/260, 8/185
Trimble, V.	21/141, 21/147	Wiemann, S.	11/317
Trümper, J.E.	8/1	Wilhelm, K.	22/81
Truran, J. W.	16/261	Wilking, B.A.	6/191
Tucholke, H.-J.	2/109	Williams, D.A.	6/49
Tuffs, R. J.	15/239	Wilson, R.N.	7/1
Turon, C.	9/69	Winter, J.	22/203
Ulrich, M.-H.	5/247	Wisotzki, L.	12/231
Ungerechts, H.	1/210	Wittmann, A.D.	4/61
Ungruhe, R.	7/207	Woitke, P.	14/185
Urry, M.	21/249	Wolf, B.	5/1
van Boekel, R.	22/155	Wolff, B.	10/219
van de Hulst, H.	9/1	Wuchterl, G.	17/129
von der Luehe, O.	2/105	Wurm, M.	22/203
von Feilitzsch, F.	22/203	Wyse, R.	22/99
Vrielmann, S.	11/285	Yorke, H.W.	2/283
Vögler, A.	17/69	Zeilinger, W.W.	11/229
Wagner, S.J.	14/209	Zensus, J A.	2/177, 9/221
Walter, F.	22/167	Ziegler, B.L.	13/211, 14/209, 18/109
Wambsganss, J.	12/149	Zinnecker, H.	6/191, 22/215
Wang, W.	22/155	Zucker, S.	15/133
Wegner, G.	7/235